The illustrated encyclopedia of the
Mineral Kingdom

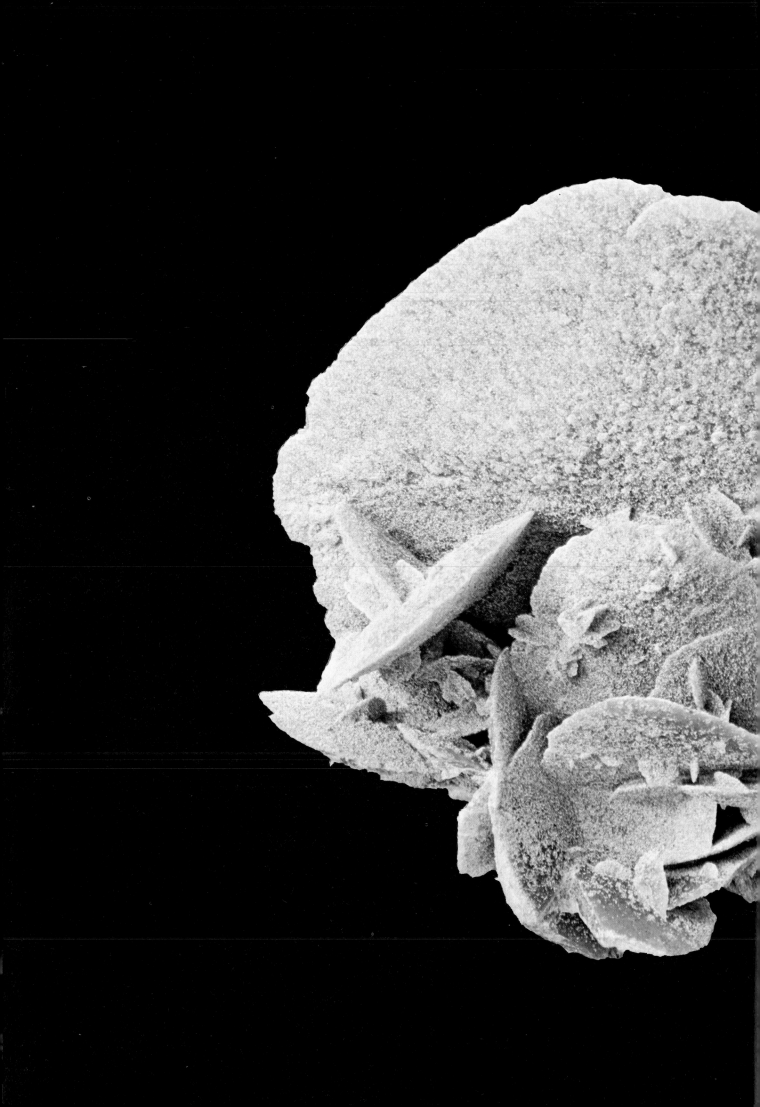

The illustrated encyclopedia of the
Mineral Kingdom

Consultant editor: Dr Alan Woolley

Hamlyn
London · New York · Sydney · Toronto

Contributors

Robert Symes is involved in the physical and chemical analysis of rocks and minerals in the Department of Mineralogy, British Museum (Natural History). He also lectures in geology.

Dr Andrew Rankin is a lecturer in mining geology at the Royal School of Mines, Imperial College, University of London. He has published several scientific papers on his main research interests, which include the genesis of mineral deposits and fluid inclusions in minerals.

Dr Andrew Clark works on the systematic mineralogy of oxide and sulphide minerals in the Department of Mineralogy, British Museum (Natural History).

Jennifer Bevan is involved in geochemical research and the running of the electron microprobe in the Department of Mineralogy, British Museum (Natural History).

Allan Jobbins is Keeper of Minerals and Gemstones at the Institute of Geological Sciences, London. He lectures in post-diploma gemmology and is an examiner for the F.G.A. Diploma of the Gemmological Association of Great Britain. He has conducted geological/gemmological surveys in Brazil, Guyana and Cambodia and carries out research in gemmological mineralogy and gem-testing.

Arthur Notholt works in the Minerals Resources Division, Institute of Geological Sciences, with particular responsibility for chemical and fertilizer raw materials on which he has written a number of papers. His particular interest is in phosphate rock, in which connection he represents the Institute of Geological Sciences on a number of international agencies.

Dr Bob King is the Curator of the Collections in the Department of Geology, University of Leicester. His main research interests include topographical mineralogy, stratabound ore deposits and hydrocarbon mineralization, and he lectures on economic mineralogy and curatorial techniques. He owns one of the finest private collections in Britain.

Dr Alan Woolley is in charge of the rock collections in the Department of Mineralogy, British Museum (Natural History). He has travelled extensively in the course of his research into rocks and minerals.

Published by The Hamlyn Publishing Group Limited
London · New York · Sydney · Toronto
Astronaut House, Feltham, Middlesex, England

Phototypeset by Tradespools Ltd, Frome, Somerset, England.
Colour separations by Culver Graphics Ltd, Bucks, England.
Printed in Italy by New Interlitho Ltd, Milan.

Contents

Robert Symes

Introduction

Millions of people have seen photographs of the Earth from space and studied the pictures of the rocky wastes of the surfaces of the Moon and Mars taken from orbiting and soft–landing spacecraft. Therefore, no detailed study of the minerals and rocks forming the Earth should start without considering the Earth in its cosmic setting. It is within this cosmic environment that our galaxy, our solar system and the Earth itself has been formed. The chemical elements, the building materials for the formation of the Earth, were originally formed in a galactic setting. The rocks of the Earth are composed of one or more minerals which are themselves naturally occurring chemical compounds of the elements. We know that the relative proportions of the elements which now constitute the Earth differ from those in the original mixture from which the Earth and the planets were first formed, and the structure of the newly formed Earth some 4 600 million years ago must have been very different from the Earth today. It is necessary, therefore, to study briefly what processes have taken place to account for these differences in time and to describe the structure and composition of the Earth as it is today.

The solar system

Our solar system consists of the Sun and the material revolving around it, the planets (of which the Earth is one), moons, asteroids, comets, meteorites, dust and gas. The solar system can be thought of as a disc with the Sun in the centre, most of the material circling the Sun in the same direction and within the same plane, as though it all formed at about the same time. The ages of the oldest known lunar rocks, stony meteorites and the estimated age for the Earth's crust all show that the solar system was formed about 4 600 million years ago. Each planet moves around the Sun in the same direction, all of the orbits being almost circular and in the same plane which is close to the equatorial plane of the Sun's rotation. Thus, the entire solar system is rotating although different parts are moving at different speeds. Most planets also rotate on their own axes. The Earth takes 23 hours 56 minutes to complete a single rotation but, for most of the larger outer planets, the period of rotation is about 10 hours. Most of the planets' satellites (moons) move around their respective planets in the same direction.

A system of stars physically connected by gravitational forces and moving in space as a whole is called a galaxy. Our galaxy is just one of millions forming the universe, and at least 400 million galaxies are detectable from the Earth. Within our galaxy there are estimated to be 100 000 million stars of which our Sun is just one (Fig. 1·1).

The Earth and solar system are very small in relation to the universe. Distances are so great that they are measured in light years [1 light year is the distance light

The Great Spiral in Andromeda. The Andromeda galaxy is very similar to our own in many respects; like ours it is a spiral and has two smaller satellite galaxies.

travels in one year – almost 10 million million kilometres (6 million million miles)]. The nearest of the millions of stars in our galaxy, other than the Sun, is 4 light years away. Our galaxy is a flattened disc of gas, dust and stars 80 000 light years across, made up of two spiralling arms which probably condensed from a vast cloud of hydrogen gas, perhaps 7 000 to 10 000 million years ago. We can see our own galaxy, edge-on, as the band of brighter night sky we call the Milky Way.

Origin of the elements

Elements are primary, chemically indivisible substances. The origin of the elements, their relative abundance and distribution in the universe are known to be closely associated with the evolution and history of stars of which our Sun is one. Researches indicate that the thermonuclear reactions occurring within the ultra-hot interiors of stars lead directly to the creation of the 'lighter' elements from primordial hydrogen, which has the simplest atomic structure, and is the least dense and most abundant element in the universe. Further complex reactions, involving protons and the nuclei of the light elements, can lead to the creation of the heavier elements. These reactions are accompanied by the release of radiant energy as light. Not all stars can create the heavier elements; their size and temperatures have to be great enough. Old stars often explode or eject material into space, modifying the composition of interstellar matter. Newly formed stars are partly composed of the earlier-formed elemental debris which has been scattered into space. The elemental materials of the solar system may have originated over 5 000 million years ago in the evolution and destruction of stars.

The Sun

The Sun is a star, composed of ultra-high-temperature gases – mostly hydrogen and helium. It is immensely large, being nearly 100 times bigger (in terms of mass) than all the planets in the solar system put together. Its composition, therefore, approximates to solar system composition, and can be used in studies of elemental abundances in the solar system. The Sun's energy is derived from the thermonuclear reactions taking place at its centre, where hydrogen is converted to helium. The energy so produced is transported to the Sun's outer surface – the photosphere – and radiated into space. The temperature at the Sun's centre has been estimated at seventeen million degrees Centigrade (°C), and at the surface 5 500 °C.

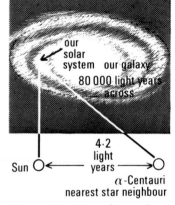

Fig. 1·1 Position of our solar system in the galaxy. The nearest star in the 100 000 million stars which make up our galaxy, α-Centauri, is 4 light years distant. The nearest galaxy to ours is the galaxy Andromeda, which is about 2·2 million light years away (*see* page 6).

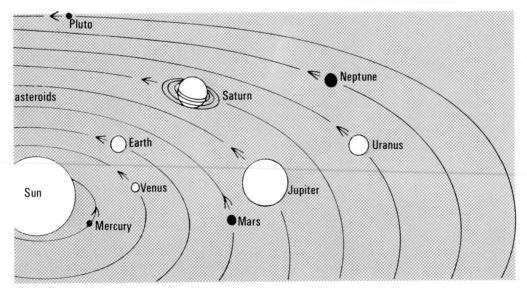

Fig. 1·2 Main components of the solar system, showing the planetary orbits and relative sizes compared to the Sun.

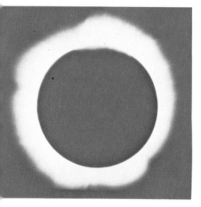

Above
Solar eclipse of March 7th, 1970, photographed from Miahautlán, Mexico. The Moon totally blocked out the Sun for 3·5 minutes enabling the nature of the photosphere to be studied.

Right
Photograph of the planet Earth taken by spacecraft Apollo 11 on the outward journey to the Moon. The spacecraft was 181 500 kilometres (112 700 miles) from Earth when this picture was taken and shows through the cloud cover most of Africa and portions of Europe and Asia.

The planets

The planets are of two main types: the four innermost (Mercury, Venus, Earth and Mars) are small and dense and are known as the terrestrial planets because they are alike in many respects (*see* Fig. 1·2), while the outer planets (Jupiter, Saturn, Uranus, Neptune and Pluto) are, with the exception of Pluto, much larger and less dense. Jupiter has a radius ten times that of Earth and has about a thousand times the volume. The outer planets are much less dense than the terrestrial planets, Saturn having a density less than that of water. The fact that most of the larger outer planets are of low density suggests that they consist mostly of substances such as hydrogen and helium, together with other hydrogen compounds such as water, ammonia and methane. The small, dense inner planets consist almost entirely of silicates and metal. This is explained by the fact that the large, outer planets, remote from the Sun, have retained large amounts of gas from the time of their original formation, while the inner planets have shed much of their gaseous component, leaving a rocky residue.

Mercury is the smallest [diameter 4 865 kilometres (3 024 miles)] and innermost planet relative to the Sun. It is dense and is probably partly formed of metallic minerals. If Mercury has an atmosphere it is likely to be transient, made up of gases such as krypton and xenon, that have come recently from the interior and are too heavy to escape into space. Mercury rotates once every fifty-nine Earth days so that at any particular place sunrise comes at intervals of about 170 Earth days. The result is that owing to the closeness of the Sun the day temperature at the surface is up to 350 °C.

Venus is small [diameter 12 100 kilometres (7 520 miles)] and is possibly similar in structure to the Earth. It has an atmosphere rich in carbon dioxide which may be as much as a hundred times as dense as the Earth's. Roughly 64 kilometres (40 miles) above the surface occurs a cloud layer of ice and water which always tends to obscure the view of the surface. Part of the surface appears to be rather smooth, probably made of loose dust which is constantly blown over the surface by winds. The surface temperature is estimated at 300 °C.

The Earth is the largest of the inner planets [diameter 12 734 kilometres (7 914 miles)]. It has a layered internal structure with a dense, possibly metallic core surrounded by layers of silicate rocks. Almost three-quarters of the outer surface is covered by water, and there is also a dense atmosphere of oxygen and nitrogen. The polar regions are currently covered by ice.

Mars is smaller than the Earth [diameter 6 760 kilometres (4 201 miles)] and rotates at a similar speed to the Earth, but has a mass of about one-tenth that of the Earth. It has an atmosphere rich in carbon dioxide and the poles are covered by icecaps. The surface of Mars can be plainly seen and the features identified. The surface is reddish and cratered like that of the Moon, and in places it is deeply eroded. Sunlight on Mars is on average less than half as strong as on Earth and the surface, therefore, is much colder, the thin cloud cover being insufficient to retain any heat. (The day surface temperature is about 25 °C but at night it drops to −80 °C.)

Jupiter is the largest planet in our solar system [diameter 142 700 kilometres (88 689 miles)] and it rotates very quickly making one complete rotation in under 10 hours. The density is low at 1·3 grams per cubic centimetre (g/cm^3) [0·047 pounds per cubic inch (lb/in^3)], little more than that of water. Jupiter has a thick, turbulent atmosphere, probably thousands of kilometres deep and composed of hydrogen and helium, which passes down into a liquid and what appears to be a solid modification of these gases due to the tremendous compression. Jupiter appears to generate some internal energy possibly as a result of material moving slowly towards its centre and so releasing gravitational energy. Some scientists believe that the present conditions on Jupiter are not very different from those on Earth about 4 000 million years ago. In Jupiter's central core, or scattered throughout its interior, there may be enough iron, silicon and other heavy elements to make a dense planet about the size of the Earth if the hydrogen and helium were to be lost into space.

Saturn is also a huge planet [diameter 120 800 kilometres (75 078 miles)] but has the lowest density of all the planets at 0·7 g/cm^3 (0·025 lb/in^3). Saturn, like Jupiter, is composed largely of hydrogen and helium, together with small amounts of other gases and a sprinkling of other heavy elements. Its total mass, however, is only one-third that of Jupiter and does not, therefore, provide enough gravitational pull to compress the gases tightly. It probably has a central body of solid hydrogen much smaller than that of Jupiter. The rings which orbit Saturn appear to be accumulations of small, ice-covered fragments.

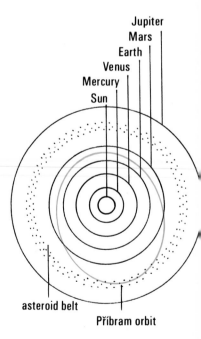

Fig. 1·3 In 1959 a meteorite fell at Příbram, Czechoslovakia. The fall was photographed from two different places, and from this the Příbram orbit was plotted as an elongated elliptical orbit which brought it from the asteroid belt.

Uranus is a large planet [diameter 47 600 kilometres (29 584 miles)] with a density of 1·6 g/cm^3 (0·058 lb/in^3). The surface temperature is very low at about −180 °C. The planet's composition is probably mostly ice with some solid ammonia, hydrogen and helium, and it owes its green appearance to a methane-rich atmosphere. Uranus is only just visible to the naked eye, when it is at its brightest.

Oblique view of the Moon's surface showing a characteristic large crater 80 kilometres (50 miles) in diameter, the whole surface pitted by smaller impact debris.

Neptune is slightly larger than Uranus [diameter 48 400 kilometres (30 080 miles)] and has a density of 2·2 g/cm³ (0·08 lb/in³). It is probably formed of frozen methane, ice and ammonia.

Pluto is the most distant planet yet discovered. Very little is known about its composition; neither its mass nor its density can be stated with certainty. The diameter is about half that of the Earth, and Pluto has an orbit which is markedly elliptical. Pluto is small, cold and dark, and not likely to support life.

The Moon is the Earth's natural satellite, and is a quarter of the diameter of the Earth and has a lower density. It has, therefore, only about one-sixth of the Earth's gravitational attraction. It has no atmosphere and a waterless surface. The Moon's surface consists of large, smooth, low areas (maria) which are meteorite impact basins in the Moon's crust that were mainly flooded by basaltic lavas about 4 000 million years ago, and rugged, mountainous regions (terrae) which are pieces of the Moon's early crust, dated at 4 600 million years. Circular craters of all sizes indicate continued bombardment by meteorites. We can see no active volcanoes on the Moon.

Several of the other planets also have satellites (moons). Jupiter has twelve known satellites – more than any other planet. In 1610, Galileo recorded his discovery of four of Jupiter's satellites, which are called Europa, Io, Ganymede and Callisto and became known as the Galilean satellites. These four satellites circle Jupiter comparatively closely although Callisto is more than 1·6 million kilometres (1 million miles) away. Ganymede, the largest of the satellites, is 145 kilometres (90 miles) greater in diameter than Mercury. The surfaces of the Jovian satellites appear to be partially covered with a mixture of ice and solid ammonia, nitrogen and carbon dioxide.

Meteorites

The fall of meteorites on the Earth's surface provides scientists with samples of rock types from other parts of the solar system. Most differ in composition from anything occurring in the Earth's crust, but one group of meteorites, the chondrites (*see* page 26), have compositions that support our theories of the origin of the chemical constitution of the Earth. By accurately plotting the flight paths of

Aerial view of meteor crater, Arizona [1·2 kilometres (0·75 miles) in diameter, 180 metres (600 feet) deep]. A large meteorite impact crater believed to have been formed by a large iron meteorite which probably weighed about 50 000 tonnes and vaporized on impact.

certain meteorites seen to fall, it is known that they move in elliptical orbits which originated in the asteroid belt (between Mars and Jupiter). The asteroid belt contains thousands of small bodies mostly less than 1 kilometre (0·6 mile) across although one, Ceres, the largest, is 769 kilometres (478 miles) across (Fig. 1·3).

Meteorites are classified into: 1 iron meteorites – made essentially of nickel-iron alloys (4–20 per cent nickel); 2 stony-irons – silicates plus metal; 3 stony meteorites – consisting mainly of magnesium and iron silicate minerals. Many of the stony meteorites contain small spheroidal aggregates known as chondrules and are called chondrites. A few chondrites contain water-bearing minerals and carbon compounds, and are called carbonaceous chondrites. It is believed that these probably represent some of the primitive matter of the solar system.

Comets

Many astronomers now believe that a large cloud of comet-type material surrounds the outermost planets, this material forming bodies with diameters of up to a kilometre (0·6 mile). They are non-luminous but, occasionally, this cloud of material may be disturbed by a passing star whose gravitational attraction causes a comet to be pulled from these outer regions and to pass in an elliptical orbit around the Sun. In so doing the brilliant long-tailed comets, that are seen occasionally from Earth, are formed. Some, such as Halley's comet, return to the region of the Earth at regular intervals; others pass close to the Sun and then move away from the solar system never to return.

Comets appear to consist of two parts: the head, which expands enormously as the orbit approaches the Sun; and the brilliant tail which points away from the Sun and is formed from dust and gas driven from the head by the sunlight and charged particles (solar wind) emanating from the Sun. Comets have low densities but it is not accurately known whether the head, which appears to consist of ice and frozen gases, contains a solid nucleus or diffuse dust particles. Molecules such as carbon monoxide, nitrogen, carbon dioxide and cyanide have been detected in comets. Some scientists believe the comets to be related to carbonaceous chondrite material.

It is apparent that there still is a great deal of matter diffused throughout space which has not condensed into stars or planetary bodies. Over 90 per cent of this matter is gaseous hydrogen but much of the elemental debris from stars is present as very fine dust which ranges in size from atomic and molecular particles to aggregates of particles and minute solid grains.

Meteorite specimens: *left* polished and etched slice of the Gibeon iron meteorite (south-west Africa) showing the characteristic Widmanstatten pattern; *bottom* cast of the Middlesbrough, Yorkshire stone which fell in 1881 (olivine-hypersthene chondrite) showing typical flow lines on the fusion crust, due to partial melting in flight through the atmosphere; *right* cast of the Limerick, Eire stone meteorite which fell in 1813 (olivine-bronzite chondrite).

Comet Humason, 1961. Photograph shows the well-developed head and diffuse tail, characteristic of such comets.

Formation of the solar system

There are many different theories to explain the formation of the solar system but two of these with their modifications are the most important. A generation ago it was widely held that the Earth and planets were torn from the Sun by a star that happened to pass close by; this was the 'catastrophic' theory of formation of the solar system. It is now believed that planet formation is a normal process that can happen near any growing star anywhere in the universe.

Over 5000 million years ago a large cloud of dust and gas, mostly hydrogen, floating in space, began to concentrate by mutual gravitational attraction to form the nucleus of the star that we now call the Sun. As this body increased in size, its gravitational pull became stronger, pulling more gas and dust towards it. How long the process took is not known but eventually the newly formed star was big enough to heat its centre to many millions of degrees Centigrade so that thermonuclear reactions could start within. Nuclear energy was released by the reaction converting hydrogen to helium, the reaction keeping the interior hot and sustaining the nuclear reactions previously described. Our Sun has shone as a star ever since.

The planets may have condensed out of the remnants of the original cloud. The initially formed bodies were small, and their gravitational attraction was too weak to make colliding objects hit them with large impact energy. As their mass increased, however, the gravitational pull grew stronger and smaller bodies impacted into them releasing large amounts of energy. The Earth at this time would probably have had a huge cosmic atmosphere and a small rocky core. Studies of the geochemical nature of planetary accretion show that the inner planets formed from solid materials at relatively low temperatures.

The Earth's early history

The Earth today has a layered structure, consisting of an extremely thin outer layer, the *crust*, below which is the *mantle*, much thicker and denser than the crust, and, at the centre, the *core*, part of which is probably a dense iron–nickel alloy (Fig. 1·4).

At its formation, the Earth was composed of a relatively homogeneous mixture of core and mantle material. A period of heating, probably of radioactive origin, then caused melting of the iron minerals so that the molten iron descended towards the centre of the Earth, leaving the overlying mantle mostly solid.

Careful studies have shown that meteorites first formed from low-temperature material similar to that which forms the carbonaceous chondrites. Some of the meteorite parent bodies then became heated enough to cause local melting, and then gradually to cool. It is believed that the heating source was short-lived radioactivity (radionuclides). These break down within a relatively short time, in geological terms, to stable elemental forms so that all these radionuclides decayed long ago to non-radioactive isotopes, but the characteristic decay products can still be isolated and studied in some meteorites. Similarly, it could be that the Earth was also heated in its earliest history by short-lived isotopes and then, after decay of these, a state was reached where the internal heat production was mainly from the radioactive decay of the long-lived isotopes of uranium, thorium and potassium concentrated into the upper crust. It is also possible that gravitational energy of accretion and core separation produced near total melting of the outer Earth, which would account for the loss of most of the primordial rare gases and the transfer of uranium, thorium and the rare earth elements from early condensates to the surface zone. Core separation and surface cooling must have been completed in 600 to 700 million years because some metamorphosed sedimentary

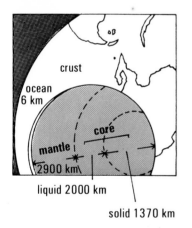

Fig. 1·4 The Earth's interior, showing the relative thicknesses of crust, mantle and core.

13

rocks in Greenland have been dated at 3 700 to 3 800 million years. Probably there was a very early period of the Earth's history when major Earth differentiation occurred.

It is probable that some localized melting of the surface layers of the Earth was caused by considerable meteorite bombardment in the period 3 800 to 4 200 million years ago as happened to the Moon. The meteorites which fell to Earth at this time may have been larger and more numerous than those of the Moon because of the greater gravitational attraction of the Earth. Part of the pattern of element distribution in the outer part of the Earth may be attributed to this early impacting and melting stage.

Formation of the Earth's crust

Initially the world-encircling crust was probably a thin basaltic layer with no distinction into parts that we now call the continents and oceanic basins. Localized fractures in this hot basaltic crust would afford passage to the surface of lava, hot gases and solutions from the hotter subcrustal layers. The action of these ascending solutions would be to produce less basic crustal rocks. Erosion and subsequent sedimentation in water on the surface of the Earth would accelerate the chemical differentiation of the crustal material. The successive addition of further sedimentary deposits and volcanic material to the initial continental fragments continues to the present day. The numerous original continental pieces may have been partly remelted and rebuilt many times before growing to a sufficient size to resist complete further engulfment. By about 3 500 million years ago, the continental masses (the cratonic shield areas) and early oceanic basins had been formed, and rocks of this age are known in all continental masses.

Formation of the hydrosphere and atmosphere

The present oceans and atmosphere are secondary features, the oceans being formed from the outgassing of the Earth's interior rocks after the initial Earth formation. The original water would probably have accreted not as a gas, but in solid hydrated minerals such as amphiboles or micas, and then been transported to the surface by volcanic activity. Studies of the origin of the atmosphere and hydrosphere of the Earth suggest an initial atmosphere rich in hydrogen and lacking free oxygen; other gases associated with the hydrogen would have been helium, nitrogen, methane, ammonia and water vapour. The gravity of the newly formed Earth was probably insufficient to hold helium and hydrogen which, therefore, diffused into space at an early stage. It may be also that the photochemical dissociation of the water vapour present in the early atmosphere by solar radiation caused the first free oxygen to be formed within the atmosphere. It is now generally agreed that most of the Earth's original atmosphere was lost in the period of early planetary heating and resultant outgassing described above. The gradual change from the reducing primary atmosphere to one containing oxygen, able to support primitive, oxygen-dependent life forms is generally believed to have occurred at about 2 000 million years ago.

The Earth today

We have seen so far that the Earth as we now know it is a small, rocky planet, fortunate in the way that it has evolved, and in the position it occupies in the solar system. In contrast with the surfaces of planets nearer the Sun which are extremely hot, and those further away which are icy wastes, the Earth is fortunate in having a modest range of surface temperature between about 60 °C (highest recorded temperature) and $-90\,°C$ (lowest recorded temperature). On the surface of the Earth there are large quantities of water (the hydrosphere) which cover three-quarters of the outer surface of the crust. The Earth also has an atmosphere rich in oxygen, a further condition that means that the Earth is able to support life as we know it.

Nevertheless we live on a thermally active Earth, much of the thermal energy being released by the concentrations of radioactive minerals in the upper layers. This is probably the principal source of internal energy responsible for volcanic activity and the forces necessary to move the crustal layers. In contrast with the Moon and Mars the Earth's surface is continually being changed as crust is created and destroyed, fractured, folded, and weathered. This internal activity results also in a relatively strong magnetic field, formed, we believe, in the outer core which is in a mobile state.

The Earth's internal structure

The greater the depth in the Earth, the greater the temperatures and pressures become, so that at the core the temperatures may be as high as $3500\,°C$ and the pressure about 300 million kilonewtons per square metre (3 million atmospheres).

Scientists cannot collect samples and study the interior of the Earth directly, so that most of the information about the internal structure has been obtained from studies of the way earthquake waves travel through the Earth. Some of the waves pass through the body of the Earth and are focused by the core, whereas others pass through the mantle but fail to penetrate the core. As a result the core acts as though it casts a shadow on the surface opposite to the origin of the earthquake and in this shadow zone no waves of this particular type are received. The radius of the core can then be calculated from the size of this shadow.

Most earthquakes originate by movement on fractures very close to the surface of the Earth, usually at depths less than 30 kilometres (19 miles). They have, however, been recorded at depths down to 700 kilometres (435 miles). The majority of earthquakes originate in well-defined earthquake zones where the Earth is being actively deformed. Earthquakes vary from mild movements to violent oscillations of the Earth's surface. When an earthquake occurs, there is often a mild foreshock, then the principal movement, and finally a milder after-shock. Analysis of this rather complex pattern of waves by sensitive instruments arranged in a worldwide network allows the important features of the interior of the Earth to be deduced.

Speaking generally, two distinct types of wave are generated by earthquakes: these are the body waves which travel through the Earth and longitudinal or L waves which travel around the Earth's surface. The body waves are also of two types: primary or P waves produce successive compressions in the direction in which they are travelling; and secondary (more correctly shear) or S waves where the medium through which the waves pass is caused to change its shape (Fig. 1·5). Primary waves travel almost twice as fast as the shear waves. It follows that the time interval between the arrival of the two types of wave at a recording station should increase with distance travelled. On their passage through the Earth, seismic waves behave like light in that they are reflected and refracted as they pass between media of different densities. The overall density of the Earth is $5\cdot5\ \mathrm{g/cm^3}$ ($0\cdot2\ \mathrm{lb/in^3}$) whereas the observed density of the surface rocks is $2\cdot5$ to $3\cdot5\ \mathrm{g/cm^3}$ ($0\cdot09$ to $0\cdot12\ \mathrm{lb/in^3}$) so that there must be considerably denser material at depth. By studying the travel-time curves for the P and S body waves passing through the Earth it is possible to infer the existence of layers of rock of varying density in which waves travel with different velocities.

If we consider the diagram shown in Fig. 1·6 and if an earthquake occurs at

Road off-set by faulting, associated with the Gediz earthquake of the 28th March, 1970 in Turkey.

point A, then at an epicentral angle of 103° from point A no further direct S waves are recorded; in fact, between 103° and 143° no direct body waves of any kind are received and this area is called the P and S wave shadow zone. Between 143° and 180° direct P waves are received once more but their travel time is longer than would be expected. Similar information for all earthquake positions in the surface layers of the Earth gives us information on the radius of the core, the nature of the outer core (S waves are not transmitted) and an impression that the inner core is solid.

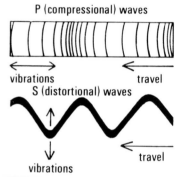

Fig. 1·5 Diagrammatic representation of P (compressional) waves and S (distortional) waves.

The core

Seismic evidence tells us that at 2900 kilometres (1800 miles) depth there is a major change in the composition or structure of the Earth – this is the core-mantle boundary. In the Earth's core, S waves are no longer propagated. P waves are refracted by the core so that they reappear only at epicentral angles greater than 143° (see Fig. 1·6), and when they do reappear they have taken longer than expected. S waves cannot be transmitted through a fluid and, therefore, it is reasonable to assume that the core, or at least its outer part, has the properties of a liquid. It has long been assumed that the core consists largely of iron or nickel-iron, a view supported by the analysis of the iron meteorites which themselves are possibly pieces of planetary bodies whose internal structure once resembled that of the Earth. The core is almost twice as dense as the mantle and is believed to be composed of nickel-iron alloys with some dissolved sulphur and silicon.

The inner core

Recent studies into the structure of the Earth's core have revealed that it is not liquid throughout, because P waves travel with slightly increased velocity through its central part which indicates that the very centre of the Earth must be in a solid condition, probably as a result of the enormous pressures there.

The mantle

At the base of the crust, the earthquake waves increase abruptly in velocity. The discontinuity between the high and low velocity layers is called the Mohorovičić discontinuity (Moho for short), and it is taken as the boundary between the crust

Ultramafic olivine nodule (peridotite) in vesicular basalt; Lanzarote, Canary Islands.

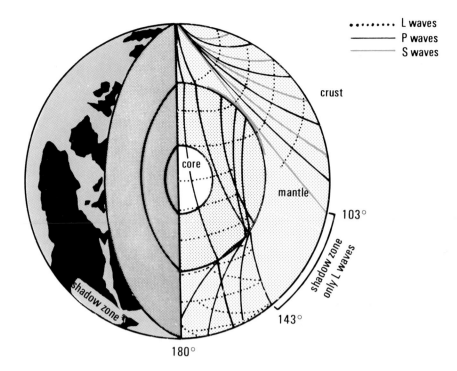

Fig. 1·6 Nature and travel paths of earthquake shock waves. P and S waves travel from the earthquake focus in all directions including through the Earth (body waves), while L waves can only travel around the surface of the planet.

and mantle. The mantle forms 83 per cent of the Earth by volume and about 58 per cent by mass.

Laboratory experiments suggest that the upper mantle consists of dense, dark rocks composed essentially of iron-magnesium silicate minerals and called peridotite perhaps together with small patches of eclogite and dunite.

It is almost certain that the overlying crust, as we now know it, developed from the mantle probably by the addition of volcanic material at the surface. Many of these lavas contain a rich variety of inclusions of rocks torn off from deeper Earth layers during the passage of the lava. Some of these inclusions are of common crustal rocks such as granodiorite, or granulite. Others, however, which are rather rare but widely distributed in volcanoes on both continental and oceanic crusts, are olivine-rich nodules. As well as olivine these nodules contain the characteristic minerals diopside, enstatite, and chrome spinel. In the relatively rare rock, kimberlite, which forms pipe-like intrusions, garnet peridotite inclusions are found. These are more variable in composition than the olivine nodules and, in addition to the minerals described above, they contain garnet, the mica, phlogopite, and occasionally diamonds.

In many fold-mountain ranges, masses of peridotite are found emplaced within the folded sedimentary rocks. The enormous stresses involved in mountain building are believed to have thrust great pieces of mantle into the upper part of the crust. The speeds of seismic waves in the upper mantle are similar to values obtained in experiments with peridotite in the laboratory.

At the greater depths in the lower mantle, minerals change to more dense forms and ultimately take on a simple, dense oxide structure produced by extreme pressures. There are two seismic discontinuities within the mantle at about 400 and 700 kilometres (250 and 435 miles) depth, and these are likely to represent the phase changes described above, where the main mineral components of the mantle are forced, by the increasing pressure, to readjust their crystal structures so that the atoms are more densely packed.

Low velocity layer

Seismic velocities increase with depth throughout most of the mantle, but it has been known for some time that seismic waves from earthquakes with foci at

17

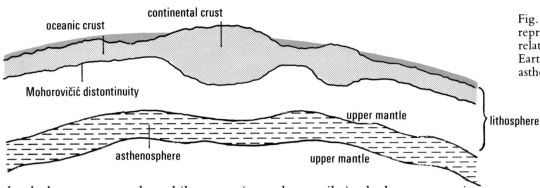

Fig. 1·7 Diagrammatic representation of the possible relationship between the Earth's crust, lithosphere and asthenosphere.

depths between 50 and 250 kilometres (30 and 155 miles) take longer to arrive at recording stations than theoretically they should. Therefore, it appears that there is a layer of lower velocity within the upper mantle, and it appears further that this low velocity layer lies at greater depth beneath the continents than under the oceanic crust. The exact nature of the layer is not fully understood; it may consist of several narrow layers or small pockets of molten material, or it may be that within this layer the peridotite of the mantle is partially molten.

This low velocity zone is important in our understanding of the surface movements on the Earth because it may represent a zone between the outer and inner parts of the Earth above which mountain building, and volcanicity can occur. The low velocity layer is sometimes called the *asthenosphere*. Above this the crust and upper mantle form a relatively rigid shell named the *lithosphere*. The lithosphere is formed of several pieces called plates, and it is the movement of these plates and the nature of their margins that controls the major features of the Earth's volcanicity, earthquakes and mountain building. It is inferred that mantle circulation provides heat and material for the continued production of pockets of magma under constructive and destructive margins (Fig. 1·8).

The idea that the major geological structural features of the outer shell of the Earth are due to the interaction of moving plates of lithosphere, is now known as the theory of *plate tectonics*. There is evidence that there are about twenty lithospheric plates, seven of which, such as the Pacific plate, are large and account for most of the Earth's surface. The plates are between 70 and 100 kilometres (43 and 62 miles) thick and move over a partially molten layer (the asthenosphere) within the mantle, carrying the continents with them. Virtually all earthquakes, volcanicity and mountain-building take place along the margins of these plates. There are four main types of plate margin: oceanic spreading ridges (constructive margins), where two plates move apart and molten material from the Earth's interior fills the gap; oceanic subduction zones (destructive margins), where two plates meet and one plate slides under the other to be destroyed; collision zones (destructive margins), where two plates carrying continents meet

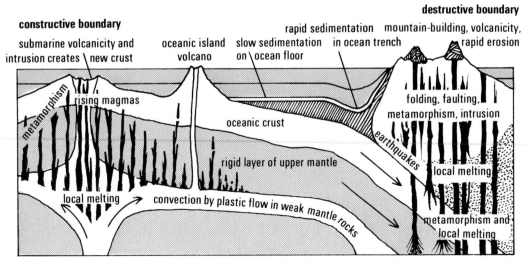

Fig. 1·8 Constructive and destructive plate margins in section.

Fig. 1·9 Simplified map of the major lithospheric plates and the nature of the plate margins; constructive margins are shown in black, destructive margins by a broken line and weakly active boundaries and those along which plates are sliding past each other in blue. The arrows show the direction of plate movement assuming Africa to be stationary.

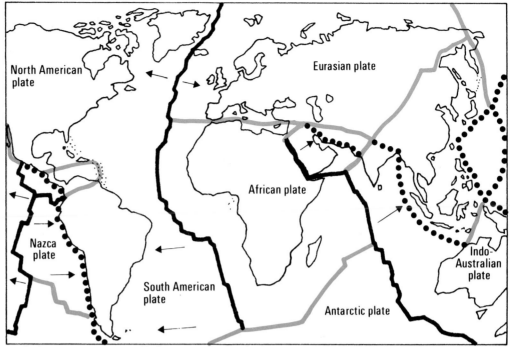

North American plate

Eurasian plate

African plate

Nazca plate

Indo-Australian plate

South American plate

Antarctic plate

Octahedral crystal of diamond in 'blue ground', a weathered form of kimberlite; South Africa.

and the continents collide; transform faults, where two plates simply glide past one another (Fig. 1·9).

Finally, from geophysical and geochemical studies, it is believed that the upper mantle is laterally heterogeneous. In the earliest geological times, before any basalt had been extracted from the mantle and before the crust was formed, the chemical composition of the undepleted upper mantle could be considered to be that of garnet peridotite. Now, however, the upper mantle must contain a mixture of depleted and undepleted peridotite.

Most magma (molten rock) formation occurs within the lower part of the crust and outer few hundred kilometres of the mantle, but there is no evidence of any world-encircling layer of magma beneath the solid crust. Basaltic magma is the most common molten rock to appear at the Earth's surface; it is the intrusive material at oceanic ridges forming new oceanic crust. At the ridges there is an upward movement of mantle material; the falling pressure combined with the concentration of volatiles, such as water, causes melting. The resultant magma has the composition of basalt.

Partial melting of rocks at even deeper levels in the upper mantle, especially beneath the continents, can give rise to ultramafic magmas which contain minerals showing evidence of a high-temperature, high-pressure origin. Good examples of this are the ultrabasic kimberlite pipes which carry diamond, a phase formed under high pressure.

Andesitic volcanism occurs along destructive margins, where oceanic crustal rocks and overlying sediments are carried down into the hot mantle. Here, andesitic and basaltic magmas are derived at depths of 50 to 70 kilometres (31 to 43 miles) and occasionally up to 150 kilometres (93 miles).

Granite represents continental crust which has been partially melted and slowly cooled. The viscous granitic magma moves from its source to the surface at relatively low temperatures often not reaching the surface but forming large intrusions within the upper continental crust. Smaller amounts of granite-like material may form by separation from basaltic magma. Granite magma often reaches the surface to form highly viscous rhyolitic lavas, and often causes violent volcanic eruptions.

The crust

The crust is the name given to the outer layer of the Earth. It is a relatively thin skin of rock composed of low density minerals that have been separating from lower levels throughout the 4 600 million years of the Earth's history. The underlying mantle is much denser than the crust and earthquake waves speed up sharply as they enter it. There are two distinct types of crust: the continental crust, which makes up the continental masses, and the oceanic crust, which forms the ocean floors. Compared with the oceanic crust, the continental crust is lower in average density but is much thicker, averaging 35 to 40 kilometres (22 to 25 miles) and beneath some of the high mountain ranges reaching thicknesses of 60 or 70 kilometres (37 to 43 miles). The oceanic crust is only on average 6 kilometres (3·7 miles) thick. The continental crust is generally older: parts of the continental crust (cratons) are 3 500 million years old and large areas are over 1 600 million years old, whereas the oceanic crust is nowhere older than 200 million years.

The surface of the Earth is, in the main, readily sampled and the rocks can be analysed so that much is known about the surface rocks and their compositions. The deepest boreholes only penetrate about 8 kilometres (5 miles) into continental crust, however, so that all the samples taken from mines and boreholes are only scratching the surface. At present, the oceanic crust is being drilled, in a phase of research associated with the IPOD programme (International Phase of Ocean Drilling) but, although deep penetration has been achieved at several of the sites, the boundary between the mantle and crust has not yet been reached and the mantle has not, therefore, been sampled directly.

Oceanic crust

Our knowledge of the structure and composition of the oceanic crust is based mainly on geophysical evidence and the studies of rock samples taken from

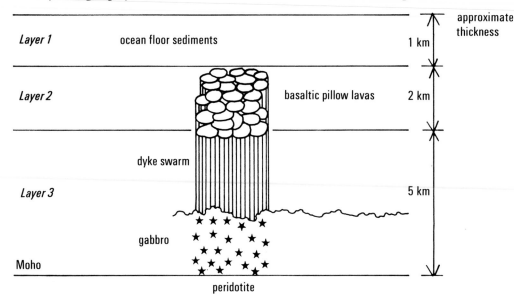

Fig. 1·10 Idealized section through the oceanic crust – a suggested model.

boreholes or specimens dredged from the ocean floor. Interpretation of seismic waves, gravity anomalies and cores have revealed a three-layered structure to the oceanic crust with an average thickness of about 6 kilometres (3·7 miles). The uppermost layer, *Layer 1*, consists of unconsolidated sediments up to 1 kilometre (0·6 mile) in thickness (Fig. 1·10). *Layer 2* is usually 1·5 to 2 kilometres (0·9 to 1·2 miles) thick and consists of consolidated sedimentary rocks, such as limestones and shales, interbedded with flows of basaltic lava. In deep oceanic environments, however, the sediment cover is very thin. Most of the lavas found in Layer 2 are pillow lavas indicating that they are erupted under water. *Layer 3* is about 5 kilometres (3 miles) thick and extends down to the Mohorovičić discontinuity. The composition of this layer is still not accurately known, but from our knowledge of the processes leading to the formation of oceanic crust, it is believed that the layer consists of an upper basic igneous dyke complex and a lower basic igneous plutonic complex, probably gabbroic in composition. Occasionally at the surface of the Earth, large, near-horizontal slabs of predominantly ultramafic (peridotite) rocks occur, perhaps the best example being rocks forming part of the Troodos Mountains of Cyprus. Here, slabs of peridotite are overlain by basic igneous rocks (gabbro) and a large dyke complex which is similar to the pattern described above for Layer 3. It is possible, therefore, that these are pieces of upper mantle and overlying oceanic crust which have been thrust to the surface by earth movements characteristic of some destructive plate margins.

Sea floor spreading is the name now applied to the mechanism which causes the lithospheric plates to move. Oceanic crust is continually being added to by the extrusion at oceanic ridges of fresh molten basaltic rocks generated in the underlying mantle. This new crust spreads away from the ridges at the rate of 1 to 10 centimetres (0·4 inch to 4 inches) per year. In every major ocean, oceanic crust is being drawn down into the mantle at destructive margins occurring where two of the Earth's lithospheric plates converge. One of the plates usually underthrusts the other so that oceanic crust plus some sediment is carried down into the mantle. Here the crustal material is remelted, forming magma which rises back to the surface. In this way, a chain of volcanic islands may be formed at the margin of the overriding plate such as the Aleutians and Japanese Islands or, perhaps, a chain of volcanoes capping high mountain ranges such as the Andes.

Continental crust

The thickness and internal structure of the continental crust are known from the study of earthquakes and the seismic waves they generate. In most areas, the

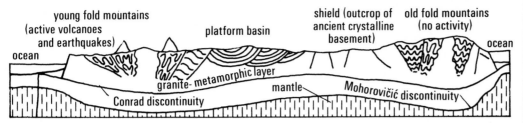

continental crust consists of a lower, more dense layer and an upper, less dense layer, the junction between the two layers, where it is developed, being called the Conrad discontinuity. The upper layer is probably granitic in overall composition, being composed of igneous and metamorphic rocks such as granite, schist and gneiss, and an overlying sequence of sedimentary rocks. The exposed granitic masses and metamorphic rocks of continental areas represent deep-seated igneous and metamorphic rocks which have been exposed by erosion. The composition of the lower layer of the continental crust is still in some doubt. Until recently, the lower continental crust was thought to be a fairly homogeneous layer. Because the seismic wave velocities are consistent with a basaltic composition, the ocean floor was believed to pass laterally below the upper continental crust. This is not now widely believed, however. Experimental studies indicate that under the pressure and temperature environment of the lower crust, rocks of basaltic composition would form either amphibolite or eclogite. Seismic studies indicate that the velocity of waves through the lower continental crust are such as would rule out eclogite. It is now widely held that a large part of this lower layer is, in fact, formed from a high-grade metamorphic rock termed granulite. Outcrops of these anhydrous rocks are found in deeply eroded metamorphic areas of all continents, and inclusions of granulite commonly occur in volcanic lavas. It is thought, therefore, that these represent parts of the lower continental crust brought to the surface (Fig. 1·11).

Structure of the continents

Every continent is a combination of ancient platform and shield areas of Pre-cambrian age, bordered or crossed by younger (less than 500 million years old) fold mountains. The structures of the rocks forming the shield areas are extremely complicated, consisting mainly of intensely folded, highly metamorphosed rocks. In some shield areas the association of granitic gneisses with metamorphosed basaltic lavas (greenstones) is characteristic. Further large areas of the continents are covered by layers of relatively undeformed sedimentary rocks, accumulated in shallow seas that covered different parts of the continental crust during past geological time. Large areas of the edges of the continental crust are still covered by shallow seas, and are known as the continental shelf.

In the relatively young fold mountain belts, the ancient crystalline basement rocks are reactivated and the overlying younger sediments extensively folded and faulted giving rise to such mountain ranges as the Alps. The deformation and dislocation of the continental crust occur along narrow 'mobile belts'. These consist of thick piles of sediments and volcanic rocks accumulated in relatively deep water. These accumulations are subsequently uplifted as mountain ranges. It is thought that this process can now be best explained by plate tectonics, the fold mountains being formed at destructive margins where slabs of oceanic and continental crust are forced together.

Abundance of the elements

In the universe, hydrogen and helium, the lightest elements, are by far the most

Fig. 1·11 Section across a typical continent showing the main structural units.

A view of part of the Ennstal Alps, Steiermark, Austria. The Alps are one of the major mountain ranges on the Earth, formed along a destructive plate margin in the Miocene geological period.

abundant. The abundance of the remaining elements decreases approximately with increasing atomic number, although there are a few irregularities such as the abundance of iron. This distribution of elements accords with what we would expect from the formation of the elements in the interior of stars; the lightest elements being the first formed. The lightest elements are not as abundant in the Earth, however, because the Earth's gravitational field is not strong enough to retain them. Segregation and concentration of the remaining elements on the Earth have occurred as a result of various geological processes; for example, the gravitational separation of minerals from magmas and the chemical differentiation of rocks by weathering at the surface. Within the surface layers of the Earth, these processes produce varying abundances of the chemical elements in different geological environments. Of all the known elements at least ninety-two occur naturally on Earth, and over seventy have been identified spectroscopically in the Sun. Of these ninety-two naturally occurring elements, eight constitute nearly 99 per cent by mass of the Earth's crust (*see* list of elements on page 113). In Table 1·1 we can see that these eight elements are oxygen, silicon, aluminium, iron, calcium, sodium, potassium and magnesium. Oxygen is obviously predominant and is volumetrically much more abundant than any other element. The Earth's crust consists almost entirely of oxygen compounds, most of them silicates of aluminium, calcium, magnesium, sodium, potassium and iron.

Table 1·1 The most abundant chemical elements in the Earth's crust

element	percentage by weight	percentage by volume
oxygen	46·60	93·77
silicon	27·72	0·86
aluminium	8·13	0·47
iron	5·00	0·43
calcium	3·63	1·03
sodium	2·83	1·32
potassium	2·59	1·83
magnesium	2·09	0·29

Trace elements

By chemically analysing rock samples from all parts of the world, it has been shown that many of the elements upon which man relies for his technology are present only in extremely small amounts; such elements are called trace elements, and their concentrations are often as low as a few parts per million. Table 1·2 shows the distribution of some trace elements in different igneous rocks.

For instance, it can be seen that chromium and nickel are concentrated in the ultrabasic rock, peridotite; the concentration of copper is nearly nine times higher in basalt than in ultrabasic or granitic rocks. On the other hand, cerium,

Table 1·2 Distribution of elements in some typical igneous rocks

major elements in terms of oxides (percentage by weight)	ultrabasic rock (peridotite)	basic rock (basalt)	acid rock (granite)
silica (SiO_2)	45·7	49·0	70·8
titania (TiO_2)	0·1	1·0	0·4
alumina (Al_2O_3)	2·7	18·2	14·6
ferric oxide (Fe_2O_3)	1·6	3·2	1·6
ferrous oxide (FeO)	5·7	6·0	1·8
manganese oxide (MnO)	0·1	0·2	0·06
magnesia (MgO)	41·5	7·6	0·9
lime (CaO)	2·0	11·2	2·0
soda (Na_2O)	0·2	2·6	3·5
potash (K_2O)	0·03	0·9	4·2
	——	——	——
	99·63	99·9	99·86
trace elements (in parts per million)			
nickel	2000	130	4
chromium	1600	170	4
zinc	50	105	39
copper	10	87	10
cerium	less than 1	48	92
tin	less than 1	1·5	3
lead	1	6	19
thorium	less than 1	4	17
uranium	less than 1	1	3
gold	0·009	0·008	0·003

element	percentage by weight	element	percentage by weight
oxygen	46·60	sulphur	0·05
silicon	27·72	chromium	0·02
aluminium	8·13	nickel	0·008
iron	5·00	zinc	0·0065
calcium	3·63	copper	0·0045
sodium	2·83	lead	0·0015
potassium	2·59	tin	0·0003
magnesium	2·09	silver	0·00001
titanium	0·44	platinum	0·0000005
manganese	0·10	gold	0·0000005

Table 1·3 Abundance of some of the chemical elements in the Earth's crust

lead and thorium are obviously concentrated in granitic rocks. It can also be seen from this table that gold, which appears to be at its highest concentration in ultrabasic rocks, has a much higher figure here than that for the overall Earth's crustal abundance. Table 1·3 shows the abundance of some of the chemical elements in the crust and it can be seen from this table just how low are the overall crustal concentrations of some of the economically important metals.

Obviously, for such elements to be extracted economically, situations must be found where they have already been concentrated by geological processes to many times the average figures shown in Table 1·3. For example, in the Witwatersrand gold field of South Africa, which is believed to be a very large, deep, and ancient sedimentary deposit, gold and uraninite grains are scattered between conglomeratic quartz pebbles; gold here is concentrated by a factor of some 1750 times the crustal average. Even in the low-grade copper ores which are now worked in the large porphyry copper deposits of the world, the concentration of copper is about 0·80 per cent of the metal; this means that here it is concentrated to about 200 times the amount found in average rocks. Therefore, we must realize that it is not necessarily the primary abundance of any metal in a particular rock type that is important to the economic geologist, but rather the means of concentrating it by geological processes.

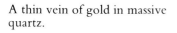

A thin vein of gold in massive quartz.

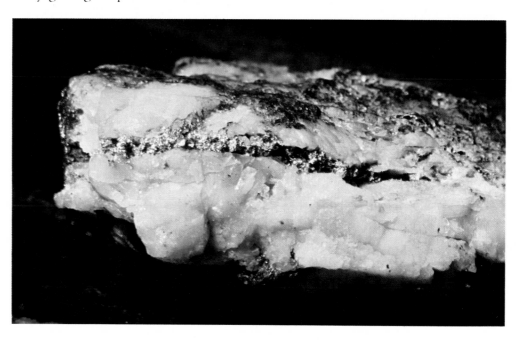

25

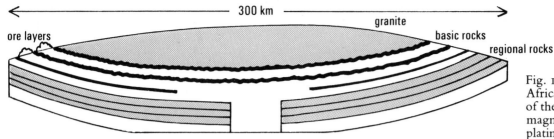

Fig. 1·12 Bushveld (South Africa) layered complex. One of the world's richest layered magnetite, chromite and platinum deposits.

Geochemical classification of the elements

The distribution of the elements within the Earth as a whole is inferred from their chemical properties and from the study of their relative distribution in minerals in bodies of extraterrestrial origin, such as meteorites, which contain both silicates and metallic phases. Elements were originally classified on the basis of their distribution between the phases of meteorites because of the analogies that meteorites provide with the Earth. This theory is based on the hypothesis that meteorites are fragments of disrupted planets, the different types of meteorite representing different parts of the parental bodies. We have already seen that we believe the irons were derived from the core of a planet; stony-irons could represent a part of the planet similar to the Earth's core/mantle boundary and the chondrites are believed to have a similar composition to that of Earth as a whole. It is now believed that the chondritic meteorites represent fragments of material from planets which resembled the early Earth, but which did not undergo complete differentiation into core, mantle and crust. Examination of the minerals present in the chondrites should, therefore, help in a study of the distribution of the elements within the Earth. The chondritic meteorites contain three major phases: magnesium-iron silicates; metallic nickel-iron; and troilite (FeS). If the Earth has a composition similar to that of the chondrites then there must be a

Table 1·4 Geochemical classification of some common metals	
(a) siderophile metal elements (commonly found in the native form and in the minerals below)	
element	**occurrence**
gold	calaverite ($AuTe_2$), sylvanite ($AuAgTe_4$)
platinum	sperrylite ($PtAs_2$)
palladium	arsenopalladinite (Pd_2As)
ruthenium	laurite (RuS_2)
(b) chalcophile	
copper	covelline (CuS), chalcosine (Cu_2S), chalcopyrite ($CuFeS_2$)
zinc	sphalerite (ZnS)
lead	galena (PbS)
antimony	stibnite (Sb_2S_3)
arsenic	arsenopyrite ($FeAsS$)
mercury	cinnabar (HgS)
(c) lithophile	
tin	cassiterite (SnO_2)
tungsten	wolframite ($FeWO_4$)
uranium	uraninite (UO_2)
beryllium	beryl [$Be_3Al_2(SiO_3)_6$]

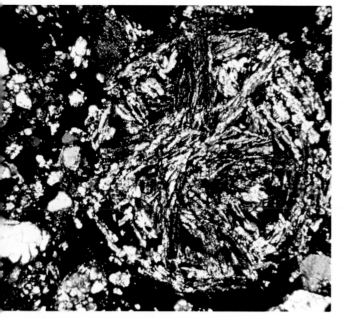

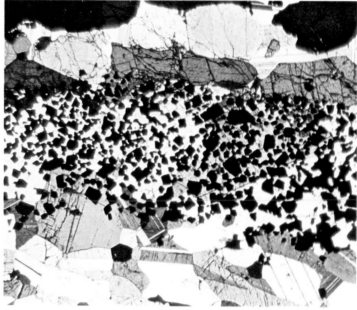

large quantity of sulphide present somewhere within the Earth, possibly in the core or lower mantle. The elements are classified according to their tendency to occur in one of these three different phases: siderophile elements have an affinity for metallic iron; chalcophile elements have an affinity for sulphide; lithophile elements have an affinity for silicate or oxide. Siderophile elements have little affinity for oxygen and sulphur, and they readily alloy with iron and are likely to be preferentially enriched in the Earth's core. Where they do occur at the Earth's surface, they may be in uncombined or native form. Gold is classified as a siderophile element, and it is likely that a proportion of the Earth's gold and other siderophile elements are, in fact, concentrated in the core and inaccessible to us. Chalcophile elements show a strong affinity with sulphur, and commonly occur as metallic sulphide ore deposits and as minor sulphide minerals in many rocks. It is possible that a sulphide layer may be present within the mantle or core, and these elements may also be highly concentrated at depth. The lithophile elements have a strong affinity with oxygen, are common in silicates, and constitute the rock-forming minerals dominant in the Earth's mantle and crust. Some chalcophile, siderophile and lithophile metals are listed in Table 1·4.

Availability of the elements

Characteristic minerals and their associated elemental concentrations can be found in all three major geological environments: igneous; sedimentary; and metamorphic.

In the igneous (magmatic) environment, mineral concentrations are formed by such processes as the accumulation of heavy minerals crystallizing early in a magma chamber; for example, chromite, magnetite and platinum; and in pegmatites by the crystallization of hot, relatively low viscosity melts to give characteristically the growth of large crystals. Typical minerals concentrated in this way are beryl, lithium micas and minerals carrying concentrations of the rare earth elements. There are also those important ore minerals formed by the high-temperature watery solutions of hydrothermal deposits; for example, metallic sulphides.

In the sedimentary environment, concentrations occur by, for example, the 'placer' deposits of gold and diamonds, the residual deposits of lateritic weathering such as bauxite, the major ore of aluminium, and minerals such as gypsum and halite formed by the evaporation of sea-water.

In the metamorphic environment, minerals are formed by the action of heat and pressure at some depth within the Earth; for example, graphite, kyanite and garnet.

To the economic mineralogist, it is the siderophile and chalcophile elements that are normally the major ore minerals and supply the needs of man's technology. Their abundance in the upper layers of the Earth must result from the geological processes outlined above. To the mining companies, however, a distinction. must be drawn between the abundance of an element and its availability. This availability depends largely on whether or not the element forms minerals in which it is a major constituent, and this depends in turn upon its atomic structure. Even some elements of very high abundance in the Earth's crust may not occur in easily worked ore minerals. Perhaps the best example is aluminium which, although the third most abundant element in the Earth's crust and a major constituent of virtually all igneous rocks and sedimentary clays and shales, nevertheless has only one ore, namely bauxite. Similarly, magnesium is extracted from sea-water, in which it is present in very low abundance, even though there are vast deposits of the mineral olivine (magnesium iron silicate) which contain up to 30 per cent magnesium, but from which it is uneconomic to obtain the element. Although they may be present in the crust in quite considerable amounts, some elements do not form minerals of their own but are dispersed throughout common minerals and never occur in any great concentration, generally forming less than 1 per cent in minerals of other elements. Examples of these are rubidium, which is dispersed in potassium minerals, and gallium which is dispersed in aluminium minerals. Other elements such as titanium and zirconium form specific minerals but these are then dispersed in small amounts in some very common rocks.

Mineral chemistry

A mineral has a characteristic chemical composition which can be expressed by a chemical formula that indicates the elements present and the proportion in which they are combined. The chemistry of an element is determined by its atomic structure which dictates whether an element readily forms mineral compounds such as oxides, silicates, phosphates, and so on.

The basic unit in all crystal structures is the atom which consists of electrons (negatively charged), protons (positively charged), and neutrons (electrically neutral). A popular model of the structure of the atom is due to the Danish physicist, Bohr, and involves the concept of the planetary atom in which electrons are visualized as circling the nucleus in orbits or energy levels at a distance from the nucleus dependent upon the energies of the electrons (Fig. 1·13). The nucleus, except in a hydrogen atom, is made up of protons and neutrons. Each proton carries a unit positive charge, the neutron as the name implies is electrically neutral, and each electron carries a unit negative charge so that the atom as a whole is electrically neutral. The mass of the atom is concentrated in the nucleus because the mass of an electron is only $\frac{1}{1850}$ that of the lightest nucleus. The simplest atom is that of hydrogen, in which the nucleus is a single proton with one electron moving around it. Atoms of the other naturally occurring elements have from two electrons (helium) to ninety-two electrons (uranium) moving in orbits about their nucleus. The fundamental difference between atoms of the different elements lies in the electrical charge of the nucleus. This positive charge is the same as the number of protons, and this number, equal to the number of electrons, is called the atomic number. The forces that bind together the various atoms composing crystalline mineral solids are largely electrical in nature, and their intensity determines the physical and chemical properties of the minerals.

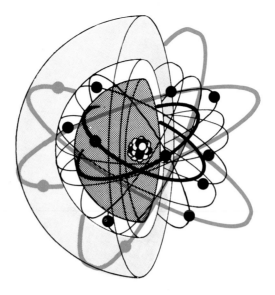

Fig. 1·13 Diagrammatic representation of the structure of the atom. The nucleus consists of positively charged protons (black) and uncharged neutrons (white). Round it circle negatively charged electrons. Their orbits are grouped into 'shells' shown by the spheres in the diagram.

We have now studied the evolution of the elements and the nature of their atomic configuration which controls their coordination to form the characteristic minerals of the Earth and the solar system. The combination of groups of minerals to form the rocks of the Earth throughout geological time and the present-day structure of the Earth have been described along with the continuing geological processes leading to the differentiation and concentration of the elements in the upper layers of the Earth. All these processes depend upon two major energy sources and it is important to remember that these control the pattern of the Earth's history.

The vast supply of energy that the Earth receives at the surface from the Sun is only a minute fraction of the Sun's output. This is the energy that provides almost the entire driving force for the geological cycles involving the interaction of the Earth's hydrosphere, atmosphere and land surface – that is, the processes of weathering, erosion, sediment transport and deposition. The energy for the internal components of the geological and geochemical cycles – the processes associated with magmatism, metamorphism and orogeny – is provided in part by the Earth's original heat plus the radiogenic heat produced from the breakdown of unstable radioactive isotopes, especially those of uranium, thorium and potassium concentrated in the crust (Fig. 1·14).

It is within these cycles that the elements are redistributed in the surface layers of the Earth and the characteristic minerals described in this book are formed.

Fig. 1·14 The rock cycle.

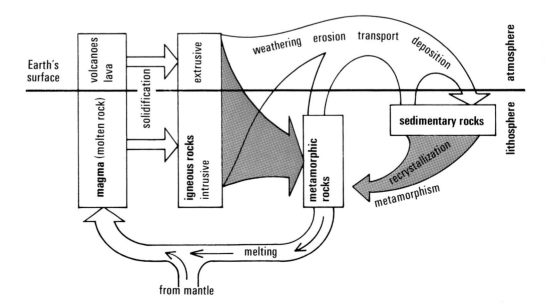

Dr Andrew Rankin

Minerals, rocks and their geological environment

Definition of a mineral

Although the word, mineral, can have several different current meanings, most practising mineralogists define a mineral as a naturally occurring, inorganically produced solid possessing a definite chemical composition, or limited range of compositions, and, usually, a systematic three-dimensional atomic order.

Solids possessing a regular, ordered atomic structure are said to be crystalline. Therefore, with very few exceptions, minerals are crystalline. Naturally occurring ice is a mineral, but in its more common molten state, water, it ceases to be a mineral. Amber, the naturally occurring fossil tree resin, much prized as a decorative ornament, does not qualify as a mineral for two main reasons: it is non-crystalline, and it is not inorganically formed. The term, *mineraloid*, is used to describe naturally occurring substances, such as water and amber, coal and petroleum (the so-called mineral fuels) which belong to the mineral kingdom but are non-crystalline and not necessarily formed by inorganic processes.

Synthetic minerals

With modern technology, it is possible to synthesize most minerals, including diamond, in the laboratory. Almost perfect crystals of quartz are produced in this way for use in electronic components, and all ruby crystals now employed in lasers are also artificially grown under laboratory conditions. Such crystals are not true minerals. They do not occur naturally and are, therefore, referred to as synthetic minerals (synthetic quartz and so on).

Biogenic minerals

Calcite ($CaCO_3$) is the most common carbonate mineral and is found in a variety of different geological environments. Calcium carbonate in the form of calcite is also the main constituent of the shells of several common shellfishes, such as oysters. In this instance, it does not qualify as a true mineral because it is biologically formed, not inorganically produced. It is, therefore, referred to as biogenic calcite to distinguish it from the true mineral calcite.

Ore mineral

Erosion of rocks by the action of rivers is beautifully demonstrated in the Grand Canyon in Arizona. Horizontal beds can be seen in the deeply eroded sedimentary rocks within the canyon.

A mineral utilized as a source of a particular metal or chemical substance is often referred to as an ore mineral. Profitability is an important factor and some authorities restrict the definition of an ore mineral to those minerals from which a metal can profitably be extracted.

Mineral names

All minerals are given names. These should always be used when referring to a particular mineral. Simply quoting the chemical formula or composition is not specific enough. This is because two or more different minerals can have identical chemical compositions but different atomic structures and physical properties. This phenomenon, known as *polymorphism,* is exemplified by the two native, naturally occurring forms (polymorphs) of carbon, graphite and diamond.

It is not possible to deal fully with the fascinating subject of the origin of mineral names. Some are named after places, anglesite (from Anglesey, an island off the north-west coast of Wales); some after a particular property displayed by the mineral, hematite (from the Greek word for blood, *haima*, which alludes to the blood-red colour of the powdered mineral); and others after people, scheelite (named after the Swedish chemist, K W Scheele who first discovered tungsten in the 1700s).

Amber.

Separate names are sometimes given to a single mineral species because of a particular physical property, form or habit displayed by that mineral. The mineral quartz is often referred to as amethyst when coloured purple, citrine when yellow, and morion when coloured dark brown or black. Likewise, transparent crystals of the mineral corundum when coloured blue are called sapphire, and when red, ruby. These names represent varieties of the two minerals quartz and corundum.

Many minerals have fixed chemical compositions. These can be expressed by simple chemical formulae; for example, quartz, SiO_2, galena, PbS, graphite, C. Others, such as olivine, $(Mg,Fe)_2SiO_4$, have a limited range of compositions. In this example, the relative proportions of magnesium and iron are variable, as depicted in the chemical formula where Mg and Fe are grouped together in brackets and separated by a comma. Ideally, olivine could have any composition from pure magnesium silicate to pure iron silicate. These end members of the olivine series are given the respective names, forsterite and fayalite. Olivines of intermediate composition, for example $(Mg_{0.87}Fe_{0.13})_2SiO_4$, can be considered as a homogeneous crystalline mixture, or *solid solution* of forsterite (Fo) and fayalite (Fa) and we can express the composition of any olivine in terms of the relative proportions of these two end members, that is $Fo_{87}Fa_{13}$.

Another notable example of a mineral series is found in the feldspar group of minerals. Plagioclase feldspars range in composition from pure albite ($NaAlSi_3O_8$) to anorthite ($CaAl_2Si_2O_8$) and the general formula of plagioclase can be written $(Na,Ca)(Al,Si)AlSi_2O_8$. Any intermediate composition can be expressed in terms of the relative proportions of albite and anorthite, that is, $Ab_{60}An_{40}$ represents $(Na_{0.6}Ca_{0.4})(Al_{0.4}Si_{0.6})AlSi_2O_8$. It should be noted that allowances are made in our definition for minerals such as olivine and plagioclase feldspar which exhibit a range of compositions. *See also* Properties and Study of Minerals.

Definition of rocks

Minerals occur everywhere in the Earth's crust. They are the principal components of the materials we call rocks which constitute the outer solid portions of the Earth. Rocks may simply be defined as naturally occurring multigranular aggregates of minerals (and/or mineraloids). Some rocks, such as quartzite (pure quartz) and marble (pure calcite), are monomineralic aggregates, but most, like granite, are composed of a variety of different minerals. Rocks need not necessarily be hard and resistant. Loose sand and wet clay are just as much rocks to the geologist as the compacted sands called sandstones and the hardened clays called shale.

Above
A rock eroded by sand-loaded wind.

Above right
Sandstorm in Monument Valley, Utah.

Not all mineral aggregates are rocks. Scale is important. Rocks must have a relatively widespread occurrence and must constitute a significant portion of the Earth's crust. An occasional chance occurrence of a group of minerals in small quantities does not constitute a rock. There is, however, no restriction on the size of individual minerals that make up a rock. They can range in size from less than 0 001 millimetre (0·00004 inch, clay minerals in shales) up to several metres (crystals of beryl in pegmatites).

There are three main groups of rocks, classified according to their origin: igneous rocks; sedimentary rocks; and metamorphic rocks. Their origin, composition and geological occurrence will be dealt with below.

Sedimentary rocks
Weathering

The Earth's surface is continually being sculptured and modified by two main agents of *erosion,* wind and water. The mechanical action of wind, rain, rivers, moving ice and waves have, over prolonged periods of geological time, worn away whole mountain chains and carved deep valleys and canyons in the land's surface. In geology, the words, erosion and weathering, are not synonymous.

Weathering is the process by which rocks and their constituent minerals are broken down in situ by mechanical (ice, water, wind) and/or chemical (from the action of percolating ground waters) means. The physical and chemical removal of weathered material is referred to as erosion.

Mechanical and chemical weathering are usually inseparable but, under different climatic conditions, one or other may predominate. The solution of vast amounts of limestone by carbon-dioxide-charged ground water percolating through cracks and crevices to produce extensive cave systems is an example where chemical weathering predominates.

All rocks, irrespective of origin, are affected by weathering. Different mineral phases within a rock, however, show varying degrees of resistance to weathering. For the common rock-forming minerals, the general order of decreasing resistance to chemical weathering is: quartz, feldspars, ferromagnesian minerals. The ferromagnesian minerals (amphiboles, pyroxenes, olivines and micas) are converted into clay minerals (notably illites and montmorillonites), carbonates, silica, and oxides of iron. Feldspars are converted to clay minerals (notably kaolinite) and silica. Quartz often remains unchanged.

The materials weathered from rocks are usually transported by water or wind away from the site of weathering to be deposited elsewhere as sediments. Under suitable conditions part, or all, of the weathered material remains in situ to form a soil or laterite. Soils are mixtures of the mineral products of weathering, such as sand and clay, and organic material derived from the decay of plant and animal matter. Laterites are essentially clays rich in aluminium and iron hydroxides with minor amounts of silica. They are often red, brown or yellow in colour due to the presence of iron hydroxides, and are formed under humid, tropical conditions. Laterites rich in aluminium hydroxides are called bauxites which are, at the present time, the only economically important source of aluminium.

Transportation and sedimentation

Sedimentation is the accumulation of sediments derived from the weathering of pre-existing rocks to a site of deposition. Rivers are by far the most important agents in the mechanical transport of sedimentary material, carrying it towards the open sea or inland lakes. During the course of its journey, the material is roughly sorted, that is, deposited according to size. Coarse-grained rock and mineral fragments form river bed gravels. Resistant minerals, mostly quartz, the next largest size fraction, form river sands. The finest-sized particles, notably clay minerals, are deposited as river muds.

The ability of wind to transport mechanically considerable amounts of material is evident from the recurrence of sand- and dust-storms in modern-day desert environments. The fine-grained fraction of desert sands can travel considerable distances before being fixed by moisture in vegetated regions outside the desert area as loess. Deposits such as loess and desert sands, formed by wind action, are referred to as *aeolian* deposits.

The movement or flow of icesheets and glaciers can also transport weathered rock material. This can range in size from huge rock boulders to exceedingly fine clay minerals, and is deposited as boulder clay or till when the glacier melts.

Deposition or sedimentation is the intermediate stage in the formation of sedimentary rocks. It follows weathering and precedes lithification. Not all sedimentary rocks, however, are formed by mechanical accumulation. Many owe their origin to the accumulation of the remains of plants and animals (organic matter) while others are formed as a direct result of chemical precipitation of material from the waters of lakes and seas.

Lithification and diagenesis

Unconsolidated, loosely compacted sediments are, under our definition, still rocks, but we will confine our discussion mainly to those sediments which have undergone conversion into a coherent, hardened aggregate. Their consolidation is known as lithification. Three main diagenetic processes are involved: compaction, cementation, and recrystallization. Diagenesis refers to all those near-surface physical and chemical changes that take place within a sediment after burial. It is a low-temperature phenomenon, but at higher temperatures it grades into *metamorphism*.

Above
Typical limestone cavern produced by solution of rock by carbon-dioxide-charged ground water. Stalactites (on roof) and stalagmites (on floor) are well developed; Bihor, Rumania.

Above right
Erosion of rocks by the action of ice. Ice 'smoothing' on rocks below the Briksdal glacier, Norway.

Considerable quantities of water are present in the pore spaces in unconsolidated sediments. Pressure caused by overlying younger sediments causes most of this water to be squeezed out. The pores close and the sediment is compacted. Cementation involves the precipitation of material such as silica and calcium carbonate from pore solutions within pore spaces and along grain boundaries. Recrystallization of minerals may take place during diagenesis. Sometimes, pre-existing minerals are replaced by new minerals. Dolomitization, the conversion or replacement of the calcium carbonate of limestone by dolomite, $(Ca, Mg)(CO_3)_2$, is an important example of diagenetic replacement.

Texture and structure of sedimentary rocks

Texture refers to the shape, size and arrangement of the grains within a rock. Size is an important factor in deposits formed by mechanical accumulation (clastic sediments) because it forms one of the bases for their classification. In chemically precipitated or organically formed deposits size is less important because of post-depositional (diagenetic) processes that commonly cause coarsening of the constituent particles. Size categories for clastic sediments are: *coarse*, particles greater than 2 millimetres (0·8 inch); *medium*, particles 2 to 0·0625 millimetres (0·08 to 0·0025 inch); *fine*, particles less than 0·0625 millimetre (0·0025 inch). Shape of particles and fragments is also an important factor in the classification of sedimentary rocks. Three basic shapes are recognized: angular, subangular, and rounded. In chemically formed sediments, small, pinhead-sized, rounded concretions of precipitated material sometimes occur. These are called ooliths, or pisoliths if their size is larger (about the size of a pea).

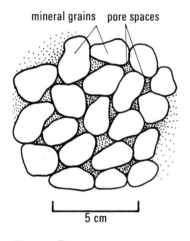

mineral grains   pore spaces

5 cm

Fig. 2·1 Pore spaces.

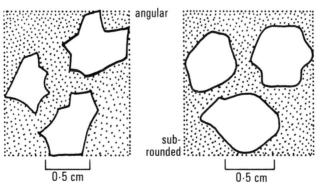

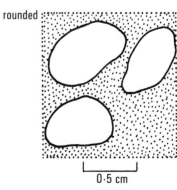

angular

rounded

sub-rounded

0·5 cm   0·5 cm   0·5 cm

Fig. 2·2 Grain shapes.

35

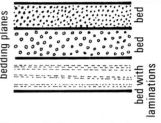

Current bedded conglomerates overlying current-bedded grits and silts; Suguta Trough area, Kenya.

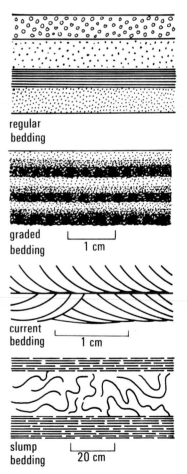

Fig. 2·3 Sedimentary beds, bedding planes and laminations.

Structure refers to the larger scale features displayed by the rock. Most sedimentary rocks show bedding or layering, expressed by variations in texture and mineralogy. A bed represents a single episode of sedimentation. It is bordered top and bottom by a bedding plane. These mark pauses in time between each successive episode of sedimentation. A single bed may show laminations. These are similar to bedding, but are smaller in scale. They represent minor variations in the nature of material deposited during one single episode of deposition. Four main forms of bedding may be recognized (Fig. 2·4).

Regular bedding The beds are separated by parallel bedding planes. Regular bedding is not confined to horizontal beds. Folded and tilted beds caused by post-depositional earth movements may also show regular bedding.

Graded bedding The grain size varies from coarse at the bottom to fine at the top. This type of bedding results from rapid sedimentation of material in water. Coarser grains settle more rapidly than the finer-grained material.

Current bedding The beds show a distinctive cross-lamination. These laminations are inclined and stacked up in the form of a wedge in the direction of the prevailing wind in aeolian deposits or in the direction of water currents in deposits formed on the sea bed and in river mouths. The top surface is commonly truncated by local erosion and overlain by the next current-bedded deposit.

Slump bedding This is a distorted, folded structure produced by mass movement, slumping and sliding of wet, unconsolidated sediments, usually down a slope on the sea-floor.

The surface of bedding planes may show minor structures such as mud cracks and ripple marks. These were formed in much the same way as modern-day mud cracks in the sun-baked sediments of dry lakes and rivers, and modern-day ripple marks found on shallow sea-floors and beaches at low tide.

Classes of sedimentary rocks

Two main groups of sedimentary rocks may be distinguished: the clastic or detrital rocks – those formed by the mechanical accumulation of material; and the chemical-organic rocks – those formed by accumulation of chemically precipitated or biologically derived material. The detrital sediments are subdivided according to grain size and the chemical-organic sediments according to composition (Table 2·1).

Fig. 2·4 Types of bedding.

A glacial moraine in north Scotland. The pebbles and boulders are composed largely of granite, schist and quartz left behind by the glacier.

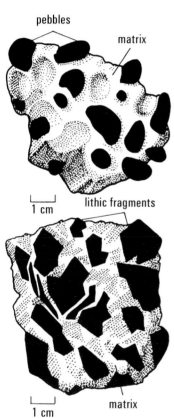

Fig. 2·5 Conglomerate (*above*) and breccia (*below*).

Fig. 2·6 Striated pebble.

Naturally, most detrital sediments contain chemically precipitated matter or material of organic origin, and most chemical-organic sediments carry some detrital material. We shall now consider the more important rock types within each of these two main groups.

Coarse-grained detrital rocks

Conglomerates are consolidated gravel, pebble or boulder beds, and consist of rounded pebbles, cobbles and boulders set in a fine-grained matrix. The matrix usually consists of sand or silt and is commonly cemented by calcite or quartz. In contrast, breccias contain angular rock fragments similarly set in a fine-grained, sandy or silty matrix. The angularity of rock fragments in breccias suggests that their material could not have travelled very far from its source (Fig. 2·5).

Boulder clay or till is unconsolidated material deposited by glaciers. It contains rounded to angular rock fragments, varying widely in size from large boulders to sand, set in a fine-grained clayey matrix. The pebbles are often scratched or striated due to abrasion during ice transport (Fig. 2·6). Consolidated till is called tillite.

Table 2·1 Classification of sedimentary rocks

clastic rocks	
coarse-grained	gravel, conglomerate, breccia
medium-grained★	sandstone, arkose, greywacke
fine-grained★	clay, siltstone, shale
chemical-organic rocks	
carbonate★ (calcareous)	limestones, chalk, dolomite
siliceous	flint, chert
ferruginous	ironstone
aluminous	bauxite, laterite
phosphatic	phosphate rock
saline	rock salt, gypsum, anhydrite
carbonaceous	coal

★most common sedimentary rocks

37

The nature of the pebbles, blocks and angular fragments in all three of these deposits is variable, and depends largely on the composition of the rocks in the source area.

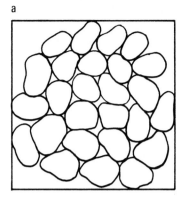

Medium-grained detrital rocks

The three most important rocks in this subgroup are sandstone, arkose and greywacke. Sandstone is composed almost entirely of well-sorted, subangular to rounded quartz grains. Lesser amounts of feldspar and mica and other minerals, such as olivine, rutile, zircon and magnetite, may also occur. The quartz grains are often cemented by silica (most commonly), calcite or iron oxides and hydroxides; the latter giving rise to the reddish-brown to yellow colour of many sandstones. The green colour of certain types of sandstones (greensands) is caused by the presence of a mineral called glauconite. Sandstones are formed mainly either in shallow waters or in desert environments. Desert sandstones typically contain almost spherical wind-polished grains. Medium-grained, well-sorted detrital rocks containing sharply angular grains are called gritstones.

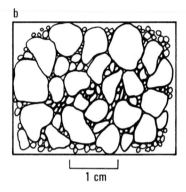

Greywackes consist of poorly sorted, angular fragments of minerals (mainly feldspar and quartz) set in a very much finer-grained matrix, commonly composed of chlorite. This mineral sometimes imparts a greenish colour to the rock, but normally the rock is grey in colour, hence the name greywacke (from the German, *grauwacke*). These sediments are commonly formed by rapid deposition in deep sea waters.

1 cm

Fig. 2·7 Sorting of grains.

Arkoses are somewhat coarser grained than greywackes and sandstones. They are composed essentially of moderately well-sorted angular grains of quartz and feldspar, with lesser amounts of mica. Mineralogically, they resemble granite and are, in fact, commonly formed by the rapid accumulation in rivers and alluvial fans, of material derived from this rock.

Fine-grained detrital rocks

Siltstones (compacted silt) contain grains varying in size between 0·0625 and 0·004 millimetre (0·0025 and 0·00015 inch). They are formed by accumulation of fine-grained sediment on the bottom of seas, rivers and lakes. The minerals are too small to be identified without the use of a microscope.

Clay, shale and mudstone are much finer grained, with grain sizes less than 0·004 millimetre (0·00015 inch). They are formed by the compaction of mud. Mudstone is structureless and shale is laminated. These rocks consist predominantly of clay minerals and vary in colour from white to blackish brown. The black colour in some shales is due to organic matter and the red colour to oxides and hydroxides of iron. Two important clay-like deposits, loess and bauxite have already been mentioned.

Chemical-organic sediments

These sediments are subdivided according to composition. The carbonate sediments are most common and include limestone and chalk. They are composed essentially of calcium carbonate provided by: the accumulation of the calcareous skeletons of organisms; and/or the direct chemical precipitation of this material from water (usually sea water).

Limestones usually contain at least some detrital sandy or clayey debris. As these components increase, limestones grade into calcareous shales, sandstones and mudstones. Formation of limestone generally takes place in warm, shallow seas, though freshwater limestones also occur. Those formed predominantly from the remains of certain organisms are named accordingly: shelly limestone, coral limestone, and so on. Chalk is a special type of pure-white limestone formed by the accumulation of tiny skeletons of certain types of micro-organisms.

Oolitic limestones are essentially chemical precipitates composed of closely packed calcite ooliths set in a matrix of calcite. Ooliths are tiny, spherical bodies usually about 1 millimetre (0·04 inch) across and look similar to fish roe.

Dolomite is commonly found together with limestone and its diagenetic origin has been discussed previously.

Flint and chert are composed of cryptocrystalline silica and are usually found as beds and nodules in limestones and chalk.

Phosphatic sedimentary rocks are rich in phosphorus and comprise an important source of this element for the fertilizer industry. The phosphate is derived from the teeth, bones and excrementa of vertebrates, though much is probably chemically precipitated.

Ferruginous and carbonaceous sediments are exemplified by *ironstone* and *coal* respectively. Ironstones are variable in origin and mineralogy but are characterized by a high proportion of iron-bearing minerals such as siderite, hematite, chamosite, pyrite and magnetite. Fossils and detrital material are commonly present and, with decreasing iron content, ironstones grade into mudstones, sandstones, limestones and other sedimentary rock types. Coals are bedded rocks formed from the accumulation and decay of vegetable matter in swampy environments rather similar to modern-day equatorial forests. After burial, heat and pressure turns the vegetable matter into hardened, compacted and reconstituted coal seams.

Saline rocks are formed by the evaporation of brines in lakes and lagoons under warm, arid climatic conditions. Marine evaporites, formed from sea-water brines, are the most common, and the principal minerals are the chlorides and sulphates of sodium, potassium, magnesium and calcium.

The mineralogy of non-marine evaporites, that is, those found in inland lakes, depends on the composition of the brines from which they are derived and the environment of deposition. Borate minerals, such as kernite and borax, dominate in some, whilst others may contain nitrates, such as nitratine, alkali carbonates and sulphates (trona, natron). Halite or rock salt is the most common evaporate mineral in both marine and non-marine deposits.

Igneous rocks

Magmas

Approximately 80 per cent of the Earth's crust is composed of rocks formed by crystallization from hot (usually in the range of 700 °C to 1 200 °C) molten masses of rock called magmas. Magmas are essentially silicate melts containing small amounts of dissolved water and other volatile components such as chlorine, boron, fluorine and sulphur; and the rocks they produce on cooling are called igneous rocks (derived from the Latin word for fire, *ignis*). These rocks are composed mainly of varying proportions of silica-bearing minerals, principally quartz, feldspars, amphiboles, pyroxenes and micas.

Geologists believe that most magmas originate at considerable depths below the surface of the Earth and migrate upwards along planes of weakness and partings within the Earth's crust. Magmas that reach the Earth's surface and extrude over it are called lavas and the rocks formed by consolidation of these lava are called extrusive or volcanic rocks. Magmas that intrude the Earth's crust, but fail to reach the surface, crystallize at depth to form intrusive igneous rocks.

Volcanic activity

Considerable quantities of lava have issued forth on to different parts of the Earth's surface periodically over the whole of geological time and, even today, lava is continuing to issue from active volcanoes in certain areas of the world. Volcanic activity is not restricted to the continents. Large portions of the oceanic crust, composed of a dark grey to black, fine-grained igneous rock called basalt, are formed from lavas that have extruded on to the sea-floor during periods of submarine volcanic activity.

Volcanologists have recognized two different types of volcanic activity: *fissure type* and *central eruptive type*. In fissure eruptions, lavas, fed by a magma chamber at depth, issue relatively quietly from linear cracks or fissures. These lavas can cover a very large area before solidifying, and considerable thicknesses of rock (mainly basalt) can be built up from repeated eruptions. These vast accumulations of solidified lava are referred to as flood or plateau basalts and are exemplified by the Deccan Traps of India which cover an area of about 500 000 square kilometres

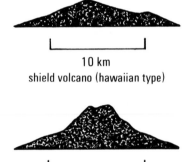

10 km
shield volcano (hawaiian type)

1 km
stratovolcano

Fig. 2·8 Central eruptive volcanoes.

Volcanism in action. The Halemaumau fire pit in Kilauea crater, Hawaii, erupting in 1961.

Steam vent in congealed ropy basaltic lava, Hawaii.

bedded ash and agglomerate

agglomerate

vent

1 km

fragments torn from vent

Fig. 2·9 Pyroclastic rocks.

(193 000 square miles). Ocean floor basalts are also the products of this type of volcanism. Central eruptive type volcanism produces volcanoes built around a central vent. Two main groups of volcanoes are recognized: shield volcanoes and stratovolcanoes (Fig. 2·8).

Shield volcanoes are so called because their shape in cross-section is reminiscent of a shield. They result from the periodic extrusion of highly fluid and mobile lavas of basic composition, that is, mainly basalts (low in silica SiO_2). These can flow for some considerable distance before solidifying. The Hawaiian volcanoes such as Mauna Loa, Kilauea and Mauna Kea are classic examples of shield volcanoes.

The lavas produced by stratovolcanoes, such as Vesuvius in south-west Italy and Mount Fuji in Japan, are generally richer in silica (acidic) and hence more viscous and less mobile than those produced by shield volcanoes. The lavas cannot travel such great distances and, in consequence, stratovolcanoes are much less extensive and their cones dip more steeply than shield volcanoes.

Often the volatile components of the viscous lavas of stratovolcanoes cannot escape freely. Pressure, caused by these volatile components, builds up and the volcano erupts explosively. Fragments of lava, sometimes together with material ripped from the throat of the volcano, are hurled into the air (Fig. 2·9). These

41

quickly solidify in contact with the cold air and fall to the ground as ash and cinders. Rocks formed by consolidation of material explosively ejected from volcanoes are called *pyroclastic rocks* (*pŷr*, Greek for fire, *klastós*, Greek for broken). Some geologists classify these under the heading of sedimentary rocks because they form *beds* and *layers* and show many textural and structural features typical of this class of rocks. Because of their undoubted igneous origin, however, they are considered here as a particular type of igneous rock.

In lavas where gases are released during consolidation, small bubbles may be frozen into the rock as vesicles. This is most marked in viscous acidic lavas where vesicles may be so abundant that the rock takes the form of a light spongy mass which is given the name pumice.

At some stage after lava has solidified, the vesicles may be filled by a variety of secondary minerals, especially calcite and zeolites. The infilled vesicles are referred to as amygdales and the lavas containing them, amygdaloidal lavas (Fig. 2·10).

Volatiles released from lavas during eruption may escape through gas vents or fumaroles at the surface of the volcano. These volatiles often emerge at temperatures in excess of 500 °C and as a consequence of their rapid cooling on contact with cold air or rock, deposit solid material (volcanic sublimates) in and around the vent. Large deposits of native sulphur may form in this way. Other common sublimates include brightly coloured (red to yellow) arsenic sulphides, realgar and orpiment, and the chlorides of calcium, magnesium, iron and lead.

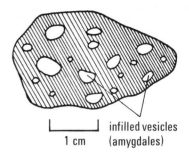

Fig. 2·10 Fragment of amygdaloidal lava.

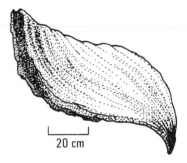

Fig. 2·11 Volcanic bomb.

Pyroclastic rocks

It is convenient at this point to consider the characteristics of pyroclastic rocks before dealing with those typical of the vast majority of igneous rocks which crystallized at depth from a molten magma.

Volcanic fragments (ejecta) expelled during explosive volcanic activity range in size from fine ash [fragments less than 2 millimetres (0·08 inch) in diameter] through medium-sized lapilli [fragments between 2 and 64 millimetres (0·08 to 2·5 inches) in diameter] to blocks [angular fragments with dimensions greater than 64 millimetres (2·5 inches)]. If the blocks are rounded or ellipsoidal they are called bombs. These are ejected clots of partly congealed lava which have been streamlined and moulded into their characteristic shape during flight (Fig. 2·11).

Pyroclastic rocks formed by consolidation of ash fragments are called tuffs (lapilli tuffs if they contain a high proportion of lapilli) and agglomerates or volcanic breccias if they contain a large number of bombs and blocks. Tuffs are further subdivided into lithic tuffs if the fine material is composed predominantly of rock fragments, vitric tuffs if they contain abundant glassy fragments, and crystal tuffs if broken crystals are commonly present.

Igneous intrusions

Consolidation of magma below surface gives rise to bodies of intrusive igneous rocks showing considerable variations in their size and shape. Minor intrusions such as dykes, sills and plugs (Fig. 2·12) are small in size and generally occupy planes of weakness in the enclosing crustal rocks.

Dykes are often vertical or steeply inclined, sheet-like bodies which cut across the pre-existing rock strata. Sills are also sheet-like bodies, but are often flat-lying and develop along partings or bedding planes in the pre-existing strata. Plugs are small, cylindrically shaped, near-vertical bodies.

When intrusive bodies attain dimensions of several hundred cubic kilometres or more, they are referred to as major intrusions. Their outcrop at the surface, exposed because of erosion of overlying rocks, is usually roughly circular or oval in shape, and in cross-section they usually possess steep, outwardly dipping

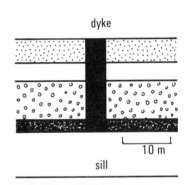

dyke

sill

plug (roughly circular in plan)

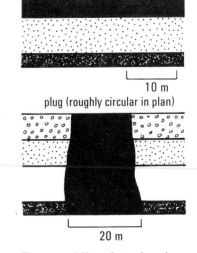

Fig. 2·12 Minor intrusions in cross-section.

contacts. These are called bosses or stocks if their diameter is between about 10 and 20 kilometres (6 and 12 miles). Larger intrusions, 100 kilometres (60 miles) or more in diameter, are called batholiths. Granite, by far the most common rock of major intrusions, commonly forms huge batholiths occupying several thousands of square kilometres of the Earth's surface.

Batholiths, stocks and plutons, like dykes, are described as *discordant* bodies because the magma from which they have formed has cut through overlying strata. *Concordant* intrusive bodies, in contrast, lie between beds of strata. Sills fall into this category as, on a larger scale (major intrusions), do lopoliths and laccoliths (Fig. 2·13).

Grain size, texture and structure of igneous rocks

The size of the constituent minerals (grain size) in an igneous rock depends to a large extent on the rate of cooling of the magma. Extrusive magmas (lavas) and minor intrusions cool rapidly on contact with cold air or surrounding rocks. As a result, the minerals quickly precipitate as a mass of tiny crystals, usually less than 0·1 millimetre (0·004 inch) in size. Sometimes cooling can be so rapid in some lavas that the magma fails to crystallize but quenches to form a rock almost entirely composed of glass which is called obsidian or tachylyte.

In contrast, the cooling of a large, deep-seated intrusive body of magma usually takes place over a protracted period of time (except at the margins which are chilled rapidly on contact with the colder surrounding country rocks). Minerals crystallize much more slowly and, therefore, form much larger crystals.

Igneous rocks, excluding the glassy types, are said to be fine grained if the constituent minerals are less than 1 millimetre across (0·04 inch), medium grained if the minerals are between 1 millimetre and 5 millimetres (0·04–0·2 inch), and coarse grained if the minerals are greater than 5 millimetres and easily distinguished by the naked eye. If the constituent minerals exceed several centimetres in size, the rock is said to be pegmatitic. Pegmatites are generally formed from the volatile-rich fractions of large bodies of intrusive magmas and often contain a variety of large and unusual minerals such as beryl, rare earth minerals, lepidolite, and so on. The most common pegmatites are those associated with granites, the granite pegmatites.

Black obsidian (volcanic glass) showing characteristic conchoidal fracture; California.

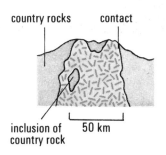

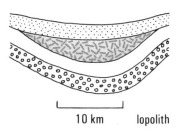

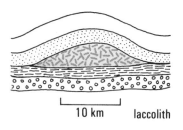

Fig. 2·13 Major intrusions in cross-section.

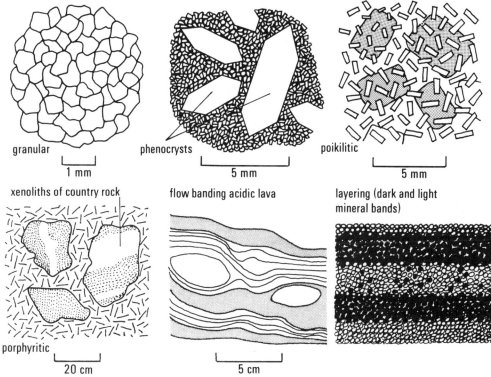

Fig. 2·14 Textures in igneous rocks.

Rocks containing roughly equidimensional minerals are said to have a granular texture. Porphyritic texture refers to rocks containing large, early formed crystals (phenocrysts) set in a finer ground matrix, and poikilitic texture to rocks containing large grains of one crystal enclosing smaller grains of other minerals.

Intrusive and extrusive magmas can incorporate fragments of the country rocks through which they pass. Sometimes these are assimilated or digested by the magma, but on other occasions the magma crystallizes and traps these fragments as xenoliths (or xenocrysts if extraneous single crystals are involved).

On a larger scale, igneous rocks, especially the coarse-grained basic intrusive rocks such as gabbro, may show a distinct layering or banding of different minerals (layered igneous intrusions) caused by differential precipitation and accumulation of minerals from the melt. Flow texture refers to the alignment of early formed tabular crystals caused by fluid flow within the magma.

Pumice showing typical vesiculated appearance; California.

Chemical and mineralogical composition

Igneous rocks exhibit a rather limited range of chemical compositions. About 99 per cent of the total bulk of nearly all igneous rocks is made up of eight major elements: oxygen, silicon, aluminium, iron, magnesium, calcium, sodium, and potassium. The other 1 per cent is made up by a very much larger number of trace elements which include phosphorus, titanium, fluorine, and hydrogen.

It is usual to express the chemical composition of igneous rocks (and, indeed, sedimentary and metamorphic rocks) in terms of the percentage oxides. Silica (SiO_2) is the most abundant component in the vast majority of igneous rocks, varying from about 40 per cent to 75 per cent. The silica percentage is important because it is used as a basis in classification schemes for igneous rocks. Acid rocks contain more than 66 per cent silica, intermediate rocks between 52 and 66 per cent silica, basic rocks between 45 and 52 per cent and ultrabasic rocks less than 45 per cent.

Fig. 2·15 Classification of igneous rocks.

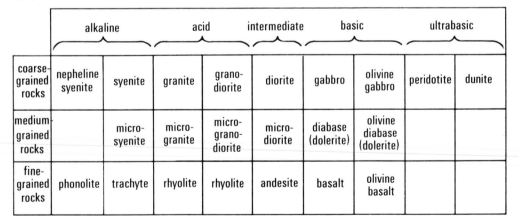

	alkaline		acid		intermediate	basic		ultrabasic	
coarse-grained rocks	nepheline syenite	syenite	granite	grano-diorite	diorite	gabbro	olivine gabbro	peridotite	dunite
medium-grained rocks		micro-syenite	micro-granite	micro-grano-diorite	micro-diorite	diabase (dolerite)	olivine diabase (dolerite)		
fine-grained rocks	phonolite	trachyte	rhyolite	rhyolite	andesite	basalt	olivine basalt		

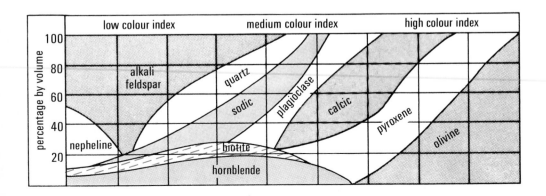

44

Most igneous rocks are composed of varying proportions of a relatively small number of essential minerals (all silica bearing). These are conveniently subdivided into: the light-coloured minerals – quartz, feldspars, feldspathoids (chemically related to the feldspars) and muscovite; and the dark-coloured, mafic or ferromagnesian minerals – pyroxenes, amphiboles, olivine and biotite. The volumetric proportion of dark-coloured minerals within a rock of igneous origin is called its colour index. The lower the index, the more light-coloured minerals it contains. In addition to the main rock-forming minerals, small amounts of other non-essential or accessory minerals such as apatite, zircon, spinel group minerals, sphene, rutile and topaz, may also occur.

It is these essential and accessory minerals that contain the major and trace elements listed previously. When the silica content is low, dark-coloured, mafic minerals are abundant, when it is high, light-coloured minerals predominate.

Classification of igneous rocks

The classification and naming of the main igneous rock types (excluding pyroclastic rocks) are based on the grain size, composition and relative proportions of constituent minerals. Important mineralogical criteria are:

the presence or absence of quartz; quartz is an essential mineral in acid rocks, only an accessory mineral in intermediate and basic rocks, and is usually absent in ultrabasic rocks;

the types of feldspars/feldspathoids present; the feldspars are a complex group of sodium, potassium and calcium alumino-silicates showing considerable variations in composition. There are two groups: (a) the potassium or potash feldspars (orthoclase, microcline, sanidine); (b) the plagioclase feldspars which range in composition from pure calcium feldspar (anorthite) through feldspars containing significant amounts of both calcium and sodium (oligoclase, andesine, labradorite) to pure sodium feldspar (albite). Potash feldspars and plagioclases rich in sodium are collectively referred to as alkali feldspars.

Alkali feldspars are essential minerals in acid rocks, but absent, or present only in minor amounts, in intermediate, basic and ultrabasic rocks. Calcium-rich plagioclases are typically present in basic rocks and andesine in intermediate rocks. Feldspars are normally absent from ultrabasic rocks. The feldspathoids, a group of minerals chemically similar to feldspars, are found in significant amounts only in a special group of igneous rocks, the alkaline rocks. These rocks also contain commonly alkali feldspars;

the types and relative proportions of ferromagnesian minerals; olivine as an essential mineral is restricted to basic and ultrabasic rocks. Pyroxene and amphiboles (particularly hornblende) are common as accessory minerals in acid, alkaline and intermediate rocks, but are more important components of basic and ultrabasic rocks. Biotite, in contrast, a common accessory in most igneous rock types, is more common in acid rocks.

A simplified classification of the more important igneous rock types is shown in Fig. 2·15. The mineral content refers to the most common major mineral phases typically present in each rock type. The coarse-grained rocks, usually referred to as plutonic rocks, are those which occur in deep-seated major intrusions. The fine- and medium-grained varieties represent the volcanic rocks and minor intrusions. Basalt, for example, commonly occurs as dykes and sills as well as in lava flows.

Only certain members of the alkaline rock family, intermediate in composition, have been included in the classification. Alkali-rich basic and acid rocks are in most cases prefixed by the term alkali (such as alkali-granites, alkali-gabbros, and so on). Alkali-rich peridotites (those rich in mica) are called mica-peridotites.

Lamprophyres are medium-grained alkali-rich rocks of basic to intermediate composition.

It must be emphasized that the division between the different types of igneous rocks is somewhat arbitrary and generalized. There are several fairly rare igneous rocks of unusual mineralogical composition that do not fit neatly into the classification scheme. Two notable examples are carbonatite (an igneous rock composed essentially of calcite or dolomite but often containing rare and unusual mineral species), and kimberlite (a complex, brecciated rock, containing a high percentage of ferromagnesian minerals, which is the only primary source of diamond).

The coarse-grained, ultrabasic rocks, dunite, peridotite and pyroxenite (composed essentially of pyroxene) have no fine-grained extrusive equivalents. It is thought that some of these rock types, which often occur as bands in layered basic intrusions, represent accumulations of early formed crystals in a magma chamber. Others are thought to have originated in the mantle and were emplaced into the crust as a crystal mush. These rocks are sometimes altered on cooling to serpentinite, a secondary rock composed essentially of the mineral, serpentine. Serpentinite could be considered to be a special type of metamorphic rock but is commonly grouped together with rocks of igneous origin.

Magma origin and differentiation

Granites are the most common plutonic rocks and basalt the most common extrusive rock. The origin of magmas of granitic composition and magmas of basaltic composition, and the relationship between them are, therefore, of profound importance in understanding how magmas originate. It is generally considered that granites can be produced in two ways: either from continental crustal rocks, or from primary magmas of basaltic composition.

The continental crust is much thicker than the oceanic crust. At considerable

Granite, showing large phenocrysts of pink orthoclase feldspar set in a coarse matrix of quartz (colourless to grey), feldspar (milky white to pink) and biotite (black); Shap, Cumbria.

depths, where temperatures are high enough to cause melting of continental crustal rocks, magmas of granitic composition can probably form. These rise to higher, cooler levels in the crust where they solidify to form batholiths and other igneous bodies.

Basaltic magmas are believed to originate in the upper mantle or lowest parts of the crust. If these magmas remain at depth for a significant period of time, they cool slowly and early mineral phases begin to crystallize from the melt. The composition of these early formed crystals is not the same as the composition of the remaining melt, however. This melt will be substantially different in composition from the original basaltic magma (in general, as crystallization proceeds, the residual melt becomes more acidic in composition). If this melt is drawn off in some way from the early formed crystals (crystal differentiation) it may consolidate as a separate body of igneous rock chemically dissimilar from the original melt. A large variety of different igneous rocks can be produced by the differentiation process. Granite is one of these.

Metamorphic rocks
Metamorphism

Metamorphism is the process by which the mineral assemblages, structures and textures of pre-existing crustal rocks (parent rocks) are modified by the effects of heat and pressure within the Earth's crust. These changes generally involve recrystallization and formation of new mineral phases (*metamorphic minerals*) many of which are only found in metamorphic rocks. During metamorphism the rocks remain essentially solid, and recrystallization takes place in the solid state by interactions with co-existing pore fluids. Surface weathering and changes involving partial melting of deep crustal rocks are excluded from consideration under the heading of metamorphism.

Pre-existing rocks are metamorphosed because their mineral assemblages (and textures) are usually unstable at the temperatures and pressures prevailing within the Earth's crust. Consequently, the minerals recrystallize to form a new assemblage of minerals stable under these new conditions.

Chemical changes usually accompany metamorphism, but when rock compositions, although not necessarily mineralogy, remain unchanged, the metamorphism is said to be isochemical. Metasomatism refers to the process whereby rocks are altered primarily by the addition or subtraction of material. This is affected by movement of aqueous fluids through rocks, usually under moderately high temperatures and pressures.

The metamorphic process involving the mechanical breakdown (grinding, crushing and deformation) of a rock, due to directed pressures or stresses caused by Earth movements (folding and faulting) where temperatures are moderately low is known as *cataclaysis*. Rocks formed predominantly by this process are called *cataclastic rocks*. Rocks metamorphosed principally as a result of high pressure with little increase in temperature are called dynamically metamorphosed rocks, and slates are perhaps the best example.

The majority of metamorphic rocks are produced by a combination of recrystallization (isochemical or metasomatic) and mechanical breakdown. Temperature is, perhaps, the most important factor controlling the types of chemical and physical changes that take place during metamorphism. In general, the higher the temperature, the greater the change, but above about 700 – 800 °C the rocks start to melt and igneous processes take over.

Heat for metamorphism is provided by intruding magmas, as in the case of contact metamorphic rocks, by friction, and as a consequence of the fact that temperature increases with depth in the Earth. In regions where the geothermal gradient is high, even those rocks buried at moderately shallow depths are subject to quite high temperatures (and pressures). These are accordingly metamorphosed on a large scale to form regional metamorphic rocks.

Contact and regional metamorphic rocks are the two main classes of metamorphic rocks. These will be considered in detail later, but first it is necessary to describe the common structures and textures of the metamorphic rocks.

Metamorphic textures

The textures, that is, the form and relationship of the mineral grains, of metamorphic rocks depend to some extent on the conditions of metamorphism, but they are also influenced by the mineralogy of the parent rock. The textures described below may be recognized.

Granular The minerals are roughly equidimensional.

Hornfelsic The minerals are again equidimensional with no preferred orientation. Rocks with a hornfelsic texture (hornfels) are fine grained, hard and compact. This texture is found only in contact metamorphic rocks.

Slaty Rocks showing a slaty texture (slates) are very fine grained and have a distinct parallel parting or cleavage caused by strong parallel orientation of flaky minerals (mainly micas and chlorite). Slaty rocks cleave readily into thin sheets. Most slates are derived from laminated, fine-grained sediments, but the cleavage direction is generally quite different from the direction of these laminations. Cleavage of this type (fracture cleavage) is a metamorphic phenomenon caused by regional stresses developed during metamorphism. It overprints the primary sedimentary structure of the parent rock.

Phyllitic Rocks displaying a phyllitic texture (phyllites) cleave or part in a similar way to slates, but the grain size of the minerals is somewhat coarser, though rarely distinguishable to the naked eye. The fracture cleavage surfaces of phyllites have a characteristic lustrous or sparkly, silver or green sheen caused by alignment of flaky minerals – mainly chlorite and mica.

Schistose Rocks with schistose textures (schists) are characteristically coarse grained. The minerals are not equidimensional, usually platy or elongate, and show a preferred parallel orientation, layering or parting (schistosity). Schistosity is again caused by regional stresses.

Gneissose Coarse-grained rocks, such as gneisses, showing a marked but irregular layering or parting are said to have a gneissose texture. The constituent minerals are large enough to be recognized with the naked eye.

As mentioned briefly above, metamorphic rocks can display two types of structures. The primary or relict structure inherited from the parent rock (for example, bedding in sediments and amygdales in lavas) and the overprinted metamorphic structure. The most notable large-scale metamorphic structures are various types of cleavage and folding. On a smaller scale, microstructures and orientation of grains may be observed with the aid of a microscope or hand-lens.

Contact metamorphic rocks

Contact metamorphism is a direct consequence of the intrusion of a hot igneous body. In contrast to regional metamorphic rocks, contact metamorphic rocks are formed in the absence of pronounced differential stresses, and temperature is the controlling factor. Heat from the intrusion 'bakes' the surrounding rocks to produce an altered rim or contact metamorphic aureole of metamorphic rock.

Temperature decreases away from the intrusion, so that the outer rocks in the aureole are less intensely metamorphosed and, therefore, contain a mineral assemblage different from that of the innermost rocks. Concentric zones of

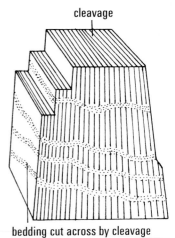

cleavage

bedding cut across by cleavage

Fig. 2·16 Slate showing relationships between bedding and cleavage.

preferred orientation of platy minerals (schistose)

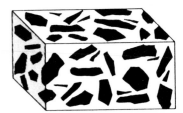

no preferred orientation of platy minerals

Fig. 2·17 Orientation of minerals in metamorphic rocks.

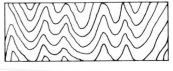

Fig. 2·18 Some different styles of folding.

Above
Schist with garnet
porphyroblasts; Vermont.

Right
Water-worn outcrop of mica
schist showing typical layering,
Lugard's Falls, Kenya.

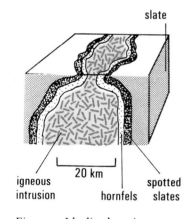

Fig. 2·19 Idealized section
across a contact metamorphic
aureole.

metamorphic rocks containing similar mineral assemblages, formed under similar temperature conditions are commonly developed in the aureoles. A typical sequence, derived from fine-grained sediments, is shown in Fig. 2·19. The spots in spotted slates are fairly small, up to about 5 millimetres (0·2 inch) across, and may be composed either of a single mineral or a fine-grained mass of different minerals.

In the hornfels, porphyroblasts (large, well-shaped crystals set in a fine-grained matrix) and poikiloblasts (similar to porphyroblasts but containing numerous small inclusions of other minerals) may be present. These may reach 4 or 5 centimetres (1·5 to 2 inches) in size. Typical porphyroblastic and poikiloblastic minerals in contact metamorphic hornfelses are cordierite, andalusite and silli-manite. Kyanite, the other polymorph of aluminium silicate, though common in high-grade regional metamorphic rocks, is never observed in contact metamor-phic rocks. Pyroxene may also develop as porphyroblasts in hornfels, but only in the highest temperature (innermost) zones of the aureole.

We have only considered contact metamorphic rocks derived from fine-grained sediments. Basic igneous rocks such as basalt and diabase give rise to hornfels generally containing pyroxene and plagioclase feldspar. In medium- to coarse-grained sedimentary rocks such as greywackes, and impure sandstones, aluminium-bearing minerals, such as feldspar, are changed to micas and garnets. Hornblende, epidote and diopside develop when carbonate minerals such as calcite are present as impurities in the parent rocks. Contact metamorphosed pure sandstones recrystallize to form a harder, more compact monominerallic (quartz) metamorphic rock called *quartzite*.

Pure limestones may also recrystallize during contact metamorphism to a compact monominerallic (calcite) rock called marble. Impure limestones containing sand, chert or clayey material form calc-silicate rocks. These are again composed essentially of recrystallized calcite, but contain additional silicate minerals, partly derived from the original silicates. These minerals, which include garnets, olivine, serpentine, wollastonite, tremolite, diopside (and, if the limestone originally contained significant quantities of magnesium, non-silicate magnesium-bearing minerals) often form large crystals clustered in bands and nodules.

Skarns are contact metasomatized limestones found in the aureoles of certain types of intrusive bodies, notably granite. Silicon, magnesium, iron and other components migrate outward from the cooling magma and react with the limestone to form rocks similar in mineralogy to calc-silicate rocks, but with the additional presence of garnet, scapolite and iron-rich pyroxene. Metalliferous ore deposits containing sulphides of lead, zinc, copper and iron are commonly found in association with skarns.

Regional metamorphic rocks

These rocks are by far the most common varieties of metamorphic rocks. They are formed in parts of the Earth's crust where temperatures are moderate to high and stresses are developed on a regional scale. Their formation appears, in most cases, to have accompanied mountain-building processes, and belts of regional metamorphic rocks, exposed at surface by erosion, can occupy several thousands of square kilometres.

Regional metamorphic rocks are very varied in texture and mineralogy. This is partly because of the wide variety of parent rock types from which they can form and partly because of the variable temperatures, pressures and stress conditions prevailing during metamorphism. Schists, gneisses, phyllites, slates and granular rocks (amphibolites and granulites) are the most common regional rock types, derived from parent rocks containing a range of different minerals. Pure sandstones and limestones, however, are regionally metamorphosed to quartzite and marble respectively.

Metamorphic grade refers to the degree or intensity of metamorphism that has affected a rock. Parent rocks of the same composition form different mineral assemblages depending on the grade of metamorphism. High-grade rocks are intensely recrystallized and are generally produced at high temperatures. In low-grade metamorphic rocks, recrystallization is less marked and relict structures are commonly preserved. Low-grade metamorphism takes place at low temperatures. At intermediate temperatures, metamorphism can be said to be medium grade. *See* Table 2·2.

Regional metamorphic rocks, regardless of parentage, formed under the same general conditions of metamorphism (temperature, pressure and stress), are said to belong to the same metamorphic facies. This forms the basis for a more useful classification, usually adopted by petrologists because it takes into account the influence of pressure and stress conditions during metamorphism. It is not possible to give a comprehensive account of the different types of metamorphic facies within the scope of this book. It is only possible to illustrate the effect of increasing grade of metamorphism with reference to fine-grained detrital sediments (mudstones and shales) and basic igneous rocks (basalts and diabase).

Under low-grade metamorphism, mudstones and shales form phyllites and slates. The original clay minerals are reconstituted to form platy minerals, mainly chlorite and muscovite. Other minerals recrystallize to form quartz and iron oxides. At higher temperatures (medium-grade metamorphism), pronounced recrystallization produces a coarser-grained, banded schist. These commonly contain alternating mica-rich and mica-poor layers. Porphyroblasts of minerals, such as garnet and staurolite, grow. Under high-grade metamorphic

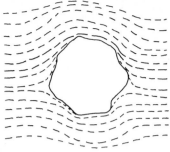

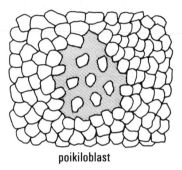

porphyroblast distorting ground mass

poikiloblast

Fig. 2·20

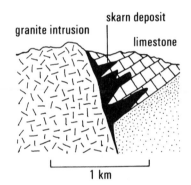

granite intrusion

skarn deposit

limestone

1 km

Fig. 2·21 Skarn deposits.

Miners splitting slate along its cleavage planes; Brazil.

conditions, when temperatures are even higher, gneisses are formed. At these elevated temperatures, the rocks become plastic. Recrystallization is more pronounced, large crystals may form and the banding becomes coarser and less well developed than in the lower-grade equivalents (slates, phyllites and schists). Gneisses, derived from muds and shales, may contain quartz, mica, garnet, kyanite or sillimanite, and feldspar. At temperatures in excess of about 700 °C partial melting occurs and the metamorphic process passes over into magmatism.

Low-grade regional metamorphism of basic igneous rocks such as basalt, dolerite and diabase, involving relatively low temperatures but quite high stresses (strong deformation) produces greenschists. These typically contain chlorite, albite and epidote and sometimes minor amounts of calcite and actinolite. As the name implies, greenschists commonly develop a schistose texture; the alignment of chlorite flakes sometimes imparts a distinctive greenish sheen to the rock.

At medium- to high-grades of metamorphism, greenschists pass into amphibolites. New minerals, such as garnet and hornblende or diopside, develop as medium- to coarse-grained crystals or even porphyroblasts. Schistosity, or a distinctive alignment/banding of dark-coloured minerals, may be present. At the highest metamorphic grades, granulites are formed. These usually have a coarse- to medium-grained, granular texture and are characterized by the presence of plagioclase, hypersthene and diopside. Charnockites are the most widespread granulite-type rocks formed from basic igneous rocks. Hypersthene is the characteristic mineral present.

Eclogites are coarse-grained rocks composed essentially of pyrope garnet and a particular type of pyroxene (omphacite). Though often classified as metamorphic

Table 2·2 Some metamorphic minerals in some regional metamorphic rocks

parent rock	low grade	medium grade	high grade
shale/mudstone	chlorite	quartz	quartz
	muscovite	muscovite	orthoclase
	albite	biotite	plagioclase
	quartz	garnet	garnet
		feldspar	sillimanite or
		kyanite	kyanite
		staurolite	
		sillimanite	
		cordierite	
		andalusite	
calcareous mudstone	calcite	calcite	plagioclase
	epidote	epidote	calcite
	tremolite	tremolite	diopside
	dolomite	diopside	quartz
	quartz	idocrase	
		forsterite	
		grossular	
		quartz	
basic rocks	albite	albite	plagioclase
	epidote	epidote	diopside
	chlorite	hornblende	hypersthene
	calcite	plagioclase	garnet
	actinolite	garnet	olivine

rocks, their origin is rather obscure. Many geologists believe that they originate from the base of the Earth's crust or mantle and are brought to the surface by magmatic or tectonic processes.

It should be emphasized that the mineralogy of the rocks described above has been somewhat generalized. Most of these rocks may also contain other minerals, and those typically present at a particular grade, for example, biotite and quartz, might also exist in higher-grade rocks. Table 2·2 lists a selection of the more important minerals that may be found in metamorphic rocks of a particular parentage and grade. Not all these minerals are likely to occur in a single rock at the same time. Minor and accessory minerals are not included in the list and it is reemphasized that minerals developed at one grade may appear in others.

The most notable metamorphic minerals present are often referred to in the name of specific regional metamorphic rock types, garnet-mica-schist, hornblende-gneiss, pyroxene-granulite, and so on. A similar procedure is adopted when naming certain specific contact metamorphic rock types, for example, andalusite-cordierite-hornfels, pyroxene-hornfels. Marble, quartzite, amphibolite, eclogite and charnockite are not usually prefixed by mineral names because their major mineralogy is more or less constant.

Hydrothermal vein deposits

We have seen that minerals occur predominantly in igneous, sedimentary and metamorphic rocks. A wide variety of different minerals (including a large number of ore minerals), however, rarely encountered in these rock types, commonly occurs in hydrothermal vein deposits. These deposits, which are too small in size and too variable in composition and texture to be classified as rocks, are formed by the precipitation of material from hot (about 50 °C to 500 °C) aqueous solutions in cracks and fissures in the surrounding country rocks. Most veins are steeply dipping or vertical, but they can also occur as flat-lying concordant bodies. Their width, though highly variable, seldom exceeds a few metres (those smaller than about 1 centimetre (0·12 inch) are called veinlets) and their lateral and vertical extent is usually no more than a kilometre or so (Fig. 2·22).

Hydrothermal veins often show a distinctive crustiform texture. Material is deposited on the walls of the fissure, then covered over by successive layers of minerals formed from hydrothermal fluids episodically introduced into the vein. Cavities or vugs, often lined with beautifully formed crystals, may be present. Many hydrothermal vein deposits are associated with igneous intrusions of granitic composition. The veins often contain economic concentrations of ore minerals of a number of different metals, which include tin (as cassiterite and stannite), tungsten (as wolframite and scheelite) and copper (as chalcopyrite and other copper sulphides and oxides), admixed with unwanted, non-ore minerals (gangue minerals), notably quartz. These deposits are formed by precipitation of material from the hot, residual, mineral-bearing fluids released by cooling magma during consolidation.

Other hydrothermal vein deposits in sedimentary or metamorphic rocks do not appear to be associated with any igneous rock types. Extensive vein and replacement deposits commonly occur in limestones. These commonly contain ore minerals of lead and zinc (galena and sphalerite) and gangue minerals which include fluorite, calcite, barytes and witherite. Non-magmatic hydrothermal fluids, such as ground water and pore fluids heated at depth, appear to be responsible for most of these lead-zinc deposits in limestone. Hydrothermal fluids can also originate during metamorphism. Such fluids are believed to be responsible for some of the gold/quartz veins which occur in ancient metamorphosed volcanic rocks (greenstones).

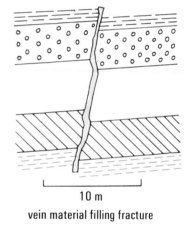

10 m

vein material filling fracture

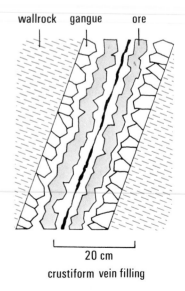

wallrock   gangue   ore

20 cm

crustiform vein filling

Fig. 2·22 Hydrothermal veins

Table 2·3 Some common mineral-rock associations

(a) Igneous rock associations

These may be found as accessory minerals, as segregation bodies, as cross-cutting veins, or mineralized joint surfaces.

(i) **acid rocks**: granites, granite pegmatites, rhyolites, etc.

elements	antimony, bismuth
sulphides, etc	molybdenite, pyrite, pyrrhotine, arsenopyrite
oxides, etc	magnetite, cassiterite, quartz, wolframite, scheelite, uraninite, columbite-tantalite
halides	fluorite, cryolite
phosphates	apatite, amblygonite, monazite
silicates	feldspars, biotite, muscovite, tourmaline, spodumene, lepido-lite, zircon, allanite, epidote, topaz, scapolite, hornblende

(ii) **intermediate rocks**: syenites, nepheline syenites, tonalites, diorites, etc

elements	silver, gold
sulphides, etc	cobaltite, niccolite, pyrite, chalcopyrite, molybdenite, galena, sphalerite
oxides, etc	opal, magnetite, ilmenite, corundum, hübnerite
phosphates	turquoise
silicates	feldspars, apophyllite, prehnite, datolite, analcime, zircon, sodalite, nepheline, cancrinite, amphiboles

(iii) **basic rocks**: basalts, dolerites, gabbros, etc.

elements	copper, silver, iron
sulphides, etc	pyrite, chalcopyrite, pyrrhotine, pentlandite
oxides, etc	magnetite, ilmenite, hematite
phosphates	apatite
silicates	apophyllite, prehnite, datolite, olivine, augite, labradorite, pectolite, pyroxenes

(iv) **ultrabasic rocks**: peridotites, serpentinites, anorthosites, etc

elements	platinum, diamond
sulphides, etc	pyrite, pyrrhotine, niccolite
oxides, etc	magnetite, ilmenite, chromite, brucite
carbonates	aragonite, magnesite
silicates	labradorite, pyrope garnet, olivine, talc, pyroxenes

(v) **solfataric deposits**: that is, associated with steam jets and fumaroles in volcanic terrain

elements	sulphur
sulphides, etc	pyrite, cinnabar, stibnite, covelline
oxides, etc	hematite, magnetite, chalcedony, opal
halides	halite, sylvite, salammoniac
sulphates	gypsum, alunite
silicates	zeolites

(b) Hydrothermal deposits

Minerals which occur usually in cross-cutting bodies such as veins, in igneous, sedimentary and metamorphic rocks.

(i) hypothermal vein mineralization (300–500 °C)

elements	gold, silver
sulphides, etc	arsenopyrite, chalcopyrite, bornite, chalcocite, enargite, sphalerite, galena, pyrite, pyrrhotine
oxides, etc	cassiterite, quartz, magnetite, hematite, uraninite, rutile, anatase, brookite
halides	fluorite
phosphates	fluorapatite
carbonates	calcite, dolomite, rhodochrosite
sulphates	baryte
silicates	rhodonite, epidote, allanite, tourmaline, topaz, datolite, axinite

(ii) mesothermal vein mineralization (200–300 °C)

elements	gold, silver
sulphides, etc	galena, sphalerite, pyrite, chalcopyrite, chalcocite, bornite, pyrargyrite, proustite, acanthite
oxides, etc	hematite, uraninite, goethite, quartz
halides	fluorite
phosphates	apatite
carbonates	calcite, siderite, dolomite, magnesite, rhodochrosite, witherite
sulphates	baryte
silicates	chlorites, epidote, clay minerals

(iii) epithermal vein mineralization (50–200 °C)

elements	gold, silver, arsenic, bismuth, antimony
sulphides, etc	chalcopyrite, covelline, sphalerite, galena, tetrahedrite, pyrite, marcasite, stibnite, cinnabar, etc
oxides, etc	hematite, goethite, chalcedony, opal
halides	fluorite
phosphates	apatite
carbonates	calcite
sulphates	baryte
silicates	clay minerals, zeolites

(c) Sedimentary associations

(i) Resistates – minerals resisting chemical and mechanical breakdown

elements	gold, platinum, diamond
oxides, etc	magnetite, ilmenite, quartz, corundum, columbite-tantalite, cassiterite, rutile, spinel, chromite
phosphates	monazite
silicates	feldspars, garnet, tourmaline, staurolite, zircon, thorite, topaz, kyanite

(ii) hydrolysates – minerals which have been produced by the decay of silicates

oxides	bauxite, quartz (chalcedony), opal
silicates	clay minerals, glauconite, chamosite

(iii) **oxidates – minerals produced by the oxidation of pre-existing minerals**

oxides   goethite, limonite, pyrolusite, psilomelane, hematite

(iv) **reduzates – minerals produced by reduction mechanisms, often in the presence of bacteria**

elements   sulphur
sulphides   pyrite, marcasite
carbonates siderite

(v) **precipitates – minerals produced by the chemical precipitation from saturated solutions**

phosphates apatite
carbonates calcite, aragonite, dolomite

(vi) **evaporites – minerals produced by the evaporation of a marine or lacustrine saline basin**

halides   halite, sylvite, carnallite, etc
borates   kernite, borax, colemanite
nitrates   nitratine
carbonates calcite, aragonite, dolomite, trona
sulphates   gypsum, anhydrite, celestine

(d) Metamorphic associations

Regional metamorphism embraces large areas of the Earth's crust (*see* above). Contact or thermal metamorphism is restricted to areas adjacent to igneous intrusions. Restricted zones within the areas may be skarns (*see* page 50) and comprise ore bodies.

(i) **high-grade metamorphism**

elements   graphite
sulphides   pyrite, chalcopyrite, pyrrhotine
oxides   magnetite, ilmenite, corundum, spinel, rutile, quartz
carbonates calcite
silicates   feldspars, sillimanite, kyanite, forsterite, staurolite, cordierite, garnets, wollastonite, rhodonite, scapolite, tourmaline, sphene

(ii) **medium-grade metamorphism**

elements   graphite
sulphides   pyrite, chalcopyrite
oxides   magnetite, hematite, ilmenite, spinel, quartz
carbonates calcite, dolomite
silicates   feldspars, garnets, kyanite, andalusite, anthophyllite, cordierite, epidote, scapolite, idocrase, hornblende

(iii) **low-grade metamorphism**

sulphides   pyrite, chalcopyrite
oxides   magnetite, hematite, brucite, anatase, brookite, rutile, quartz
carbonates calcite, dolomite, magnesite, siderite
silicates   albite, talc, chlorite, biotite, tremolite-actinolite, epidote, prehnite, spessartine

Dr Andrew Clark

Crystals

Growth of crystals

Large, well-formed, natural crystals are something of a rarity, since the conditions provided by nature are often far from the ideal required for them to grow to a good size. As a consequence, the localities at which fine crystals occur become well known and such specimens are eagerly sought after by collectors.

The word crystal derives from a Greek word meaning 'clear ice' since it was supposed by the ancients that the quartz crystals familiar to them had formed by the intense freezing of water. Crystals may form in three main ways: solidification of molten material; crystallization from cooling aqueous solution; and deposition from vapour. More than one of these processes may have been active in producing crystals in a particular geological setting.

Crystals commence growing when atoms coalesce or 'nucleate' forming a small 'seed', which develops through the addition of further atoms to the outside of this seed. In general, a near perfect crystal will only result if it is able to grow freely without meeting any obstruction, and if the chemical environment around the growing crystal remains stable.

The first of the processes listed above operates in the formation of igneous rocks. This group of rocks results from the solidification of molten magma, the largest crystals generally occurring in plutonic rocks which cool slowly far below the Earth's surface, thereby illustrating the important fact that crystals tend to grow to large sizes when cooling is slow. Usually, the first crystals to have formed in an igneous rock are those which show the most regular crystal outlines, minerals formed later having to fit into the more limited space available between the existing crystals. Observers are often puzzled by some minerals which occur as large fine crystals, apparently formed inside a solid rock, but again it must be emphasized that in such situations the large crystals generally formed before the surrounding rock.

The finest crystals found are usually those which crystallized from solution. A number of minerals, such as halite and sylvine, dissolve readily in water and form large deposits by evaporation. Many others are less soluble but, nevertheless, find their way dissolved in mineralizing fluids, often at high temperatures and pressures, into fissures and cavities in rocks where they often crystallize in very spectacular forms.

The third of these crystallization processes, that of mineral deposition from a vapour, can be illustrated by reference to one of the most important minerals formed in this way, sulphur. Large crystals are sometimes found at the openings of volcanic fumaroles where escaping gases, encountering cooler rocks, deposit solid particles.

Cubic galena crystals, from Kansas, associated with pyrite.

Crystal geometry

The most obvious feature of a crystal is that it is a solid bounded by flat faces. Only in the present century has the technique of X-ray diffraction proved that the geometry of the disposition of a crystal's faces is directly related to its internal atomic arrangement. However, for several hundred years mineralogists had suspected something of this nature and as long ago as 1669 a Dane, Nicolaus Steno, found that the angles between corresponding faces of quartz crystals were always the same, however misshapen the crystals appeared to be. This is now known as the *law of constancy of interfacial angles*, the interfacial angle being defined as the angle between the perpendiculars to a pair of faces. The law holds for crystals of the same mineral provided that their compositions are identical (Fig. 3·1).

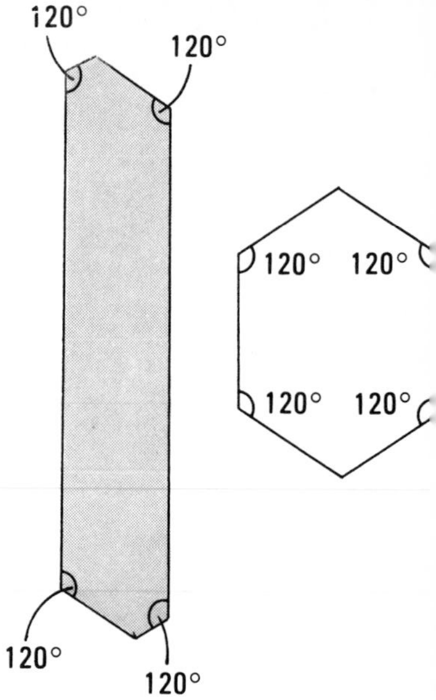

Fig. 3·1 Constant interfacial angles.

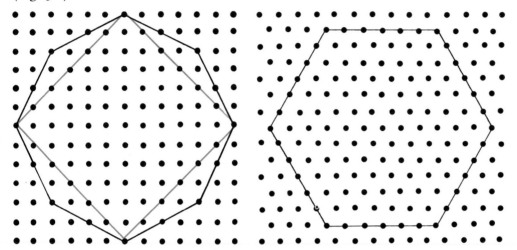

Fig. 3·2 Cubic and hexagonal arrays of lattice points.

Crystal faces are parallel to planes of atoms in the crystal structure and the relationship between a crystal's faces can therefore reveal much about the underlying structure of the mineral. For example, Fig. 3·2 shows how atomic planes in a cubic type of lattice can be translated into crystals showing four- and eight-sided outlines, and a hexagonal lattice into a crystal with hexagonal cross-section. It follows that the measurement of angles between the principal faces of a crystal contributes greatly towards assigning a mineral to its correct symmetry class, and can also be used to calculate other fundamental crystallographic parameters.

Interfacial angles can be measured very accurately on instruments known as goniometers. For small crystals a reflecting goniometer is required, which enables a crystal to be rotated at the centre of a graduated circular scale and the angle

Fig. 3·3 The contact goniometer.

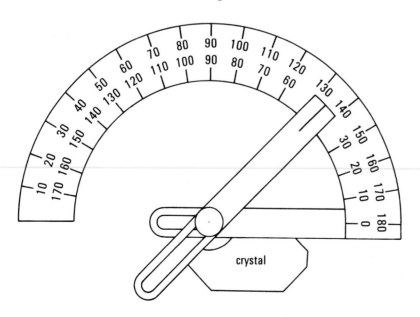

crystal

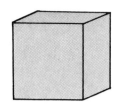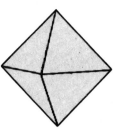

Fig. 3·4 Combinations of cubic and octahedral forms.

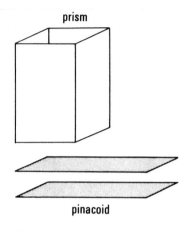

prism

pinacoid

Fig. 3·5 Open crystal forms.

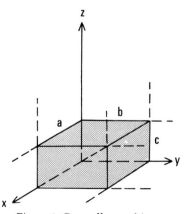

Fig. 3·6 Crystallographic axes and unit cell edges.

between two faces found by aligning each respectively with a telescopic sight. Great skill is required to obtain a precise orientation of the crystal, generally restricting the technique to well-equipped laboratories. However, for larger crystals a simple contact goniometer can be used very effectively. This consists of an arm pivoted at the centre of a graduated semicircle (Fig. 3·3). The two edges shown in the diagram can be brought into contact with pairs of crystal faces and the angles read directly from the scale.

In different crystals of the same mineral the law of constancy of interfacial angle can be applied to corresponding faces. The overall form of the crystals may be extremely variable, however, since different crystals of the same mineral may have developed under slightly different conditions, causing different faces to become prominent during crystal development. This is expressed by saying that many minerals crystallize with more than one *habit*. Often slight differences in the chemical composition of the liquid from which a mineral crystallizes can cause this to happen.

A crystal whose faces are all identical is called a *closed form*. The cube and the octahedron are both closed forms in the cubic system, the cube being composed of square faces and the octahedron of equilateral triangles. Usually minerals crystallize in combinations of different forms and Fig. 3·4 gives a simple example of combinations of cubic and octahedral forms. The different colours in the diagram serve to illustrate the different forms present in each crystal. Unfortunately, no mineral provides the information in this convenient way. The main problem in morphological crystallography is to recognize the various combinations of forms present in natural crystals. Identical crystal faces which do not totally enclose space are known as *open forms* and they can only be present in crystals in combination with other forms (Fig. 3·5).

Crystal systems and symmetry

Every crystal consists of a three-dimensional array of atoms or atomic groups arranged in a regular pattern which repeats itself over and over again throughout the crystal. The smallest complete unit of this pattern is known as the *unit cell* and the whole crystal can be visualized as comprising a very large number of these unit cells stacked together. The *crystallographic axes* are the directions defined by the unit cell edges as shown in Fig. 3·6. Lengths along these axes can be related to the lengths of the cell edges. A crystallographic axis must not be confused with a zone axis; the latter is the direction through the centre of the crystal taken by a series of prominent parallel edges and is not necessarily parallel to a crystallographic axis.

All crystals fall into one of six crystal systems depending on the geometrical relationship between the crystallographic axes. Some texts list seven systems, separating the trigonal (or rhombohedral) group from the hexagonal system. The difference between these groups will become apparent in the later section on symmetry. The main systems are illustrated in Fig. 3·7 and may be described as follows.

Cubic or isometric system This system includes all those minerals which can be referred to three equal axes at right angles to one another.

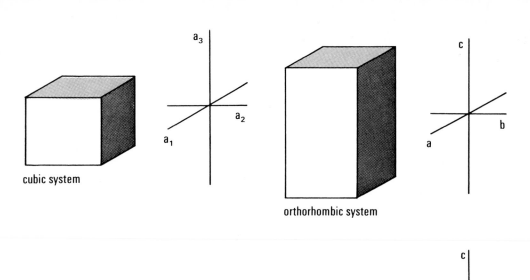

cubic system

orthorhombic system

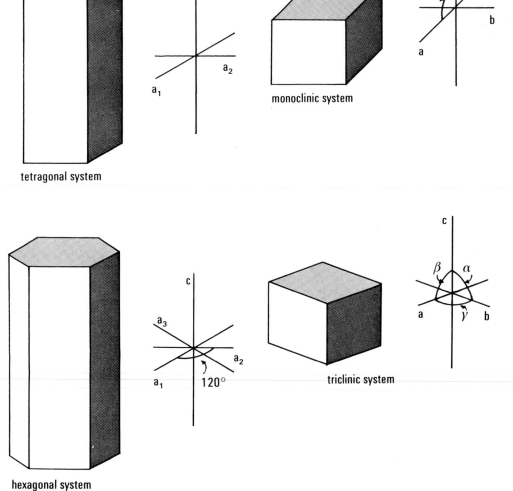

tetragonal system

monoclinic system

triclinic system

hexagonal system

Tetragonal system All crystals in this system can be referred to two equal horizontal axes and a third vertical axis of different length, all three being at right angles to one another.

Hexagonal system This differs from the others in that crystals belonging to the system have four axes, three equal horizontal axes at angles of 120° to each other and a vertical axis perpendicular to the plane of the horizontal axes. The *trigonal* (or *rhombohedral*) division is very important since it contains a number of common minerals, quartz and calcite being the most important.

Orthorhombic system This contains all those crystals which can be referred to three unequal axes at right angles to one another.

Fig. 3·8 Combinations of tetrahedron and cube.

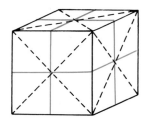

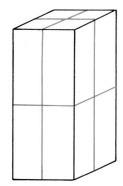

Fig. 3·9 Planes of symmetry in a cubic and orthorhombic figure.

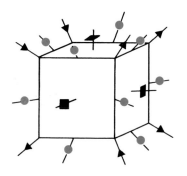

● two-fold axis

▲ three-fold axis

■ four-fold axis

Fig. 3·10 Symmetry axes in a cube.

Monoclinic system This includes all crystals which can be referred to three unequal axes with one axis normal to the plane containing the other two axes which are not at right angles.

Triclinic system The final system is the least symmetrical of all the classes, including those crystals which can be referred to three unequal axes, none of them at right angles to one another.

Symmetry A crystal can be assigned to its correct class by a careful study of the symmetrical properties of its faces. Three categories of symmetry can be recognized in crystals.

Centre of symmetry A crystal possesses a centre of symmetry if it has each face in an ideal crystal duplicated by a similar parallel face on the opposite side of the crystal. The cube, octahedron and combinations of the two forms (Fig. 3·4) possess a centre of symmetry, whereas a tetrahedron (Fig. 3·8) does not, although all belong within the cubic system.

Planes of symmetry These are planes which divide a crystal into portions which are mirror images of each other. A cube has nine such planes, whereas a solid with orthorhombic characteristics has only three (Fig. 3·9).

Axes of symmetry These are axes about which a crystal can be rotated to bring it successively into two or more identical positions during a complete rotation. The axes are referred to as two-fold, three-fold, four-fold, or six-fold, according to the number of possible symmetrical positions taken by the crystal during one rotation. A cube has three four-fold, four three-fold, and six two-fold axes, as illustrated in Fig. 3·10.

The crystal systems can now also be defined in terms of essential symmetry elements as shown in Table 3·1.

The important difference between the hexagonal system, with a six-fold axis, and the trigonal system, with a three-fold axis, should be particularly noted. The cubic system, minerals from which show a large number of symmetry axes and planes, is said to be a *high* symmetry system, while the triclinic system, with at most only a centre of symmetry, is said to be of *low* symmetry. The list above is arranged in order of decreasing symmetry and in Fig. 3·11 a number of idealized crystal forms are shown from each of the classes. These drawings should be examined carefully to ascertain the positions of the symmetry elements in each crystal. With some practice the reader can then attempt to find the symmetry elements present in a number of crystal models or well-formed natural crystals, proceeding then to deduce their crystal classes.

Crystal axes and indices

The normal convention for labelling crystallographic axes is to refer to the vertical axis as *c*, the left to right axis as *b* and the front to back axis as *a*. This can

Table 3·1	
cubic system	four three-fold axes
tetragonal system	one four-fold axis
hexagonal system	one six-fold axis
trigonal system	one three-fold axis
orthorhombic system	three two-fold axes *or* one two-fold axis at the intersection of two perpendicular planes of symmetry
monoclinic system	one two-fold axis
triclinic system	a centre of symmetry *or* no symmetry

be seen in Fig. 3·7, the system being slightly modified for the higher symmetry classes.

Cubic system The three axes are equal in length, so all are labelled *a*.

Tetragonal system The *a* and *b* axes are equal in length, so the convention adopted is to refer to both horizontal axes as *a*.

Hexagonal system The three equal horizontal axes are all labelled *a*.

When the axes are not at right angles, the convention adopted is to refer to the angle between *b* and *c* as α (alpha) between *a* and *c* as β (beta), and between *a* and *b* as γ (gamma). For the monoclinic system, therefore, the unit cell angle is β.

Crystal faces are usually located by using the system devised by W. H. Miller in 1839 and universally known as *Miller indices*. It is based on the principle that each crystal face cuts at least one of the crystal axes at some point. Consider the

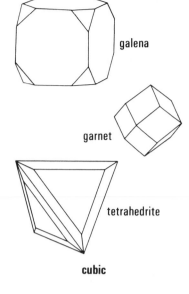

galena

garnet

tetrahedrite

cubic

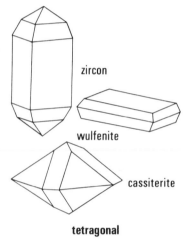

zircon

wulfenite

cassiterite

tetragonal

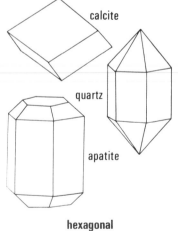

calcite

quartz

apatite

hexagonal

Fig. 3·11 Typical crystals from the six classes.

A phantom quartz crystal produced by an overgrowth of crystals in a parallel position.

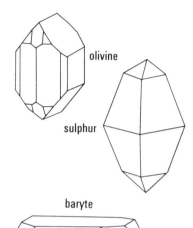

olivine

sulphur

baryte

orthorhombic

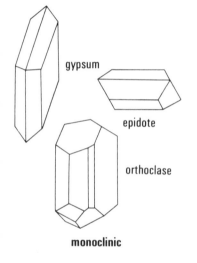

gypsum

epidote

orthoclase

monoclinic

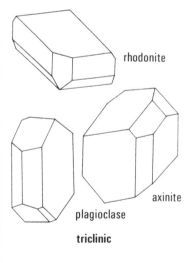

rhodonite

axinite

plagioclase

triclinic

A reticulated group of cerussite crystals from Tsumeb, south-west Africa.

axes shown in Fig. 3·12 where *a*, *b* and *c* are the lengths of the unit cell edges. The plane PQR intercepts the axes at distances from the origin of $3a$, $3b$ and $2c$, so it can be designated by the parameters –

$$3a : 3b : 2c.$$

Miller's system utilizes the reciprocal of the numerical part of this expression –

$$\tfrac{1}{3}a : \tfrac{1}{3}b : \tfrac{1}{2}c.$$

Expressed in integers this becomes –

$$2a : 2b : 3c.$$

The face is therefore designated (223). Similarly face P'Q'R', intercepting the axes at $6a$, $6b$, $4c$, is parallel to PQR and, on converting its parameters into Miller indices, is also designated (223). Therefore, *parallel faces always have the same indices*. A face with intercepts on the negative side of an axis such as PQR''

would have intercepts at $3a$, $3b$, $-3c$ and is designated ($11\bar{1}$). Finally, a face parallel to one of the axes intercepts that axis at infinity. Since the reciprocal of infinity is zero, the indices of a face parallel to the a and b axes is (oo1). Using this system readers should then verify for themselves the indices of the faces of a cube and octahedron given in Fig. 3·13. The general symbol for a face in the hexagonal system is ($hkil$) since the system has three horizontal axes (hki) and a vertical axis (l). The indices can always be verified from the fact that

$$h + k + i = 0.$$

Some possible faces in the hexagonal system are ($21\bar{3}0$), ($10\bar{1}1$) and $3\bar{1}\bar{2}1$.

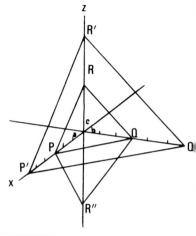

Fig. 3·12 Crystallographic axes, intercepts, and planes.

Defects in crystals

Some minerals commonly form easily recognizable and simple crystal forms. Fluorite and galena in the cubic system are good examples. The majority of minerals, however, form complex combinations of forms, or simple but distorted forms due to the irregular development of certain faces. In some minerals these irregularities occur so frequently that they are a valuable aid to identification. A few of these defects are outlined here.

Striations A number of minerals show characteristic fine parallel lines covering certain faces. They are usually caused by two different crystal forms attempting to develop at the same time. Three minerals particularly noted for this effect are pyrite, quartz and tourmaline, illustrated in Fig. 3·14. In the case of pyrite, the striations are the result of combinations of the cube and pyritohedron forms on the surface. Note particularly the directions of the striations on the adjacent faces of the cube which can be related to the edges of the ideal pyritohedron shown in the figure. In quartz the striations are always across the prism face whereas in tourmaline they proceed along the length of the crystal. These striations serve as very useful diagnostic features for a number of minerals as mentioned in the chapter, The Mineral Kingdom.

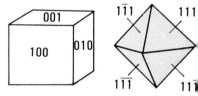

Fig. 3·13 Indices of the faces of a cube and octahedron.

Hopper crystals This effect is particularly shown by the mineral halite (common salt), as illustrated in Fig. 3·15. The mineral is cubic but it sometimes tends to grow from solution faster along the cube edges than in the centre of the faces. This results in the production of cavities in each face. The barrel-shaped crystals in the pyromorphite – mimetite series also develop their distinctive habit in this way.

Etched crystals As mentioned earlier, crystals grow from solution under very precise conditions. Sometimes a change in a growing crystal's environment can reverse the crystallization process and dissolution commences. Evidence of dissolution conditions is provided by etch-figures on crystal faces, these having the appearance of small, regularly shaped pits. Crystal etching is very easy to reproduce in the laboratory; anyone can try immersing a small crystal of calcite in dilute hydrocholoric acid. The shape of etch-figures, whether naturally or

Fig. 3·14 Striated crystals.

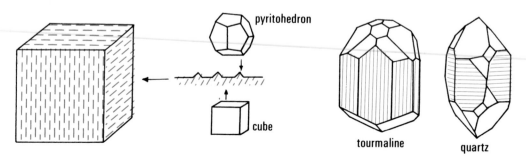

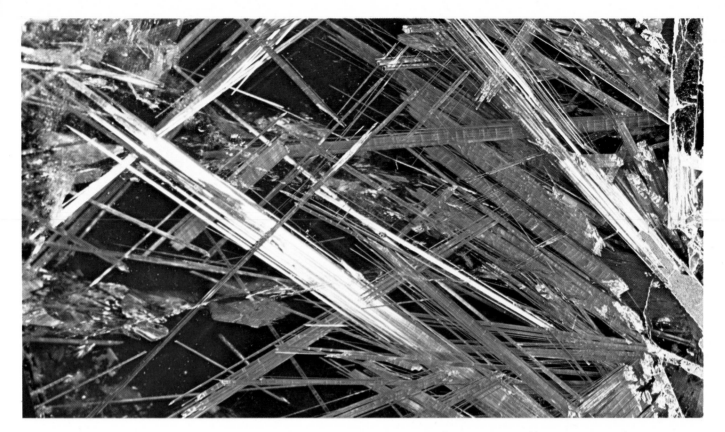

Acicular rutile crystals in quartz, from Brazil.

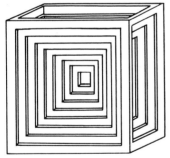

Fig. 3·15 Idealized hopper crystal of halite.

artificially produced, is always related to the crystal's structure and, as a consequence, the figures must be the same on identical crystal faces. In certain cases this property can be used to ascertain the symmetry of a distorted crystal.

Inclusions in crystals Growing crystals often trap foreign material as they develop. This material may take the form of bubbles of liquid or gas, or it may be small crystals already developed, which happen to be enveloped by a larger crystal. Milky coloured varieties of normally transparent minerals are usually found to be filled with minute gas or fluid bubbles. Well-known crystalline inclusions include needles of rutile in quartz and fine plates of hematite in quartz, the latter imparting a blood-red colour to the mineral. These are examples of crystalline inclusions which are not necessarily orientated on crystallographic planes in the host mineral. If hematite flakes are present in quartz in more or less parallel positions the variety known as aventurine quartz is produced. The same effect occurs in sunstone or aventurine feldspar, the shimmering optical effect being produced by orientated inclusions of hematite, ilmenite, or limonite. The effect known as *chatoyancy* occurs in crystals which contain long, slender, parallel inclusions or cavities. When tourmaline is polished perpendicular to these inclusions, stones known as *cat's-eyes* are produced. Similarly, ruby, sapphire, garnet and quartz furnish *starstones*, the inclusions here being produced by exsolution in the host crystals (*see also* page 184).

Twinning in crystals

Groups of associated crystals are known as *aggregates* and the type of aggregation is often typical of a particular mineral species. This topic is dealt with more fully on page 74 but here mention should be made of two types of aggregation which are directly related to a mineral's crystal structure. Groups of crystals are said to be aggregated in *parallel growths* if the crystals are *separate* individuals but with one or more of their crystallographic axes parallel.

Very often crystals develop in such a way as to show two separate portions

65

joined at a common crystallographic plane. These are known as *twinned crystals*, a fairly straightforward example being the geniculate twin of rutile (Fig. 3·16) so called on account of its resemblance to a leg bent at the knee. The plane common to each part of the crystal is known as the *twin plane* – in the rutile illustrated the plane is (011). The *twin axis* is the axis about which a 180° rotation of one half of the twin would convert the twin into a single crystal.

The two portions of a twinned crystal are usually of comparable size, indicating that development as a twin commences early in the life of the seed crystal and is not a late stage in the crystallization process. An exception is the deformation under pressure of certain crystals which sometimes produces an effect known as secondary twinning. Fortunately for mineralogists, twins are formed only on certain crystallographic planes characteristic to each species, resulting in a uniformity in the appearance of the same twin in different specimens of a mineral. It follows from this that the angles between twin segments must remain constant for a given mineral. This regularity of twinning properties for a mineral on a particular crystallographic plane is spoken of as a 'twin law'.

A characteristic feature of twinned crystals is that they usually possess *re-entrant* angles at the junction of two portions of a twin. These are acute or obtuse angles between adjacent faces, a feature which never occurs in a single crystal.

Several types of twinning can occur in minerals. *Contact* twins are the simplest form of twin as shown by rutile (Fig. 3·16) and calcite (Fig. 3·17), the latter twinned on the basal pinacoid {0001}. Two separate crystal portions are symmetrically positioned relative to the twin plane, implying that if the crystal were split along this plane, two separate single crystals would result.

Penetration twins are crystals in which the two parts are intergrown and they represent the most common type of twinning encountered in mineralogy. Some examples are given in Fig. 3·18, the main feature being that in each case the twin cannot be divided into two separate halves.

In certain circumstances the same twin law can operate more than once in a crystal. Thus, a crystal containing three parts related by the same law is known as

Above left
Hopper development of an octahedral crystal of cuprite.

Above
A fine twinned orthoclase crystal from Italy.

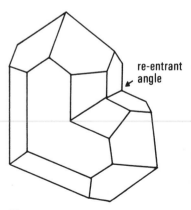

re-entrant angle

Fig. 3·16 Geniculate twin of rutile.

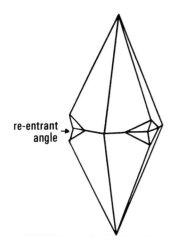

re-entrant angle →

Fig. 3.17 Calcite scalenohedron twinned on basal pinacoid.

A red scalenohedron of calcite showing twinning on the basal plane.

A single crystal and a swallowtail twin of gypsum.

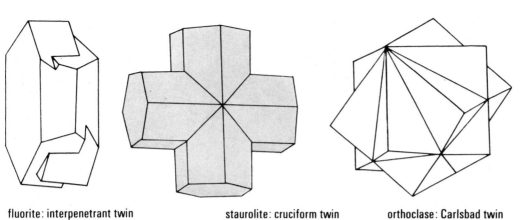

Fig. 3·18 Penetration twins.

fluorite: interpenetrant twin   staurolite: cruciform twin   orthoclase: Carlsbad twin

a *trilling*. In the same way *fourlings, fivelings*, and so on, sometimes occur. Repeated twinning on parallel planes is referred to as *polysynthetic*, often resulting in the thin twinned lamellae typical of the plagioclase feldspars. Some examples of repeated twinning are given in Fig. 3·19.

X-ray diffraction by crystals

In 1912 it was shown by Laue and his co-workers that the spacings of atomic layers in crystals were appropriate for X-rays to be diffracted from these layers. They placed a photographic plate behind a crystal of sphalerite which was then irradiated by an X-ray beam. Subsequently the plate was found to show a central dark spot formed by the X-ray beam passing directly through the crystal, but in addition a large number of small spots arranged symmetrically about the centre showed that some X-rays were being reflected off atomic planes in the structure. This profound discovery was followed shortly afterwards by the first structural analysis of a crystal by X-ray diffraction techniques. The crystal studied was halite and the scientists involved were the British physicists, father and son, W. H. and W. L. Bragg.

The younger Bragg gave his name to the law governing the reflection of X-rays by atomic planes in crystals. By referring to Fig. 3·20 it can be shown that the path difference between X-rays reflected from successive atomic layers is

$$2d \sin \theta,$$

where d is the distance between the layers and θ is the angle of X-ray incidence. If this path difference is equivalent to an integral number of X-ray wavelengths, then reinforcement of the reflected X-rays occurs giving rise to the spots observed on a photographic plate or film.

A number of methods for the X-ray investigation of crystals are now used, the most important being the following.

Rotation or oscillation method The crystal is mounted with one of its crystallographic axes vertical and surrounded by a cylindrical film. A pencil of X-rays,

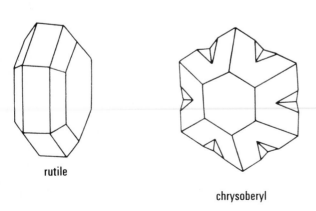

rutile

chrysoberyl   plagioclase

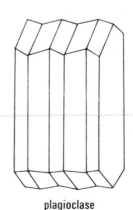

Fig. 3·19 Repeated and cyclic twins.

Fig. 3·20 Reflection of
X-rays from atomic planes.

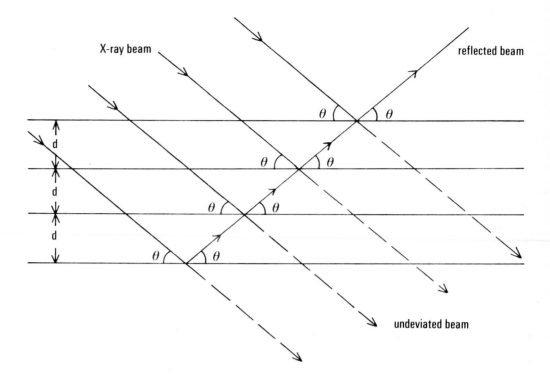

incident on the crystal perpendicular to the cylinder axis, is diffracted producing
horizontal lines of spots, the positions of which are governed by the unit cell type
and size.

Powder method This technique is very widely used for the identification of unknown minerals. One important advantage over single crystal diffraction techniques is that an orientated crystal is not required, opening up the possibility of
working on many minerals which do not form usable single crystals. A small
quantity of the mineral is crushed into a powder and cemented into a narrow
cylindrical filament. The photographic pattern is formed on a strip film as shown
in Fig. 3·21. Because there are now a very large number of randomly orientated
diffracting crystals in the powder, the diffracted rays are not composed of single
beams producing a series of spots on the film, but of conical beams producing a
series of curved lines. The spacings of these lines represent a 'fingerprint' for
each mineral – by matching a pattern with a set of standards, rapid mineral
identification is possible.

Fig. 3·21 Diagram of a powder
camera and film.

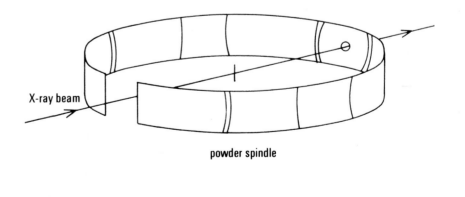

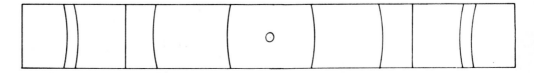

Dr Andrew Clark

Properties and study of minerals

Physical properties

Most minerals, as we saw in the last chapter, have clearly identifiable crystal structures. They also have fixed chemical compositions, or compositions that vary only within clearly defined limits. These specific crystallographic and chemical features produce in each mineral a characteristic set of physical properties. Some minerals are of great industrial and scientific importance on account of certain of these properties; for instance, the hardness of diamond makes it the ideal abrasive and cutting material as well as the most durable of gemstones, and the piezoelectric properties of quartz make it a suitable substance for use as an electronic oscillator.

Knowledge of a mineral's physical and chemical properties help us to understand its mode of formation and that of the rock in which it is found. Since a number of these properties are easier and quicker to measure than direct chemical or structural determinations, they form, in addition, a very useful way of rapidly identifiying minerals, as will be outlined in the pages that follow. In the next chapter the most important properties of each reasonably common mineral species are listed.

The physical properties are considered in the following pages in a typical sequence appropriate to the identification of an unknown mineral. The first steps are to study a mineral's colour, habit and association, which require mainly a keen eye and a little experience. Properties, such as hardness and specific gravity, which require some equipment are then dealt with, and several properties, such as fluorescence and radioactivity, which apply to only a few species, are mentioned. Optical properties, which are extremely important in identification, are dealt with last.

Colour

The colour of a mineral is one of its most obvious features, but it should be treated with care as a means of identification. Colour is the result of a mineral's capacity to absorb certain wavelengths of light and reflect or transmit others, the unabsorbed light giving rise to the colour seen by the observer. Thus, dark-coloured minerals absorb most of the light falling on them, whereas red minerals reflect or transmit predominantly the red part of the spectrum and absorb the remainder. The property of a mineral to absorb light depends on a number of factors which vary from one species to another.

The most reliable factor is the property of certain chemical groups to show characteristic and fairly constant colours. Thus, secondary copper minerals are invariably coloured green and blue, whereas secondary uranium minerals are

A rhomb of calcite ('Iceland spar'), showing double refraction.

A blue, mamillated crust of azurite on green malachite, both being typically coloured copper minerals.

generally bright yellow and yellow-green. Minerals containing aluminium, sodium, potassium, and so on, are mainly light coloured or colourless, whereas those containing the transition metals as their major constituents are usually deeply coloured. Certain, usually colourless, minerals may show colour effects owing to the presence in their structures of small amounts of particular elements. Amethyst and rose quartz are thought to be coloured by traces of titanium or manganese, whereas pure quartz is colourless. Some other characteristically coloured minerals are listed in Table 4·1, together with the metal present, either as a major or minor constituent, giving rise to the colour.

It is not always realized that certain minerals which are generally seen as dark-coloured mineral specimens, may, in their pure state, be totally different. Cassiterite (SnO_2) and sphalerite (ZnS) without any impurities are almost colourless. However, specimens of both these minerals invariably contain some iron which imparts the characteristic colour to each species.

A mineral's colour may, in certain instances, be governed by its atomic bonding rather than its chemical composition. Graphite, which is black and opaque, and diamond, which is colourless, are both chemically identical as pure carbon, but have different structures.

Minute inclusions of a mineral present in a larger crystal of another mineral produce a 'false' colour, as in the reddish colour of some feldspars which is due to minute particles of exsolved hematite. Also, a thin coating of one mineral on another produces misleading colours in many instances. Minerals, such as agate, are often artificially stained to produce new colours or enhance their natural colour.

Many crystals undergo colour changes when exposed to heat, light, or other electromagnetic radiation. Smoky quartz is an example of this, as its appearance is often the result of irradiation by radioactivity from associated minerals. Heating is often used to enhance the colour of certain gemstones (*see* page 188).

Table 4·1

mineral	colour	element responsible for colour
corundum (sapphire)	blue	titanium
garnet (uvarovite)	green	chromium
beryl (emerald)	green	chromium
corundum (ruby)	red	chromium
rhodonite	pink	manganese
rhodochrosite	pink	manganese
olivine	green	iron
hematite	dark red	iron
magnetite	black	iron
erythrite	pink	cobalt
annabergite	green	nickel
malachite	green	copper
azurite	blue	copper

When examining minerals, therefore, one should be aware that colour can be either an intrinsic property or a property unconnected with the mineral's basic chemistry or structure. It can also be artificially induced by both irradiation and staining.

Streak

The term streak is used in mineralogy to describe the colour of the powder of the mineral under examination and is less subject to variations induced in minerals by foreign elements, inclusions, and surface coatings. The word is used because the most convenient way of obtaining the colour of a mineral's powder is by drawing it across the surface of a piece of unglazed porcelain (such as the reverse side of a kitchen tile) to produce a 'streak' mark. To mark a tile the mineral must obviously be softer than the tile, and the technique is most useful when applied to dark coloured minerals which often give a much lighter streak than that indicated by their body colour. Thus, hematite, although appearing almost black in many massive specimens, gives a red streak, whereas magnetite, another oxide of iron,

The streak test, using a specimen of hematite.

gives a black streak, and goethite, the hydrated iron oxide, a brown streak. The technique is less useful for identifying silicates, carbonates, and transparent and translucent minerals in general, because they all give a white streak. Note that, in mineral data, the terms 'white' and 'colourless' are interchangeable when applied to streak determination. If a mineral is too hard to mark a streak plate, then the colour of its powder may be obtained by filing a small sample or crushing it with a pair of pliers.

Asbestos, one of the best examples of fibrous minerals.

Lustre

The lustre shown by a mineral is a property related to its surface features, reflectivity and refractive index. Various categories of lustre are widely applied, the most important being the following.

Metallic lustre The brilliant, highly reflecting surface lustre of many opaque minerals, shown particularly by native metals and most sulphides.

Submetallic lustre A feebly displayed metallic lustre, characteristic of opaque or almost opaque minerals, as in many of the metallic oxides and a few sulphides (for example, cinnabar).

Adamantine lustre The lustre of diamond, shown by transparent or translucent minerals with high refractive indices (for example, zircon, cassiterite, sulphur, rutile).

Resinous lustre The term used to describe the lustre of brown or yellow coloured minerals with moderately high refractive indices, alluding to the appearance of resin.

Vitreous The lustre of broken glass. This is the characteristic lustre shown by most of the silicates, carbonates, phosphates, sulphates, halides, and light element oxides and hydroxides. The refractive index range covered is approximately $1\cdot3$–$1\cdot8$.

A number of other terms are also used, mainly relating to surface properties of the minerals. Thus *silky* lustre is shown by minerals having a fibrous structure, as in gypsum of the satin-spar variety. *Pearly* lustre is a characteristic of minerals with numerous, partly developed cleavages parallel to the surface, having the effect of dividing the mineral into very thin plates. *Greasy* and *waxy* lustres are caused by microscopically roughened surfaces on crystals. *Earthy* is the lustre of soft friable minerals with negligible reflectivity.

Crystal aggregates

When minerals occur in the form of well-developed crystals, a positive identification can often be made without resorting to any further tests. The basic relation between external form and crystal structure has been outlined in the previous chapter and need not be repeated here, except to emphasize that crystals are rarely perfectly developed or in ideal forms since crystal faces often develop at uneven rates, thus distorting the true symmetry, and occasionally even produce a misleading crystal habit.

Most mineral specimens are aggregates of a large number of imperfectly developed crystals. They may be clusters of single crystals, or compact aggregates which only show evidence of crystallinity under the microscope. The type of aggregation provides valuable information on the conditions under which a mineral was formed, as well as giving considerable help in mineral identification, since many minerals form in characteristic and recognizable aggregates or crystal types. The following are some of the terms used to describe aggregates.

Acicular Needle-like crystals (for example, natrolite).

Bladed Flattened lath-shaped crystals (for example, kyanite).

reticulated

radiating

acicular

bladed

botryoidal

fibrous

dendritic

granular

foliated

tabular

Fig. 4·1 Some mineral
aggregates and crystal habits.

Hematite (kidney ore); the famous botryoidal form from Cumberland, England.

Malachite; a polished slab showing the botryoidal growth patterns.

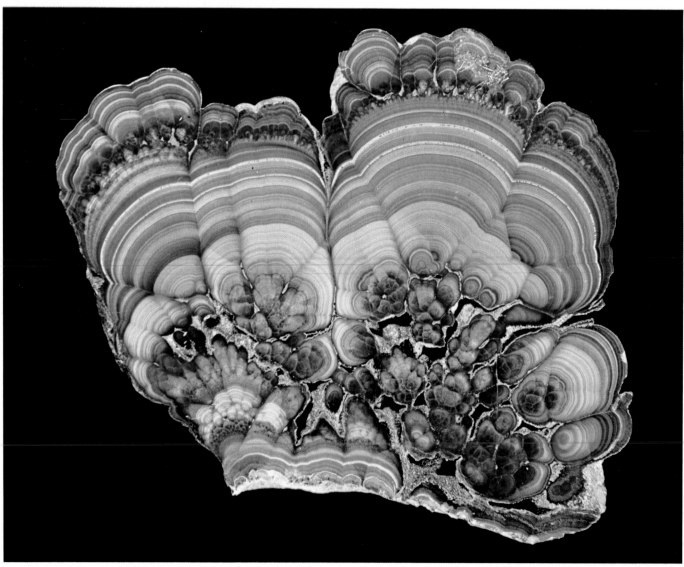

Radiating crystals of wavellite from Hot Springs, Arkansas.

Botryoidal, mamillary, reniform All these terms refer to minerals found as rounded masses without any visible crystal faces. The appearance is due to the minerals having been formed by mineralizing solutions dripping into cavities and building up layer by layer. Chalcedony, malachite, and hematite sometimes occur in this form. Botryoidal means 'grape-like' and the other terms are somewhat synonymous, mamillary generally referring to the larger scale of aggregation.

Columnar This is a common form for a mineral which has developed in an open space. In particular, *stalactitic* refers to columnar forms of minerals such as calcite and aragonite caused by the deposition of the mineral from dripping water in limestone caves.

Concretions and nodules These terms describe minerals which form in isolated aggregates, sometimes as spherical or ellipsoidal forms, as seen in flint nodules, or as irregular shapes, such as pyrite nodules. Both these minerals are found extensively in the chalk of southern England. When the concretions take on a flattened shape they are referred to as *lenticular*.

Dendritic, arborescent These are terms used to describe minerals deposited from solution in cracks and fissures of other minerals or rocks. The resulting aggregate often has the appearance of tree or moss-like forms. Manganese oxides often form in this way.

Fibrous A form commonly shown by asbestos minerals and occasionally the satin spar variety of gypsum, where the crystals aggregate as long thin strands.

Foliated, lamellar, micaceous Minerals with a platy habit commonly form these types of aggregates, developing as thin separable sheets, the best examples, of course, being the micas. If foliated masses grow from a point then roughly spherical aggregates (rosettes) are often formed. These are termed *concentric* aggregates, hematite of the 'iron rose' variety being a very good example.

Granular This form is shown by minerals which develop as massive aggregates of mineral grains without any pronounced individual form. According to the size of the grain, we speak of coarse-, medium-, and fine-grained aggregates.

Radiating, stellate Aggregates of acicular or foliated crystals show this form when they develop outwards from an initial growth point (for example, stibnite, wavellite).

Reticulated This is an aggregate of crystals in the form of a net-like mesh, often with individual crystals, crystallographically orientated to one another, or to the medium in which they are formed.

Tabular These are crystals showing broad, flat faces (for example, baryte, wollastonite). A similar platy habit, but on a smaller scale, is often referred to as *scaly*.

Tuberose These aggregates show some similarity to the botryoidal form mentioned previously, but are characterized by a very much more irregular form of rounded development (for example, certain specimens of aragonite).

Wiry, filiform These are twisted aggregates of mineral strands (for example, native silver); *capillary* forms refer to fine strands with straight development (for example, millerite).

Cleavage and fracture

Much information on the nature and crystal system of a mineral can be gained from studying broken and chipped features on the crystal faces. It goes without saying that a good quality crystal should not be unnecessarily damaged in order to study these properties, but usually there is some inconspicuous part that can be tested. If a crystal breaks along a flat plane it is said to possess a *cleavage*; if it breaks with an irregular surface, then it is said to show a *fracture*.

Cleavage will only occur along a plane in a crystal when the atomic bonding perpendicular to the plane is weaker than that parallel to the plane. The fact that a mineral has a cleavage does not mean that it is soft, for diamond has an octahedral cleavage, yet it is the hardest mineral known, while a number of soft minerals have no prominent cleavage. Cleavage always occurs parallel to a possible crystal face and so is designated by describing it in terms of the crystal face to which it is parallel, that is, cubic, octahedral, prismatic and so on, or by the Miller indices appropriate to each of these faces. According to the ease with which a crystal cleaves and the perfection of the surface obtained, the cleavage is classified as perfect, good, poor, and so on. Minerals with perfect cleavage include fluorite, graphite, calcite, mica and galena.

Careful study of the cleavage directions and their relative perfection can help in assigning a mineral to its appropriate crystal class. Thus, in the simplest case of a mineral with a cubic cleavage, if it has a cleavage in one direction it must also be possible to cleave it in two further directions at right angles to the first cleavage. Fluorite, a mineral in the cubic system, has an octahedral cleavage, because it will split easily and perfectly to produce an octahedron. Graphite, molybdenite and mica have one prominent perfect cleavage, known as a basal cleavage, and these minerals split into thin sheets. This perfect cleavage makes graphite suitable for pencil 'lead' and muscovite mica ideal for insulating sheets in electrical equipment.

Parting planes in minerals have a similar appearance to cleavages, but are developed from splitting along twin planes, or along planes of chemical alteration. They are not true cleavages and care is necessary to avoid confusion in minerals which show both types of splitting.

Fracture is shown by minerals which do not easily cleave. The type of fracture surface is usually constant for a particular species, the commonest forms being –

Conchoidal A fracture surface resembling the surface of a shell, with curved, concentric fracture lines. Quartz, flint and glass fracture in this way.

Even Fracturing with an almost flat surface.

Uneven Fracturing with a rough surface, shown by a large number of mineral species.

Hackly This term is used to describe the fracture surface of native metals. The surface is rough with sharp, jagged points.

The ease with which a mineral fractures is indicated by terms denoting the *tenacity* of the species. A very large number of minerals are *brittle*, breaking easily and cleanly into a number of fragments, but a number of the native metals are *malleable*, which means that they can be worked with a hammer without breaking. Certain soft minerals are *sectile*, meaning that they can be cut with a knife, while a few minerals are *flexible* in thin sheets and others *elastic* if they return to their original shape after bending.

Hardness

We come now to one of the most important tests used in rapid mineral identification – that of hardness determination. Hardness is related to a mineral's crystal structure and is usually measured from its resistance to scratching by other minerals of varying hardness. In 1822 Mohs, an Austrian mineralogist, proposed his well-known scale of mineral hardness, which is in use unaltered today. He chose ten common minerals of varying hardness, grading them on a scale from 1 to 10, talc (1) being the softest and diamond (10) the hardest.

1 Talc, 2 gypsum, 3 calcite, 4 fluorite, 5 apatite, 6 orthoclase, 7 quartz, 8 topaz, 9 corundum, 10 diamond.

A mineral at a certain point on the scale will scratch a mineral lower down the scale and will be scratched itself by any mineral higher in the scale. Therefore, to determine the hardness of an unknown mineral, a set of these standard minerals is required, or better still, a set of metal rods with a crystal of each mounted firmly at the end. Tests are then carried out and if, say, the mineral is scratched by quartz, but not by orthoclase, then its hardness lies between 6 and 7. With some experience it will be possible to estimate whether the mineral's hardness is nearer 6 or 7, or intermediate between the two, in which case it should then be designated $6\frac{1}{2}$.

In the absence of a set of hardness testing minerals, the following materials may be substituted to give approximate hardness indications: a penknife blade, with a hardness on Mohs' scale near $6\frac{1}{2}$; window glass, 5; fingernail, $2\frac{1}{2}$.

Mohs chose his set of standard minerals on the basis of their wide availability. His ten minerals do not form a perfectly uniform scale. If the hardness of each

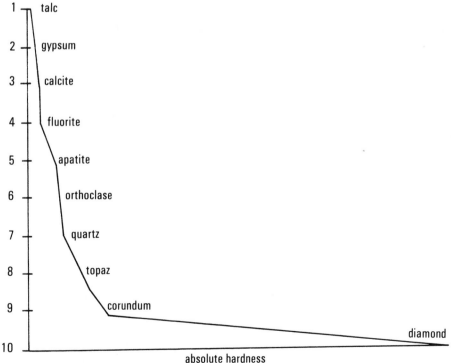

Fig. 4·2 The hardness scale for the ten standard minerals.

mineral is measured quantitatively, using a diamond indentation method, it is found that the hardness of the group increases steadily to corundum (9) although the difference between calcite and fluorite is quite small (*see* Fig. 4·2). Between corundum (9) and diamond (10) there is an extremely large jump in the measured value. For testing minerals this is of little importance, however, as there are no common minerals with hardness values lying between those of diamond and corundum, although carborundum (SiC), an artificial substance occasionally mistaken for a mineral, does lie in this range.

The crystal surface on which a hardness determination is to be made should be reasonably smooth, unbroken and without a coating of alteration products. After scratching it carefully in some inconspicuous spot, to avoid disfiguring the specimen, the mark should be cleaned and examined with a magnifying glass or binocular microscope. The reason for this is that misleading results are often obtained by attempting to scratch a harder mineral with a softer one. The softer mineral may leave a line of powder on the harder specimen, which can easily be mistaken for a scratch. Cleaning the specimen removes this mark, so one can then be sure whether the specimen has actually been scratched.

The hardness of a mineral is related to its crystal structure and tends to be greater when the structure is built up of smaller atoms or ions, and when the atomic packing density is high. Corundum (Al_2O_3) and hematite (Fe_2O_3) have the same type of hexagonal crystal structure, but corundum (9) is harder than hematite (6) owing to the small size of the aluminium ions compared to those of iron. Atomic packing and its effect on hardness are best illustrated by the polymorphs of carbon, graphite and diamond. Graphite, with its loosely bound sheet structures has very low hardness (1-2) whereas diamond has a closely packed structure and small ionic size, contributing to the mineral's extreme hardness.

Due to this relationship between hardness and structure it is to be expected that the hardness of a mineral will vary according to the crystallographic direction in which it is measured. This can be most convincingly demonstrated on a specimen of the mineral kyanite, crystals of which invariably show a well-developed {100} cleavage face. Hardness determinations made on this face give values of $4\frac{1}{2}$ along the *c*-axis, the direction of elongation of the crystal, and $6\frac{1}{2}$ along the *b*-axis, at right angles to the elongation direction.

Specific gravity

The term specific gravity is a fundamental property relating a mineral's chemical composition to its crystal structure. It is defined as the ratio of the weight of a substance to that of an equal volume of water and is, therefore, given as a dimensionless number; for example, the specific gravity of quartz is 2·65. Density is an identical quantity, but is defined as the mass of unit volume of the substance and is expressed in the appropriate units, so that the density of quartz is 2·65 grams per cubic centimetre.

For minerals which do not show appreciable chemical variation or atomic substitution, the specific gravity is a constant and for many minerals forms a very useful guide to its identity. If we measure the specific gravities of four well-known sulphate minerals, gypsum, celestine, baryte, and anglesite, we find a large difference in the values obtained, as can be seen in Table 4·2.

The specific gravities of these minerals directly reflect the relative atomic weights of the four cations, calcium, strontium, barium and lead. To the experienced mineralogist, the specific gravity of a species is important because its value can usually be measured directly and also calculated from the unit cell volume and chemical composition. If the theoretical and measured values are in agreement, then the structural work is confirmed as correct.

Table 4·2		
mineral	composition	specific gravity
gypsum	$CaSO_4$	2·32
celestine	$SrSO_4$	3·97
baryte	$BaSO_4$	4·50
anglesite	$PbSO_4$	6·38

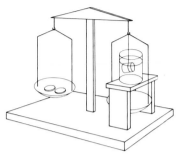

Fig. 4·3 Chemical balance.

To obtain a direct measurement of a mineral's specific gravity, it is necessary to determine either directly or indirectly the weight of the specimen in air and also the weight of water it displaces. The most convenient way of doing this is by using a chemical balance (Fig. 4.3) to measure the weight of the specimen in air and then in water, or some other liquid of known specific gravity. By applying Archimedes' principle, the loss in weight between these two measurements is equal to the weight of liquid displaced. This technique avoids the very inaccurate step of measuring directly the weight of displaced liquid.

If W_1 is the weight of the specimen in air and W_2 its weight in water, then its specific gravity will be

$$S.G. = \frac{W_1}{W_1 - W_2}.$$

If instead of water a liquid with specific gravity L is used for the determination, then

$$S.G. = \frac{W_1}{W_1 - W_2} \times L.$$

If the mineral specimen is on the small side, extreme care must be taken when making the measurements to allow for the weight of the suspension wire and the surface tension effect as the wire passes into the water. As thin a wire as possible should be used since the surface tension is directly proportional to the length of the liquid/wire contact, that is, to the circumference of the wire. Fuse wire is generally suitable for all but the finest scale of work.

Due to its high surface tension, water is not the ideal liquid for specific gravity determinations. In addition to the 'drag' exerted on the suspension wire, it tends to trap air bubbles around the specimen giving misleadingly low weighings. A drop of washing-up liquid introduced into the water will cut down the surface tension, but for the greatest accuracy, organic liquids such as toluene, bromoform, or methylene iodide can be used. These have the added advantage that they are all heavier than water and the greatest accuracy in these determinations is obtained by using a liquid in which the specimen *just* sinks.

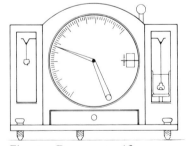

Fig. 4·4 Berman specific gravity balance.

A number of points must be borne in mind when measuring specific gravities by this type of method.

(a) Large mineral specimens are often not homogeneous, containing other intergrown and accessory minerals, and sometimes with weathered areas. Therefore, the largest specimen does not always give the most accurate results. For this reason, and also because many minerals are only available in small quantities, a microbalance is desirable for the most precise work. A balance, designed specifically for measuring densities of mineral samples up to 25 milligrams (0·0009 ounces) in weight is known as the Berman balance after its designer (Fig. 4·4). Using this instrument it is possible to examine carefully the crystals under a microscope to check their purity before taking measurements.

(b) Many minerals, such as clays and numerous secondary minerals, which occur in crusts are porous and may give misleading specific gravities since air bubbles are trapped in the specimen when immersed in the liquid, giving rise to

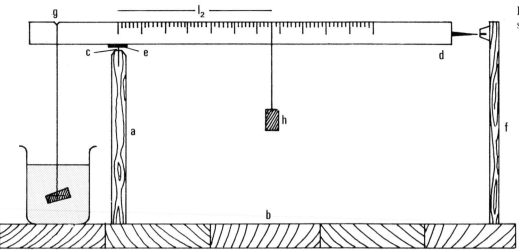

Fig. 4·5 A home-made specific gravity balance.

extra buoyancy and a specific gravity lower than the correct value. The trapped air can often be removed by boiling the liquid with the immersed mineral and allowing it to cool to room temperature while still immersed before taking the measurement.

(c) Mineral identification based on a specific gravity determination is not necessarily conclusive when the mineral has a variable chemical composition. For instance, the olivine series varies from forsterite (Mg_2SiO_4) to fayalite (Fe_2SiO_4) by a progressive replacement of magnesium by iron in the structure. The specific gravity increases from 3·2 to 4·4 through the series, although most olivines commonly found have compositions resulting in specific gravities in the range 3·3–3·4. Atomic replacement is a common feature of silicate minerals and makes specific gravity measurements on them less diagnostic. Also, a large number of silicates have specific gravities in the range 2·6–3·6, so additional tests, usually optical, are necessary for positive identifications.

Since specific gravity is such an important fundamental property it is desirable for everyone with an interest in identifying minerals to have access to a method of measurement. Balances and sets of weights are very expensive, so a description is now given of a method of constructing a simple balance (Fig. 4·5) which does not require a set of calibrated weights.

The instrument is constructed by mounting a vertical wooden post (a) with a razor blade (c) cemented into a slot at the top on a firm wooden base (b). Only about 2 millimetres (0·08 inch) of the blade needs to be exposed, just enough for it to act as a firm knife edge support for the horizontal beam (d). This beam should be made of a light material, such as a ruler or strip of hard balsa wood. To avoid errors in the determination it should be of uniform cross section, not tapering towards one end. A small strip of thin metal (e) is glued to the underside of the balance beam. This has a straight groove scored in it to engage the razor edge when the balance is centred and prevent the razor cutting into the soft wood of the beam. The beam should be pivoted as shown and marked with a scale of centimetres and millimetres starting the measurements from the groove about which the balance pivots. Finally, a second vertical post (f) is fixed to the baseboard with a marker at its upper end to align with the balance beam in a horizontal position.

The mineral specimen to be measured is suspended by a very thin wire from the shorter arm of the balance, the point of suspension being a notch (g) cut into the upper part of the beam. By placing an appropriate counterweight (h) towards the right-hand end of the graduated scale the beam is brought into a position of balance, the fine adjustment being obtained by moving the weight carefully up or down the scale until the pointers coincide exactly. The position of the counterweight on the graduated scale is recorded. A beaker of water or the appropriate immersion liquid, is then placed on the left-hand side of the balance, fully immersing the specimen. To compensate for the loss in weight of the specimen it is necessary to move the counterweight towards the left-hand side in order to rebalance the system. When the exact balance position has been found another scale reading of the counterweight position is taken, the two readings providing sufficient information to calculate the specific gravity of the mineral.

If l_1 is the scale reading with the specimen suspended in air, and l_2 the reading when the specimen is suspended in water, then its specific gravity is given by –

$$S.G. = \frac{l_1}{l_1 - l_2}$$

or if a liquid of density L is used, then

$$S.G. = \frac{l_1}{l_1 - l_2} \times L.$$

It is important to note that the size of the counterweight, often made up of collections of paper clips, must not be altered between the two readings. Again it should be emphasized that at no time in this experiment is the absolute weight of the specimen determined, just the ratio of its weight in air to its weight in a liquid.

The specific gravities of a number of common minerals will be found in Table 4·9.

There are two other methods of measuring specific gravities which should be mentioned as they are frequently used by mineralogists.

The pycnometric method. This technique uses a small glass bottle known as a specific gravity bottle, which has an accurately known volume. The bottle is fitted with a ground glass stopper containing a capillary opening for expelling excess fluid. To measure the specific gravity of some weighed mineral fragments, the bottle is filled with distilled water and weighed along with the mineral fragments. The fragments are then introduced into the bottle, displacing some of the water and the bottle reweighed. If W_1 is the weight of the mineral, W_2 the weight of the bottle full of water and the separate mineral sample, and W_3 the weight of the bottle containing both water and mineral grains, then the mineral's specific gravity is given by

$$S.G. = \frac{W_1}{W_2 - W_3}.$$

Heavy liquid method. This is a useful way of estimating the specific gravity of a mineral sample available in such small quantities that the balance and pycnometric methods are impracticable. A fragment of the mineral is floated in one of the organic liquids; if the mineral floats its specific gravity is lower than the liquid, if it sinks it is higher. By testing the fragment with a series of these heavy liquids, one can find the liquid with the lowest specific gravity in which the fragment will float. This liquid is then diluted until the grain neither sinks nor floats, but remains suspended, at which point the liquid and mineral have the same specific gravities. Finally, the specific gravity of the liquid is measured by means of a set of specific gravity floats or a Westphal balance.

Suitable liquids for use in this technique are bromoform (specific gravity 2·9), methylene iodide (specific gravity 3·3) and Clerici solution (specific gravity 4·2). The first two liquids can be diluted with acetone, while Clerici solution should only be diluted with water. Great care must be taken when handling organic liquids as many produce harmful vapours, and Clerici solution is very dangerous if brought in contact with the skin.

The heavy liquid method, as well as being useful in specific gravity determinations on small samples, is widely used for separating and purifying mixtures of minerals. If a sample containing an intergrowth or association of two minerals with different specific gravities is crushed to a powder and then shaken up with a liquid of intermediate specific gravity, the heavier mineral will sink and the lighter one will float, thereby effecting a separation of the two minerals.

Magnetic and electrical properties

Only two common minerals, magnetite and pyrrhotine, are strongly attracted by a magnetic field, a property known as *ferromagnetism*. Magnetite shows the effect much more strongly than pyrrhotine and, under the name lodestone, was used as an early form of compass. Minerals which are repelled by a magnetic field are said to be *diamagnetic* and those attracted weakly are called *paramagnetic*. Iron-bearing minerals are all paramagnetic, but to different degrees, a property which enables mineral grains to be separated electromagnetically.

Some minerals will conduct electricity, while others will not. The native metals and many sulphides are good conductors, while most remaining minerals are poor conductors. Included in the latter group are a number of minerals which show the important electrical properties of *pyroelectricity* and *piezoelectricity*. Pyroelectricity is the property of certain non-conducting substances to develop electrical charge when subject to a change in temperature. Tourmaline is a good example of a pyroelectric mineral; when heated it becomes positively charged at one end and negatively charged at the other end. Piezoelectricity is the development of electrical charges in non-conducting crystals by the application of mechanical pressure. The effect is best demonstrated with quartz crystals which are, therefore, widely used as oscillator plates in radio transmitters and receivers.

Fluorescence

A number of minerals show the effect known as fluorescence or photoluminescence when illuminated in ultraviolet light. The terms are used to describe the emission by a substance of visible light when excited by the shorter wavelength ultraviolet light. The most important mineral to show the effect is fluorite, after which the phenomenon is named. Fluorite generally shows a blue fluorescent colour, thought to be due to a small quantity of rare-earth elements substituting for calcium in the structure and acting as fluorescence activators. Other minerals which are often fluorescent are calcite (fluorescing red, pink and yellow), willemite (fluorescing green) and scheelite (fluorescing white).

A specimen of calcite and willemite from Franklin, New Jersey, illuminated in natural light (*left*) and ultraviolet light. The willemite fluoresces green and the calcite red.

As mentioned above, fluorescence can be linked to the presence of activators in the mineral structure. The concentration of these activators is critical – too little or too much can cause the effect to disappear. This can be shown by the fact that a mineral from one locality may occur in a fluorescent form, but from another locality may show no such effect. For instance, the green fluorescent willemite from Franklin, New Jersey, is unique; specimens from elsewhere do not show the same colour.

Radioactivity

Radioactivity in minerals is usually thought of in connection with minerals containing the heavy elements uranium and thorium, since these two elements show the effect most noticeably. However, potassium and rubidium are also weakly radioactive and the measurement of the amounts of the radioactive isotopes of these two elements has in recent years become very important, forming the basis of a method for dating rocks and minerals. The process of radioactivity is caused by an unstable atomic nucleus disintegrating and in so doing emitting three types of radiation: α- (alpha) particles (positively charged helium nucleii), β- (beta) radiation (electrons), and γ- (gamma) rays (electromagnetic waves).

Uranium in its natural state consists of a mixture of two isotopes, that is, atomic nucleii with the same number of protons but different numbers of neutrons, the mass numbers (the total number of neutrons and protons) being 238 and 235 in this case. ^{238}U is over 100 times more abundant than ^{235}U in any given sample, but it is the latter isotope which is required in processes involving nuclear fission. Both ^{238}U and ^{235}U decay to give ultimately two isotopes of lead ^{206}Pb and ^{207}Pb respectively. A third lead isotope, ^{208}Pb, is formed as a product of the disintegration of the major isotope of thorium ^{232}Th.

The presence of radioactive minerals in a rock can be confirmed by the signal from a Geiger counter. The instrument is able to measure the intensity of the more penetrative radiations (principally γ-rays) emitted by the decay of radioactive elements in the minerals.

The rate at which radioactive decay of an isotope takes place is constant and not affected by factors such as temperature and pressure. For all common radioactive isotopes, the decay rate is accurately known in terms of a half-life, that is, the time required for a given weight of an isotope to decay to half its initial weight. If a rock or a mineral contains measurable quantities of these radioactive isotopes and the isotopes formed from them, then the relative quantities of each will enable a date for the formation of the specimen to be calculated.

Certain isotopes of potassium and rubidium are very weakly radioactive, but because they occur in a wide range of rocks and minerals, they have become very important in determining the age of specimens. The potassium isotope ^{40}K decays to argon ^{40}Ar, and rubidium ^{87}Rb to strontium ^{87}Sr, both with half-lives which make them suitable geological time-keepers provided none of the components in the system has been chemically leached out of the rock at any time.

Optical properties of minerals

The study of the optical properties of minerals forms one of the most important methods of mineral classification and identification. Light travelling through a crystal interacts with its crystal structure in a number of very characteristic ways. The properties of colour and lustre in minerals have already been discussed and now a number of other properties are considered.

Visible light forms a small part of the spectrum of electromagnetic radiation, that with a wavelength between 4000 and 7000 angstroms (1 angstrom = 10^{-7} millimetre). Electromagnetic radiation of longer wavelengths comprise the various bands of radio-transmission frequencies. As the wavelength shortens the infrared region is encountered, passing into visible red light at a wavelength around 7000 angstroms.

Light radiation may be represented by a simple wave form (Fig. 4·6) where the wavelength is measured from the peak of one wave to the same point on the adjacent wave. The length of this wave alone governs the colour of the light and as the wavelength shortens it passes successively through the colours of the spectrum – red, orange, yellow, green, blue, indigo and violet. Just beyond the violet end of the spectrum are the rays known as ultraviolet, followed at much shorter wavelengths by X-rays and γ-rays.

Ordinary daylight and most interior artificial light is known as *white light* because it consists of a uniform mixture of all the wavelengths in the visible spectrum, the resulting colour being seen as white. To make precise optical measurements, however, a source of illumination is often required with only one wavelength, referred to as *monochromatic light*. The yellow light from a sodium lamp is frequently used as a source of this type of light.

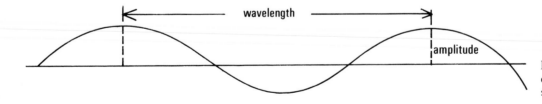

Fig. 4·6 The wave form of electromagnetic radiation such as light waves.

Reflection and refraction

When a beam of monochromatic light strikes the surface of a crystal (Fig. 4·7) part of the light is reflected from the surface and part is transmitted into it. The incident and reflected rays of light make an equal angle with the normal to the reflecting surface, that is, i=i′. The ray transmitted through the crystal does not normally maintain its original direction, since the velocity of light in air is greater than that in the crystal. This slowing down of the rays passing into the crystal causes them to make an abrupt change of direction on passing through the surface. The effect is known as refraction and the refractive index (n) of a substance is defined as the ratio of the velocity of light in air, V_a, to its velocity in the crystal V_c. If r is the angle the refracted ray makes with the normal, the

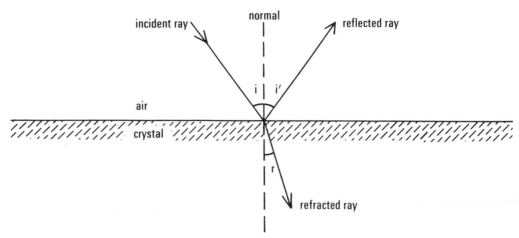

refractive index can be shown to be related to the sines of the angles of incidence and refraction as follows—

$$n = \frac{V_a}{V_c} = \frac{\sin i}{\sin r}.$$

This is a statement of the well-known Snell's law from which it can be seen that rays of light passing into minerals with high refractive indices are deflected more than rays passing into those with lower refractive indices. Note that light passing from air into a crystal is deflected towards the normal, and that light passing out of a crystal is deflected away from the normal.

If, instead of using monochromatic light, a beam of white light strikes the surface of a crystal, the refractive index is found to vary for different wavelengths. The red component of the spectrum is refracted less than rays at the violet end of the series, resulting in the splitting of the beam into a spectrum of colours, the effect being known as *dispersion*. Dispersion is not seen in light passing through substances with parallel faces, such as a sheet of glass, since the refracted rays recombine when re-emerging into air as a parallel beam of white light. Dispersion is seen at its best when a beam of light emerges through a face which is not parallel to the first face, as in a prism (Fig. 4·8).

A mineral with high dispersion has a relatively large difference between the red and violet refractive indices, hence the divergence of the beam is large. It is this high dispersive power of some minerals which gives them considerable *fire* when cut as gemstones. The cutting of gemstones is very much concerned with gaining the maximum dispersion, or fire, by skilful orientation of the faces, as will be seen on page 185.

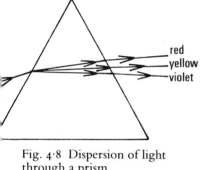

Fig. 4·8 Dispersion of light through a prism.

Total reflection As a ray of light passes from a crystal into air or a medium with a lower refractive index the transmitted light is refracted away from the

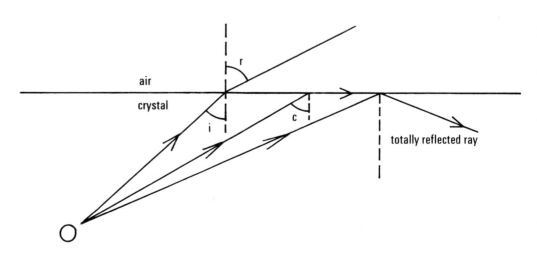

Fig. 4·9 Total reflection of light.

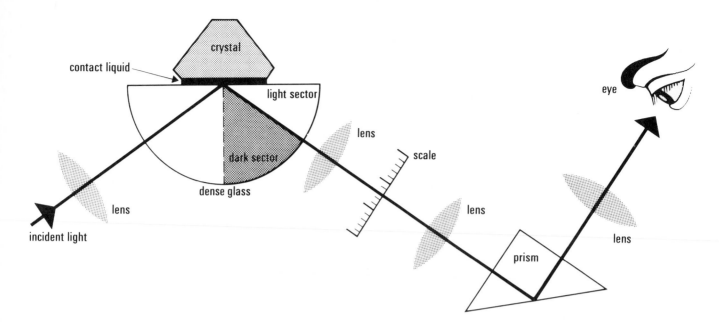

Fig. 4·10 The principle of the refractometer.

normal to the surface. As the angle of incidence increases, a situation arises where, at one particular angle, the critical angle (c), the refracted beam emerges in such a way that it passes exactly along the surface of the crystal, as shown in Fig. 4·9 by the ray refracted from OB. Further increase in the angle of incidence now means that none of the light can pass out of the crystal, so a ray such as OD is said to be *totally reflected*.

The critical angle is found from Snell's law, for the angle of the refracted beam is obviously 90°.

$$\frac{\sin i}{\sin r} = \frac{\sin c}{\sin 90°} = \frac{1}{n}$$

The term 1/n is used for the refractive index since the light is passing from a high to a low refractive index medium. Since sin 90° = 1, the equation for the critical angle becomes

$$\sin c = \frac{1}{n}.$$

The critical angle is, therefore, inversely proportional to the refractive index of the crystal. A mineral with a high refractive index has a low critical angle and more of the light passing through tends to be internally reflected. In the cutting of gemstones, total reflection is utilized in such a way as to send as much light as possible out through the front of a stone thereby producing the maximum 'sparkle'. Obviously this is easier with high refractive-index minerals like diamond or topaz, since they have lower critical angles and more light is totally reflected.

Measurement of a mineral's critical angle is the principle embodied in an optical instrument, known as the *refractometer*, used for refractive index determinations on polished minerals and gemstones with flat faces. One such instrument, known as the *Rayner refractometer*, is shown in Fig. 4·10. It contains a half cylinder section of a high refractive index glass with a polished flat surface on which is placed the crystal to be measured. A drop of liquid with refractive index intermediate between the glass and mineral is placed between them to ensure that they are in contact. The lower part of the glass cylinder is illuminated, preferably with a sodium lamp, and the light totally reflected from the mineral's surface passes into an eyepiece with a graduated scale. Depending on the value of the critical angle between the glass and mineral, a certain portion of the image in the eyepiece will be from the darker sector, shown in the diagram, and the remainder from the light sector. The boundary between the light and dark

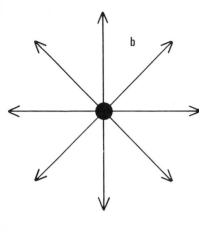

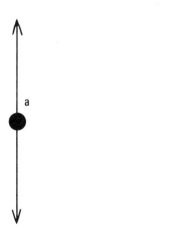

Fig. 4·11 Polarized (**a**) and unpolarized (**b**) light.

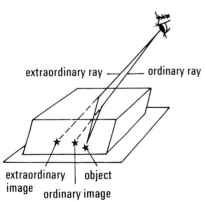

Fig. 4·12 Double refraction by calcite rhomb.

sectors falls on the eyepiece scale, calibrated to give a direct reading of the refractive index of the mineral.

The method is extremely useful for identifying gemstones and minerals with perfect cleavages, although the user must be careful to avoid stray effects, such as reading the refractive index of the contact liquid.

Refractive index by immersion methods The refractive indices of minerals are generally very characteristic for each species and form a valuable method of identification. The refractometer technique has just been mentioned, but there are two other methods suitable for a wider range of crystalline substances. Both are based on the immersion of crystals in liquids of known refractive index. If the indices of the crystal and liquid are very close then the crystal becomes invisible, as the boundary between the liquid and crystal is indistinct. If the two have very different refractive indices then the boundary between the two shows up sharply. An unknown mineral can be calibrated against a set of refractive-index liquids and by referring to tables of refractive indices (Table 4·10) the identity of the mineral can generally be narrowed down considerably.

If a microscope is available the process can be made very much more sensitive by using the Becke method. A few small fragments of the mineral are immersed in a little liquid on a glass slide and viewed through a low-power microscope objective. Owing to total reflection at the edge of each fragment, a narrow band of light surrounds each grain. If the grain has a higher refractive index than the liquid, the band of light, the so-called Becke line, occurs on the mineral side of the boundary whereas if the liquid has the higher index, the band occurs in the liquid. The Becke line will be seen to move if the microscope objective is raised slightly, the important rule being: *as the microscope objective is raised, the Becke line moves towards the substance having the higher refractive index.* In other words, if the liquid has a lower refractive index than the mineral, the Becke line moves into the mineral grain, and vice versa. By testing mineral fragments with a succession of liquids, very accurate refractive index measurements can be made.

Double refraction and polarized light Up to this point it has been assumed that light passing through minerals is refracted equally in all crystallographic directions. This is true for crystals belonging to the cubic system, as they possess a single refractive index independent of the crystal orientation. Minerals of this type, which transmit light with equal velocities in all directions are known as *isotropic.* Minerals which do not fall in the cubic crystal system transmit light with different velocities in different crystallographic directions, and are known as *anisotropic.*

To understand the behaviour of light in anisotropic substances, it is necessary to know something of the nature of propagated light waves. A beam of light is an electromagnetic radiation with vibration directions normal to the direction of propagation. Rays of light emerging vertically upwards through a pinhole (Fig. 4·11a) vibrate in any direction parallel to the page, this type of light being known as unpolarized. If, however, the vibration direction is confined to a single direction (Fig. 4·11b) then it is referred to as plane polarized light.

A ray of light passing into an anisotropic crystal is split into rays vibrating at right angles to one another, usually with different refractive indices. In calcite crystals, the difference between the two refractive indices, known as the *birefringence*, is so great (0·17) that objects viewed through a crystal appear doubled. In other anisotopic crystals, quartz for instance, the birefringence is so low (0·009) as to be detectable only by microscopic techniques. The splitting of light into two rays by anisotropic crystals is known as *double refraction.*

The double refraction produced by a rhomb of calcite is worth examining in

89

more detail. A star drawn on a sheet of paper viewed through a crystal appears as two images (Fig. 4·12). If the crystal is rotated, one star remains stationary, while the other star revolves around it. The stationary image is that produced by the so-called *ordinary rays*, plane polarized but passing through the crystals as if it were an isotropic medium. The other image is formed from the *extraordinary rays*, polarized at 90° to the ordinary rays and with variable refractive indices depending on the crystallographic direction.

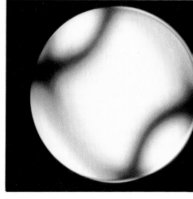

An interference figure from baryte, a biaxial mineral.

Uniaxial crystals Crystals which refract light in two component rays in this way are termed *uniaxial*, since they have one direction (the optic axis) in which no double refraction occurs. Uniaxial minerals belong to the tetragonal and hexagonal crystal classes. Wave fronts for the ordinary and extraordinary rays in uniaxial crystals are shown in Fig. 4·13. The ordinary ray behaves as if it were travelling in an isotropic crystal and consequently its velocity is unchanged whatever direction it takes. The ordinary ray wave front is therefore spherical. The extraordinary rays, however, have a velocity which varies with direction and the wave front is a rotation ellipsoid. Where the sphere and ellipsoid touch the velocities are equal and the optic axis passes through these points.

A crystal which has its ordinary ray refractive index less than that of the extraordinary ray is said to be *uniaxial negative* (Fig. 4·13a). *Uniaxial positive* minerals, on the other hand, have the ordinary refractive index greater than the extraordinary. Calcite is an example of a uniaxial negative mineral, while quartz is uniaxial positive. The refractive index of the ordinary ray is usually denoted by the Greek letter ω (omega), while the extraordinary ray is given as ε (epsilon).

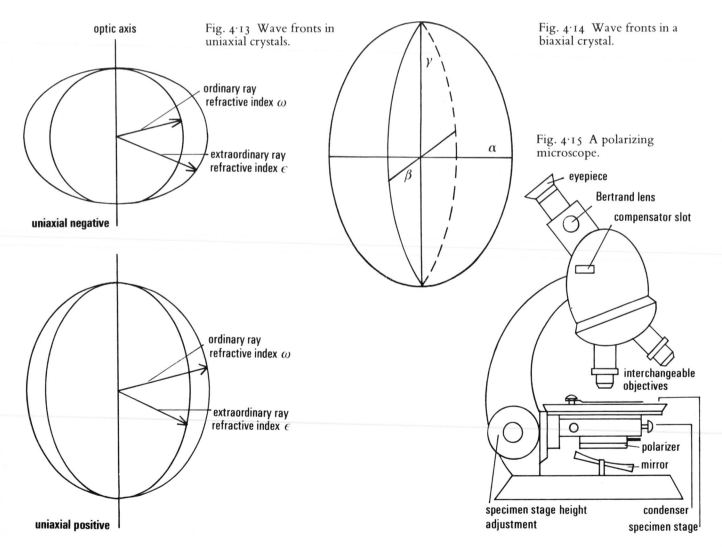

optic axis

Fig. 4·13 Wave fronts in uniaxial crystals.

ordinary ray refractive index ω

extraordinary ray refractive index ϵ

uniaxial negative

γ

β

α

Fig. 4·14 Wave fronts in a biaxial crystal.

Fig. 4·15 A polarizing microscope.

eyepiece

Bertrand lens

compensator slot

ordinary ray refractive index ω

extraordinary ray refractive index ϵ

uniaxial positive

interchangeable objectives

polarizer

mirror

specimen stage height adjustment

condenser

specimen stage

Table 4·3	crystal class	optical group
	cubic	isotropic
	hexagonal, tetragonal	uniaxial
	orthorhombic, monoclinic, triclinic	biaxial

a

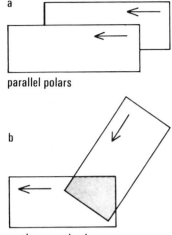

parallel polars

b

partly crossed polars

c

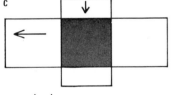

crossed polars

Fig. 4·16 Principle of polaroid.

Biaxial crystals Minerals which belong to the orthorhombic, monoclinic and triclinic crystal classes have lower crystallographic symmetry than isotropic and uniaxial minerals. Consequently their optical properties are more complicated, the wave front diagram being a triaxial ellipsoid with three axes defined by the three refractive indices α, β and γ for this group of minerals (Fig. 4·14). Such minerals have two directions with a single refractive index value, that is, two optic axes, hence the term biaxial. The angle between these axes is termed the *optic axial angle*. The three refractive indices are assigned in such a way that α is the lowest value, γ the highest, and β the intermediate value. A mineral is said to be *biaxial positive* if β is nearer to α, or *biaxial negative* if β is nearer γ.

The polarizing microscope All minerals occurring in a crystalline form must fall into either the isotropic, uniaxial, or biaxial groups, the optical and crystallographic groups being related as shown in Table 4·3.

The uniaxial and biaxial minerals can be further subdivided into positive and negative groups. By assigning a mineral to its correct group and measuring its refractive indices, an extremely powerful method of mineral identification is available.

In order to make these determinations it is usually necessary to use a polarizing microscope (Fig. 4·15). This type of instrument differs from other optical microscopes in that the user can view the specimen in polarized light. In modern microscopes the polarized light is produced by passing the illuminating beam through a sheet of polaroid. This is a very thin slice of a crystalline iodine compound, each individual crystal being in the same orientation and able to pass light vibrating in one direction only. A second polaroid disc, the analyser, is situated in the barrel of the microscope and the two can be rotated with respect to one another.

If two sheets of polaroid are in parallel orientation (Fig. 4·16a) the system passes the maximum amount of polarized light. If one of the polars is turned through an angle of 45° (Fig. 4·16b) a substantial portion of the light is absorbed. At 90° inclination (Fig. 4·16c), the polarized light produced by the lower polar is fully absorbed in the upper polar, and complete extinction of the transmitted light occurs. In this position the polars are said to be crossed.

When isotropic minerals are viewed through crossed polars they remain dark even when the microscope stage is rotated, since polarized light passing through an isotropic substance does not have its vibration direction altered. This group of minerals is therefore instantly recognizable when a few fragments on a microscope slide are examined in this way.

Anisotropic minerals possess two vibration directions at right angles and the polarized light entering a crystal (Fig. 4·17) is split into two components along the two vibration directions. The analyser, set in the crossed position, transmits the horizontal vibration component from the crystal. If the vibration directions of the crystal are parallel to the axes of the polarizer or analyser, then no light emerges from the analyser. There are four such positions for an anisotropic crystal and as the microscope stage is rotated the light passing through the crystal is extinguished four times for each complete revolution. This behaviour is an instant indication that the mineral under examination is anisotropic.

An anisotropic mineral examined in white light in a microscope with crossed polars shows, in positions of maximum illumination, characteristic colours, known as *interference colours*. The colour does not change as the microscope stage is rotated, but it does alter in intensity. These colours are graded into a series known as *Newton's interference colour scale* which is included in many optical texts. In the study of minerals the position of an interference colour on Newton's scale is determined by the thickness of the mineral grain and its birefringence. The interference colours of thicker mineral grains occur higher up the scale as do grains with higher birefringence.

To determine whether an anisotropic mineral is uniaxial or biaxial it is generally necessary to examine fragments of the mineral in convergent light, produced by inserting the condenser lens below the microscope stage. For suitably orientated mineral grains this has the effect of producing *interference figures*. These figures can be seen when a high-powered objective is used, with the polars crossed, and can be studied either by inserting the Bertrand lens, or by removing the eyepiece. Apart from outlining the difference between uniaxial and biaxial figures and the method for determining the optic sign, detailed study of interference figures is beyond the scope of this book.

A section of a uniaxial mineral cut perpendicular to its *c*-axis or optic axis gives a cross-shaped interference figure (Fig. 4·18a). Grains slightly disorientated give the same figure, but with the cross off-centre (Fig. 4·18b, c).

A typical biaxial interference figure is somewhat different as shown in Fig. 4·19, occurring in this form when the centred grain is 45° from an extinction position. There are many variants on this biaxial figure for off-centred grains, and the reader is referred to more advanced optical texts for a detailed treatment.

From what has already been described, a means is available for distinguishing between isotropic, uniaxial, and biaxial minerals. It only remains to determine the optic sign for uniaxial and biaxial minerals, which is achieved by the use of an accessory plate which will slide into a slot in the microscope tube. These plates are usually thin wedges of quartz, or thin plates of mica or gypsum which introduce a phase difference to light waves vibrating parallel and at right angles to the long dimension of the plate. Without going into too much detail, blue and yellow coloured bands appear around the edges of the dark interference bands. The positions of these coloured bands are determined by the optic sign, and are shown in Fig. 4·20 for uniaxial and biaxial positive and negative minerals, for the case in which the slow vibration direction of the accessory plate is perpendicular to its length. If the slow direction is along the length of the accessory plate the colours are reversed.

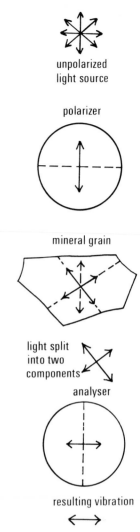

Fig. 4·17 Behaviour of polarized light in an anisotropic mineral grain.

 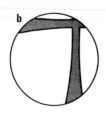 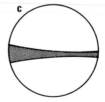

Fig. 4·18 Uniaxial interference figures.

Tables of refractive indices and other optical properties mentioned in this chapter will be found in Table 4·10.

Before concluding this section on the optical properties of minerals, several other terms often encountered in mineral descriptions should be mentioned.

Extinction angles For an anisotropic mineral grain, four extinction positions occur during a complete rotation of the microscope stage. Notice should be taken of the angle between a prominent cleavage or crystal edge and the microscope cross-wires in an extinction position. If extinction occurs parallel to a

Fig. 4·19 Biaxial interference figure.

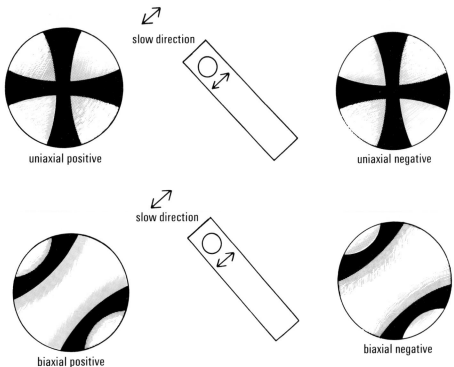

Fig. 4·20 Uniaxial and biaxial figures and determination of optic sign.

slow direction

uniaxial positive

uniaxial negative

slow direction

biaxial positive

biaxial negative

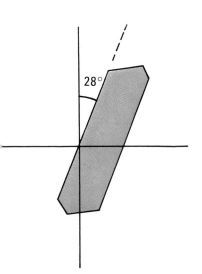

Fig. 4·21 Hornblende extinction angle.

prominent edge, then the mineral is said to have straight extinction. If it extinguishes at an angle it has *oblique extinction* as exemplified by a crystal of hornblende (Fig. 4·21) which commonly gives an extinction angle in the range of 28°. The extinction angle is of diagnostic importance in the examination of many mineral species.

Pleochroism Pleochroism is the property of numerous doubly refracting coloured minerals selectively to absorb polarized light. It cannot be observed in colourless or isotropic crystals. Pleochroic mineral grains should be examined with the microscope by rotating the polarizer without the analyser in position, and observing the colour changes in the grains. Minerals can in this way be classed as strongly or weakly pleochroic.

Chemical properties

Most minerals are *chemical compounds*, substances made up from the combination of two or more chemical elements. A few minerals, known as the native elements, occur as substances made up predominantly of one element only. An *element* is a substance which cannot be divided into a simpler substance by any chemical process. Over 100 are known to exist, although many of the very heavy elements are radioactive and unstable. Elements can be arranged in a table of *atomic numbers* and *atomic weights* as shown in Table 4·11. The atomic number corresponds to the number of electrons surrounding the nucleus in each element, varying from 1 for hydrogen to 83 for the heaviest non-radioactive element, bismuth, and to 92 for the most common radioactive element, uranium.

An element's atomic weight is the weight of an atom relative to that of a standard, often taken to be a twelfth of the weight of an atom of the isotope carbon 12. The list of atomic weights should, on this basis, be a series of integral numbers, but it is not. This important fact arises because many atoms have atomic nuclei containing different numbers of neutrons, but identical numbers of protons and associated electrons. Such elements are said to exist as different *isotopes* and the atomic weight is the average of the weights of the isotopes present in the natural element. All isotopes of a given element have identical chemical properties.

Chemical analyses and formulae

Minerals, as chemical compounds with fixed proportions of elements, can be represented by *chemical formulae*. Thus, calcite is given as $CaCO_3$ indicating that calcium, carbon and oxygen are combined in the ratio $1:1:3$. $CaCO_3$ is usually referred to as one formula unit of calcite. From the atomic weights of the individual elements, its formula weight will be 100·09. In a sample of calcite, therefore,

$\dfrac{40 \cdot 08}{100 \cdot 09} \times 100$ per cent of the sample weight will be the element calcium,

$\dfrac{12 \cdot 01}{100 \cdot 09} \times 100$ per cent will be carbon and

$\dfrac{48}{100 \cdot 09} \times 100$ per cent oxygen.

This process can be reversed to deal with an unknown mineral. Once the chemical elements it contains have been identified, their concentration can be measured and a possible formula for the mineral derived.

As an example, consider the mineral chalcopyrite which contains the elements copper, iron and sulphur. By measuring the concentration of each in a pure sample of the mineral, ideally the analysis obtained would be copper 34·63 per cent, iron 30·43 per cent, and sulphur 34·94 per cent. The formula of the mineral can then be calculated from the analysis as shown in Table 4·4.

Chalcopyrite can, therefore, be represented by the formula $CuFeS_2$. No analysis will give the atomic proportions exactly, but will deviate slightly from the ideal values, due principally to inaccuracies in the experimental technique and, occasionally, to variations in the composition of the mineral.

It is not possible to measure directly the oxygen content of a mineral, so the concentration of each element in an oxygen-bearing mineral is expressed as a percentage of the oxide of the element. An ideal analysis of the zirconium silicate, zircon, is presented in Table 4·5.

The formula of zircon is, therefore, ZrO_2SiO_2 or $ZrSiO_4$. This basic procedure is followed for calculating the formulae of all minerals, including much more complicated silicates than the example given above.

Atomic bonding in minerals

Four basic types of atomic bonding can occur in solid substances, namely ionic, covalent, metallic and residual bonding. In mineralogy the first of these is by far

Table 4·4

element	percentage by weight	atomic weight	atomic proportions	
copper	34·63	63·55	0·5449	1
iron	30·43	55·85	0·5449	1
sulphur	34·94	32·06	1·0898	2

Table 4·5

oxide	percentage by weight	molecular weight of oxide	molecular proportions	
ZrO_2	67·22	123·22	0·5455	1
SiO_2	32·78	60·09	0·5455	1

the most important, although there are examples of minerals in each group and many substances show a combination of bonding types.

Ionic compounds

An atom which has lost an electron becomes positively charged, since it has one unbalanced proton charge on the nucleus. Similarly an atom with an extra electron is negatively charged. These charged atoms are known as *ions*, positively charged ions being *cations* and negatively charged ions, *anions*. The valency of an ion is indicated by the number of electrons lost or gained, atoms tending to lose or gain electrons so as to attain the stable configuration of an inert gas. In ionic compounds anions and cations combine to form electrically neutral structures. In the mineral halite (NaCl) the sodium atoms lose one electron each to become Na^{+1}, the electrons becoming attached to the chlorine atoms, which then all become Cl^{-1}. Electrostatic attraction between these positively and negatively charged ions leads to the well-known cubic halite structure based on electrically neutral NaCl units (Fig. 4·22). In fluorite (CaF_2) the calcium atoms lose two electrons sharing one with each of a pair of fluorine atoms as shown in Fig. 4·23.

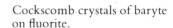
Cockscomb crystals of baryte on fluorite.

About nine-tenths of all known mineral species can be regarded as ionic compounds, the native elements and sulphides being the major exceptions. In the diagrams shown above, the sodium ion, with a radius of 0·97 angstrom, is much smaller than the chlorine ion (1·81 angstrom), and calcium (0·99 angstrom) is smaller than fluorine (1·36 angstrom). In general, cations are invariably smaller ionic units than anions. For example, oxygen, with a relatively large ionic radius of 1·40 angstrom, is present in most oxide minerals as packed structures of oxygen ions with the cations squeezed into the interstices.

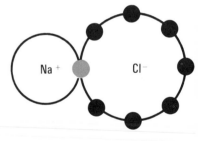
Fig. 4·22 Ionic bonding in the mineral halite (NaCl).

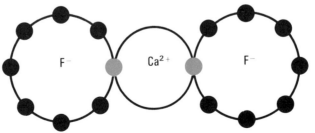

Fig. 4·23 Ionic bonding in the mineral fluorite (Ca F$_2$).

Covalent compounds

These compounds attain stable outer electron structures by sharing electrons with adjacent atoms, as in the formation of the chlorine molecule (Cl$_2$) shown in Fig. 4·24. No oppositely charged ions are involved in this type of bonding. Covalency is common in organic compounds, but in the mineral kingdom, exclusively covalent compounds are rare. Diamond is the best example, each carbon atom sharing one electron with four neighbouring atoms.

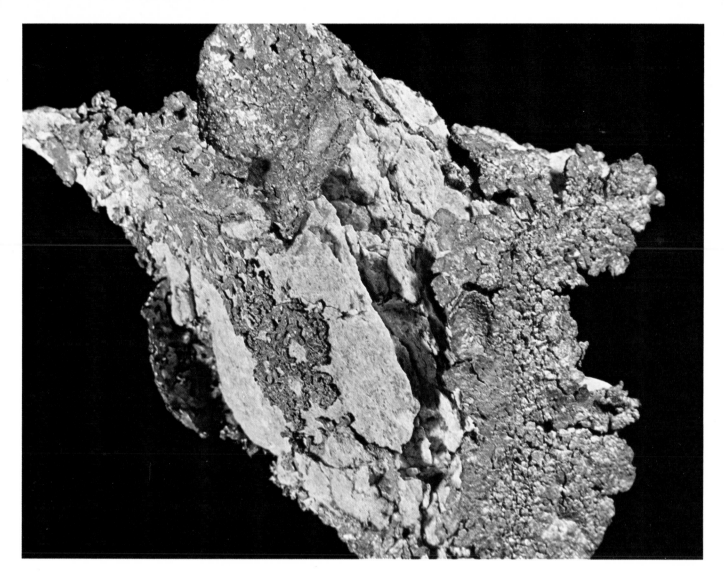

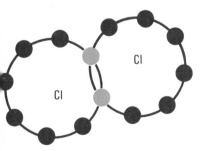

Native copper – metallic bonding in this mineral accounts for its good electrical conductivity.

Fig. 4·24 Covalent bonding in the chlorine (Cl_2) molecule.

Metallic bonding

The native metals and some sulphides possess this type of bonding, in which a number of positively charged metallic ions occupy fixed structural positions, but a 'cloud' of electrons, neutralizing the positive charge, moves about within the structure. This electron cloud is responsible for the good electrical and thermal conductivity of these minerals.

The residual (or Van der Waals) bond

This is a weak form of bonding between atoms with closed valency shells and is encountered in crystals of the inert gases. The element sulphur is partially bonded in this way, groups of eight atoms being covalently linked to form S_8 molecules. Residual bonding is responsible for the cohesion of the groups of molecules in the crystal.

Ionic sizes

When considering the ways in which groups of ions form crystal structures it is convenient to visualize ions as spherical bodies and, as mentioned previously, the size of these spheres varies from one ion to another. When crystals are made up of ions with identical radii, they tend to form highly regular structures. One ideal structure is known as *hexagonal close packing*, and another arrangement produces an overall cubic symmetry known as *cubic close packing*. Most minerals, however, are combinations of ions of different sizes and the resulting structure must consist of a geometrically stable arrangement of the various ions. This

97

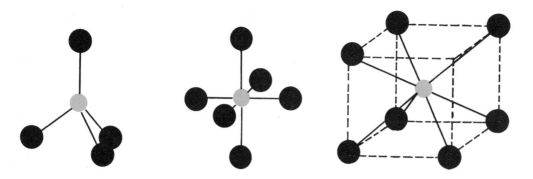

Fig. 4.25 Structures showing four-, six-, and eight-fold coordination.

requirement is reflected in the multiplicity of crystal forms found among the mineral species.

Ionic compounds must be electrically neutral, that is, the total positive and negative charge within the structure must balance. This property leads to a very useful way of checking the correctness of a mineral's formula, since the total anionic charge must equal the total cationic charge, even in chemically complicated minerals.

As mentioned previously anions tend to have larger ionic radii than the cations with which they are associated in minerals, so the structures have the appearance of anionic frameworks with the cations squeezed into the gaps. Important anions in mineralogy are O^{-2}, F^{-1}, Cl^{-1}, S^{-2}. Each cation in a mineral's structure is therefore surrounded by a group of anions; the number of anions linked to any one cation is known as its *coordination number*. Figure 4.25 shows three structure units with coordination numbers of four, six and eight. Many elements occur in only one typical coordination form, whereas others may have two or more values, aluminium being a good example. Ions of aluminium may assume fourfold or sixfold coordination and many rock-forming minerals have this element showing both types of coordination within one structure.

Complex anions

Many minerals are compounds with single element anionic radicals such as O^{-2} or F^{-1} (oxides and fluorides). An even larger number, however, have anionic groups which are compounded from a number of anions and a cation, with an overall negative charge. A number of these complex anions are – carbonates $(CO^3)^{-2}$, sulphates $(SO_4)^{-2}$, arsenates $(AsO_4)^{-3}$, phosphates $(PO_4)^{-3}$, silicates $(SiO_4)^{-4}$.

These groups are strongly bound together and behave as if they were a single anionic unit. The hydroxyl ion $(OH)^{-1}$ can also be included in this group.

Isomorphism and solid solution

Many mineral species can be arranged in groups with closely related crystal structures, but with one or more of the ions replaced by another of similar size and coordination number. One example is the baryte group of orthorhombic minerals — anhydrite ($CaSO_4$), celestine ($SrSO_4$), baryte ($BaSO_4$), anglesite ($PbSO_4$).

These minerals all have the sulphate anionic group in common. The cations are not related chemically although Ca, Sr, and Ba are in the same group of the periodic table; all four have similar ionic radii and identical coordination. Often, pairs of isomorphous minerals have both anions and cations exchanged, so that they contain no common chemical element or group. Calcite ($CaCO_3$) and soda nitre or nitratine ($NaNO_3$) are examples, both having the same type of hexagonal structure.

Isomorphous groups of minerals are structurally related but this does not mean that minerals with intermediate compositions necessarily exist between a pair of such minerals. For instance, there is no known example of lead substituting for barium in baryte. However, in certain groups atomic substitution takes place between isomorphous end-members, giving rise to a special case of isomorphism known as *solid solution*. The olivine series consists of solid solution between fayalite (Fe_2SiO_4) and forsterite (Mg_2SiO_4) with unlimited substitution of iron for magnesium. Most naturally occurring olivines fall towards the forsterite end of the series, so the olivine formula is normally written $(Mg,Fe)_2SiO_4$. The elements shown in the brackets can substitute for one another in the structure; writing it in this way implies that the ions of magnesium outnumber those of iron, since the dominant cation is placed first in the group.

Substitution between ions of similar size and coordination is the usual form of solid solution, but substitution between differently coordinated ions can also occur. Between albite and anorthite ($NaAlSi_3O_8 - CaAl_2Si_2O_8$) in the feldspar series substitution takes place between Na^{+1} and Ca^{+2}. This produces a change in the mineral's electronic charge which is balanced by an equivalent amount of Al^{+3} substituting for Si^{+4}, so that, at all compositions, electrical neutrality is maintained. This process, known as *coupled substitution* is very widespread among rock-forming minerals.

The amount of atomic substitution which can occur in a mineral depends on four factors –

The similarity in ionic radii of the substituting elements.

The appropriate coordination number or ability of the mineral to enter into coupled substitution.

The type of atomic structure – the quartz structure, for instance, permits very little substitution for Si^{+4}, whereas the spinel and garnet groups form numerous solid solution series.

The temperature at which the mineral was formed, since more atomic substitution takes place at higher temperatures. For certain mineral species, the degree of substitution possible at certain temperatures is known. The mineral, when chemically analysed, can serve as an indicator of the temperature at which it was formed, thus acting as a *geothermometer*.

Certain minerals showing extensive atomic substitution at high temperatures may have compositions and structures which are not stable at lower temperatures. In such circumstances, as a high-temperature mineral cools it may break down into two separate mineral phases stable at the lower temperatures. This process is known as *unmixing* or *exsolution*. A well-known example of unmixing is found in the feldspar known as perthite, an intergrowth of sodium-rich albite and potassium-rich microcline. At higher temperatures the mineral is a homogeneous sodi-potassic feldspar.

Interstitial solid solution

The atomic substitutions considered in the previous section arose from the replacement of ions in structural sites by ions of similar size and coordination. Another form of substitution occurs when small ions are able to find stable positions in gaps in fairly open crystal structures. In this way Na^{+1} ions can enter openings in the structures of cristobalite and tridymite, both these minerals being high-temperature forms of silica (SiO_2). The resulting imbalance in the electrical charge is restored by the coupled substitution of Al^{3+} for Si^{4+}.

Defect structures

Chemical analyses of some minerals sometimes show an excess of one element and a deficiency of another when compared to the ideal composition of the mineral. Pyrrhotine is the best known example, usually showing slightly more sulphur and less iron than the ideal mineral formula FeS. This was thought at one time to be due to extra sulphur ions occupying interstitial structural sites, but now it has been found to be due to a number of the iron sites being unoccupied, forming what is known as a defect structure.

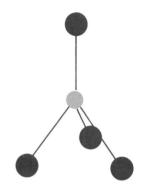

Fig. 4·26 An isolated SiO_4 tetrahedron, found in the nesosilicates.

Silicate mineral structures

Silicate minerals are very widespread, constituting about 95 per cent of the Earth's crust and are present in all common rocks apart from limestones. Their structures are worth considering in more detail in order to appreciate the differences between the major silicate mineral groups.

Fine groups of radiating crystals of hemimorphite from Mapimi, Mexico.

Fig. 4·27 Linked SiO_4 tetrahedra, found in the sorosilicates.

The basic unit in all silicate minerals is the SiO_4 tetrahedron, in which one oxygen atom is situated at each of the four corners of the tetrahedron, with the silicon atom at its centre. In the diagram (Fig. 4·26) the size of the oxygen atom has been reduced relative to that of the silicon atom in order to make the bond directions clearer. These tetrahedra can occur in the silicate structures, either as single units, or joined together in specific arrays. The classification, outlined in the following paragraphs, is based on the types of structures built up by these tetrahedra.

Nesosilicates

These are minerals built up from cations bonded with isolated SiO_4 as illustrated in Fig. 4·26. Considering the charges on the individual ions, isolated tetrahedra must be anionic $(SiO_4)^{-4}$ groups, forming minerals such as zircon $(ZrSiO_4)$, olivine [$(Mg,Fe)_2SiO_4$] and the garnets, for example, grossular $(Ca_3Al_2Si_3O_{12})$.

Sorosilicates

This group contains linked pairs of SiO_4 tetrahedra, formed by the sharing of one oxygen ion (Fig. 4·27). The anionic group is therefore $(Si_2O_7)^{-6}$, minerals in the group having a cationic charge of $+6$. An example is hemimorphite [$Zn_4Si_2O_7$ $(OH)_2.H_2O$].

Cyclosilicates (ring silicates)

In a number of silicates, the SiO_4 tetrahedra are linked into rings, which may consist of groups of three (Fig. 4·28), four, or six linked tetrahedra. For three linked tetrahedra the anionic groups are $(Si_3O_9)^{-6}$, as in the mineral benitoite $(BaTiSi_3O_9)$; four linked tetrahedra form $(Si_4O_{12})^{-8}$ groups, as in axinite [$(Ca,Mn,Fe)_3Al_2(BO_3)Si_4O_{12}(OH)$] and groups of six form $(Si_6O_{18})^{-12}$ groups as shown by beryl $(Be_3Al_2Si_6O_{18})$.

Inosilicates (chain silicates)

In this group the linkage is similar to the ring silicates, two oxygens from each tetrahedron being linked to adjacent tetrahedra, but instead of forming closed rings, the tetrahedra form straight chains of indefinite length. These chains may be single chains, forming basic $(SiO_3)^{-2}$ groups or double linked chains with $(Si_4O_{11})^{-6}$ anionic groups (Fig. 4·29). The most important single chain minerals are the pyroxenes of which diopside, from a chemical point of view, is one of the simplest with a formula $CaMg(Si_2O_6)$.

Augite is one of the aluminous pyroxenes, containing aluminium ions in two different coordination states, distributed between divalent and quadrivalent silicon sites in such a way that charge balance is maintained. Augite's formula is $(Ca,Mg,Fe,Al)_2(Si,Al)_2O_6$.

The amphibole group is the most important group of minerals forming as double chain compounds. Actinolite [$Ca_2(Mg,Fe)_5Si_8O_{22}(OH)_2$] is a typical example, the $(OH)^{-1}$ in these minerals forming an additional anionic component. Some amphiboles also show coupled substitution of aluminium as in the hornblende series [$NaCa_2(Mg,Fe,Al)_5(Si,Al)_8O_{22}(OH)_2$]. Part of the hydroxyl group in many amphiboles may be replaced by fluorine.

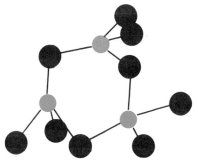

Fig. 4·28 SiO_4 tetrahedra, linked in a ring, as found in the cyclosilicates.

Phyllosilicates (sheet silicates)

These are silicate structures with a sheet-like form produced by the linkage of three of the oxygen ions to adjacent SiO_4 tetrahedra (Fig. 4·30). This results in the formation of $(Si_2O_5)^{-2}$ anionic groups and included among these silicates are many of the characteristically flaky minerals, such as kaolinite, the chlorites and the mica group.

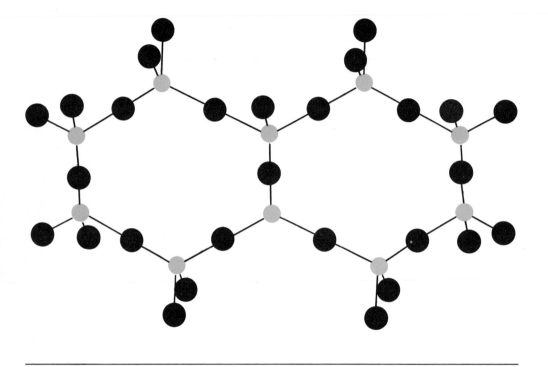

Fig. 4·29 Single and double chain silicate structures (inosilicates).

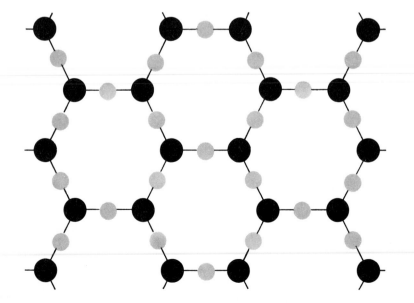

Fig. 4·30 Sheet silicate structures (phyllosilicates).

Above
Biotite, demonstrating the perfect cleavage parallel to the crystal sheets.

Above right
A typical specimen of actinolite, showing the massive bundles of needle-shaped crystals.

Among the micas, lepidolite has the formula $KLi_2Al(Si_4O_{10})(OH)_2$, although many of the common micas contain aluminium in some of the silicon sites, as in biotite $[K(Mg,Fe)_3(AlSi_3O_{10})(OH)_2]$. The atomic bonding perpendicular to the sheet structure in micas is very weak and cleavage occurs very easily along these planes giving the minerals their highly characteristic appearance.

Tectosilicates (framework silicates)

These are silicate structures built up on a three-dimensional framework of linked SiO_4 tetrahedra. Each oxygen atom in one tetrahedron is linked to an adjacent tetrahedron, leading to a basic SiO_2 structural unit. Unlike all the previous silicate structural types, the framework structures are electrically neutral in themselves and do not need the addition of further cations. There are, therefore, minerals built up exclusively of this basic unit, these being the minerals of the silica group, quartz, tridymite, and cristobalite. If other major constituents are to enter this framework silicate structure, the electrical balance must be disturbed by the replacement of some silicon by aluminium and the simultaneous entry of other cations into the structure to restore the balance. This is the basic crystal-chemical feature of the very important feldspar group of rock-forming minerals, which are found in almost all igneous rocks.

In orthoclase feldspar ($KAlSi_3O_8$), one-quarter of the silicon ions are replaced by aluminium and the charge balance restored by the presence of one potassium for each aluminium. In albite ($NaAlSi_3O_8$), sodium occurs instead of potassium. Anorthite is another important member of the feldspar group, but differs from those already mentioned in that half the silicon sites are occupied by aluminium, charge balance being restored by the presence of divalent calcium, giving a formula of $CaAl_2Si_2O_8$.

Ideal end-member feldspar compositions are very rarely found in the rocks normally encountered, but isomorphous substitution between certain group members produces feldspars of intermediate composition. Between albite and anorthite a complete range of compositions occurs, produced by the coupled substitution of NaSi for CaAl. These feldspars form the so-called plagioclase series.

Polymorphism in minerals

The tendency of certain chemically distinct minerals to fall into groups with related crystal structures has been mentioned under the heading isomorphism. *Polymorphism,* on the other hand, is the property possessed by a number of chemically identical minerals to exist in more than one structural form. It provides evidence that the structure of a mineral is not determined exclusively by its chemical composition but that its formation is subject also to further environmental factors, such as temperature and pressure.

A mineral which has two polymorphs is said to be dimorphous, native carbon being a very good example. This occurs in the forms of cubic diamond and hexagonal graphite, diamond being formed under high-pressure conditions and graphite under high-temperature conditions. A few minerals are trimorphous, in other words they exist in three polymorphous forms, the natural occurrences of titanium dioxide (TiO_2) being important examples. Rutile (tetragonal) is the most commonly found of the three oxides, usually being formed in conditions where iron is present. Anatase (tetragonal), on the other hand, is characteristically found in iron-deficient environments, while brookite (orthorhombic) often forms in the presence of niobium.

Since polymorphous minerals are formed under specific conditions of chemistry, temperature and pressure, their presence in a rock provides valuable information on the conditions under which the rock was formed.

Polymorphs of a mineral may readily change from one form to another when the right chemical-temperature-pressure conditions prevail. If the reactions are reversible they are known as *enantiotropic,* the three crystalline forms of silica (SiO_2) altering at specific temperatures as follows —

quartz $\rightleftharpoons$ tridymite $\rightleftharpoons$ cristobalite

(hexagonal)   867°C   (orthorhombic)   147°C   (cubic)

Sometimes the reactions are irreversible, one of the polymorphs being less stable than the other, and the reaction is then said to be *monotropic,* the marcasite-pyrite transition being an excellent example. Both are forms of iron disulphide (FeS_2), pyrite being the common cubic form while the rather rarer marcasite is orthorhombic. Marcasite is formed at lower temperatures than pyrite and in acid environments. Above 350°C, marcasite is converted into pyrite, but the change

A pyrite nodule from the chalk of south-east England, showing the radiating crystal development on the broken end.

A group of striated cubes of pyrite from Northern Sporades, Greece.

is irreversible since pyrite cannot be converted into marcasite in the solid state.

The higher temperature polymorph of a group of minerals generally has a more open crystal structure and hence its density will be lower, as illustrated by the quartz-cristobalite-tridymite series mentioned previously. Their specific gravities are 2·65, 2·33, and 2·27 respectively.

Pseudomorphs

Pseudomorphism is a process by which one mineral is replaced by another without any change in the external form of the original mineral. The process sometimes occurs between two polymorphous forms of a mineral, such as rutile and brookite. If the environment in which the brookite forms alters to one in which it is unstable, causing its alteration to rutile, it retains the form of the original brookite crystal. The crystal is then said to be a paramorph of rutile after brookite. Another common paramorph is calcite after aragonite.

The formation of paramorphs is essentially a structural alteration, one of the types of pseudomorphism found in mineralogy. Chemical alteration is another very important process which can change a mineral's identity. Thus a pseudomorph by replacement is obtained when, due to changes in its chemical environment, one mineral is progressively altered to another.

Alteration may take place through the loss of one chemical component as in the reduction of cuprite to native copper or in the addition of a component, as in the production of gypsum pseudomorphs after anhydrite by the action of water on the anhydrite.

$$CaSO_4 + 2H_2O = CaSO_4.2H_2O$$
anhydrite $\qquad\qquad$ gypsum

This process of replacement may not always be completed in a particular specimen and so, occasionally, partial pseudomorphism is observed. Under certain conditions, the copper mineral azurite $[Cu_3(CO_3)_2(OH)_2]$ will alter to malachite $[Cu_2CO_3(OH)_2]$. If the process is arrested before completion, crystals of intergrown bright green malachite and blue azurite may be found.

Pseudomorphic alteration of some minerals occurs by means of anionic exchange, as in the production of goethite pseudomorphs after pyrite, in which the iron sulphide is converted into a hydrated oxide.

The most spectacular form of pseudomorphism, however, occurs between certain pairs of minerals which have no chemical elements in common. From certain European deposits come cubic crystals apparently of fluorite, but appearance in this case can be deceptive, as the fluorite has in fact been completely dissolved away and replaced by quartz. In the same way native copper has completely replaced aragonite in certain specimens from a Bolivian deposit.

Two further types of pseudomorph should be mentioned. First, somewhat related to the process outlined in the last paragraph is *infiltration*; a mineral may be dissolved out of a cavity leaving a cast of its shape which is subsequently filled by deposition of mineral matter from fresh mineralizing fluids. Secondly, pseudomorphs may be produced by *incrustation,* when a coating of one mineral forms over the surface of another. Sometimes the mineral underneath the coating is dissolved away, leaving a hollow pseudomorphous cast. Siderite coatings on baryte and fluorite are good examples of this.

Non-crystalline minerals

Minerals are sometimes defined as naturally occurring inorganic substances with characteristic crystal structures, mistakenly, since there are a number of valid mineral species which have no regular crystal structure and appear isotropic when examined under a polarizing microscope. There are three processes by which non-crystalline or amorphous minerals can be formed.

A number of minerals are formed by the evaporation of gelatinous or colloidal solutions, the prime example being opal, which is chemically similar to quartz, but with the addition of up to 10 per cent of water. The very attractive internal reflections and lustre of a good quality opal are caused by the diffraction-grating effect of the minute, closely-packed silica spheres which are characteristic of opal's microstructure. Other minerals also form as gels, examples being some of the hydrated silicates of iron and manganese. They can often be recognized from the botryoidal surfaces they develop during deposition from solution.

Substances will form as non-crystalline solids (glasses) if they cool very rapidly from the molten state without allowing the melt to equilibrate during the cooling process. Obsidian, a volcanic glass, is formed in this way and is often mistaken for a mineral although it is, in fact, a rock composed of a mixture of poorly crystalline silicate phases. The only glass which can be classified as a mineral is the rare variety of silica, lechatelierite, formed by the rapid cooling of molten silica. This process sometimes occurs when lightning strikes sand producing a rapid heating and cooling, and forming tubes of silica glass known as fulgarites.

The final group of amorphous minerals comprises those which were originally crystalline, but have had their structures destroyed by radioactive bombardment. Such minerals are known as *metamict*. They often have a similar appearance to obsidian and other dark-coloured glasses, showing their characteristic lustre and conchoidal fracture. The radioactive bombardment is the result of the decay of uranium and thorium, either contained in the mineral or from an adjacent mineral, the decay process being accompanied by the emission of α-particles. Emission from radioactive elements in a mineral, commonly produces metamict 'haloes' in surrounding minerals, while at the same time the radioactive mineral itself suffers complete destruction of its structure. Metamict minerals can be reconverted to a crystalline form by moderate heating, but whether the structures attained correspond to those which existed before radiation damage occurred is questionable.

Right
Pyrite pseudomorphs after hopper crystals of halite.

Below
A fine specimen of opal from Querétara, Mexico.

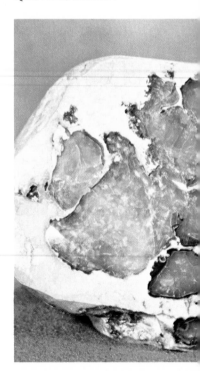

Chemical tests applied to minerals

In certain circumstances chemical tests are desirable to confirm an identification based on a mineral's physical properties, or to distinguish between two or more possibilities. Advanced quantitative analysis techniques will not be dealt with here.

The presence of a number of metals can be confirmed by reducing a mineral to a metallic residue or bead, on a charcoal block. Strong heating in a blowpipe flame is required to effect the reduction process. Characteristic colours associated with a number of metallic radicals are shown in Table 4·6.

When minerals are fused in a Bunsen burner flame the presence of certain elements is indicated by distinctively coloured flames. The most convenient way of observing these flames is to moisten a clean platinum wire with hydrochloric (HCl), nitric (HNO_3), or sulphuric acid (H_2SO_4), dip it in a little of the powdered mineral, and then place the wire in the flame. Characteristic flame colours, if seen, indicate the presence of certain elements as shown in Table 4·7.

The yellow sodium flame is very persistent and often obscures the violet potassium flame, since the two elements often occur together in minerals. A blue glass is useful for observing this type of flame, as it will absorb the sodium coloration and enable one to see the violet colour, if present. When using a platinum wire in this way, care should be taken if the elements iron, lead, and

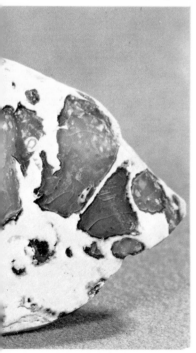

antimony are suspected, since these elements can form alloys with platinum and will lead to deterioration of the wire.

The third qualitative chemical test which can be very easily carried out is the borax bead test. A small loop is made at the end of a clean platinum wire which is then heated and dipped into powdered borax. A colourless bead will form in the loop which, while hot, is brought in contact with a little of the powdered mineral. The bead is then reheated, first in an oxidizing flame from a Bunsen burner (the intense colourless flame) and then in the less intense, yellow, reducing flame. Characteristic bead colours formed in the two types of flame are shown in Table 4·8.

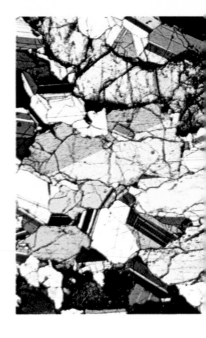

A thin section of an olivine gabbro, showing the typical lamellar twinning in the plagioclase feldspars.

Table 4·6 Charcoal block test

antimony	brittle, white-grey bead
bismuth	brittle, silver-white to reddish-white bead
cobalt	feebly magnetic residue
copper	red, malleable, spongy mass
gold	soft, yellow bead
iron	magnetic residue
lead	white, malleable bead which readily tarnishes and marks paper
nickel	feebly magnetic residue
silver	silver-white malleable bead
tin	white malleable bead which does not mark paper

Table 4·7 Flame test

barium	yellow-green
boron	yellow-green with H_2SO_4; very brief
calcium	brick red
copper	bright green with HNO_3, sky blue with HCl
lithium	red
potassium	violet
sodium	yellow
strontium	red

Table 4·8 Borax bead test

element	oxidizing flame	reducing flame
cobalt	deep blue	deep blue
chromium	yellow-green	emerald-green
copper	blue-green	opaque red
iron	yellow when hot, colourless when cold	green
manganese	red-violet	colourless
molybdenum	colourless	brown
nickel	red-brown	opaque grey
manganese	red-violet	colourless
molybdenum	colourless	brown
nickel	red-brown	opaque grey
titanium	colourless	violet
uranium	yellow	pale green
vanadium	colourless	green
tungsten	colourless	yellow

Tests for anionic radical

A few of the anionic groups in minerals can be identified by noting their behaviour in acids. All carbonate minerals are soluble in hydrochloric acid, evolving carbon dioxide (CO_2) with effervescence. Some carbonates only dissolve on heating, but others, such as calcite and aragonite, do so readily in cold acid. Many silicates decompose in hydrochloric acid forming gelatinous masses of silica. Some sulphides dissolve in this acid giving off hydrogen sulphide (H_2S) gas with its characteristic smell.

Chemical tests for other anionic and cationic groups will be found in many standard mineralogical and chemical texts.

The study of rocks and minerals

As mentioned at the beginning of this chapter, minerals are naturally occurring, inorganic compounds, usually with an identifiable crystal structure. With over ninety elements in the Earth's crust available to form chemical compounds, it is, perhaps, surprising that the number of established mineral species number only around 2500, although new species are being reported at a rate of about fifty a year. Many of these 2500 are rather rare and unlikely to be found except in very specific geological environments. Indeed, the number of widespread and reasonably common minerals, likely to be encountered by the average collector, is approximately 300.

Newly discovered, or very rare minerals are most usefully examined in well-equipped mineralogical laboratories. Here, spectrographic techniques identify the chemical elements present, while X-ray diffraction and optical goniometry are used to determine crystal symmetry. Mineral formulae can be derived from chemical or microchemical analyses or from one of the spectrochemical techniques, such as electron probe microanalysis, and optical properties from microscopic techniques.

The recognition and description of minerals in rocks is part of the science of petrography. The study involves the examination of very thin rock slices under a polarizing microscope, thereby revealing the crystalline rock fabric. A knowledge of mineralogy and petrology is necessary to identify the constituent minerals, interpret their relationships to one another, and so deduce the history of the rock.

The various physical and chemical properties of minerals considered in this chapter have placed particular emphasis on those properties which help in mineral identification. There is no set procedure for identifying a completely unknown mineral – experience and familiarity with a wide range of species are two of the greatest assets, as is a knowledge of the geological environments in which a particular mineral is likely to be found.

An inspection of a mineral's colour, habit and association usually suggests a small number of possibilities. The art of mineral identification lies in the selection and skill in using the appropriate diagnostic tests to distinguish between these possibilities as rapidly and unambiguously as possible. No single technique mentioned in the previous pages is likely on its own to give positive identifications on a whole range of specimens. Optical and microscopic techniques are most applicable to transparent minerals – refractive indices, together with other optical parameters, and position on the hardness scale should be the major diagnostic properties for these minerals. A number, however, have refractive indices which are too high to be determined in the normal range of immersion liquids and for these minerals specific gravity can be more informative. Opaque minerals are generally identified most rapidly through determinations of hardness, streak and specific gravity.

Table 4·9 Specific gravities of some common minerals

chabazite	2·08–2·16	biotite	2·8–3·2
opal	2·1	lepidolite	2·8–3·3
graphite	2·09–2·23	wad	2·8–4·4
sulphur	2·07	talc	2·8
stilbite	2·1–2·2	dolomite	2·85
stichtite	2·18	pollucite	2·9
halite	2·17	prehnite	2·9
montmorillonite	2·2–2·7	wollastonite	2·91
analcime	2·3	aragonite	2·95
tridymite	2·27	tremolite	2·98
gibbsite	2·3–2·4	anhydrite	2·98
gypsum	2·32	euclase	3·0–3·1
cristobalite	2·33	spodumene	3·0–3·2
sodalite	2·35	ankerite	3·02
wavellite	2·36	actinolite	3·05
lazurite	2·4	erythrite	3·06
brucite	2·39	lazulite	3·08
vermiculite	2·4	andalusite	3·1–3·2
apophyllite	2·3–2·4	apatite	3·1–3·2
glauconite	2·5–2·9	fluorite	3·18
microcline	2·55	enstatite	3·18
chalcedony	2·6	hornblende	3·2
orthoclase	2·57	olivine	3·2–3·4
nepheline	2·55–2·65	axinite	3·2–3·3
serpentine	2·5–2·6	diopside	3·28
kaolinite	2·6–2·7	goethite	3·2–4·3
turquoise	2·6–2·8	idocrase	3·3–3·6
albite	2·61	jadeite	3·33
antigorite	2·62	augite	3·34
quartz	2·65	epidote	3·4–3·6
beryl	2·65–2·90	sphene	3·4–3·56
scapolite	2·6–2·8	grossular	3·4–3·6
labradorite	2·70	sinhalite	3·49
calcite	2·71	realgar	3·5
phlogopite	2·75	orpiment	3·5
anorthite	2·76	diamond	3·50–3·53
muscovite	2·8–2·9	topaz	3·5–3·6

Specific gravities of some common minerals

pyrope	3·5–3·8	millerite	5·5
spinel	3·55	chalcocite	5·5–5·8
periclase	3·56	proustite	5·57
kyanite	3·6	arsenic	5·7
benitoite	3·65	bournonite	5·83
staurolite	3·7–3·8	pyrargyrite	5·85
rhodonite	3·7	crocoite	5·99
azurite	3·77	arsenopyrite	6·07
chrysoberyl	3·75	scheelite	6·10
anatase	3·90	cuprite	6·14
sphalerite	3·9–4·1	descloizite	6·2
siderite	3·96	boulangerite	6·23
celestine	3·97	cobaltite	6·33
brochantite	3·97	anglesite	6·38
corundum	4·0–4·1	skutterudite	6·5
perovskite	4·01	wulfenite	6·5–7·0
malachite	4·05	pitchblende	6·5–8·5
chalcopyrite	4·1–4·3	cerussite	6·55
brookite	4·14	antimony	6·7
rutile	4·23	bismuthinite	6·78
witherite	4·29	vanadinite	6·88
manganite	4·33	cassiterite	6·99
chromite	4·5–4·8	argentite	7·2–7·4
baryte	4·50	mendipite	7·24
pyrrhotine	4·6	mimetite	7·24
tetrahedrite-tennantite	4·6–5·1	wolframite	7·27
stibnite	4·63	löllingite	7·4
zircon	4·7	tapiolite	7·9
ilmenite	4·72	galena	7·58
molybdenite	4·6–4·7	nickeline	7·8
marcasite	4·89	cinnabar	8·09
pyrite	5·02	copper	8·95
bornite	5·07	silver	10–11
magnetite	5·18	sperrylite	10·6
columbite-tantalite	5·2–8·0	mercury	13·6
hematite	5·26	platinum	14–19
		gold	15–19

Table 4.10 Optical properties of some common minerals

mineral	optical character	refractive indices	mineral	optical character	refractive indices
fluorite	isotropic	1·43	siderite	uniaxial negative	1·63–1·87
opal	isotropic	1·45	andalusite	biaxial negative	1·63–1·64
sodalite	isotropic	1·48	baryte	biaxial positive	1·64–1·65
chabazite	uniaxial negative	1·48–1·49	spodumene	biaxial positive	1·66–1·68
analcime	isotropic	1·49	olivine	biaxial positive	1·66–1·70
natrolite	biaxial positive	1·48–1·49	rhodonite	biaxial positive	1·73–1·74
stilbite	biaxial negative	1·48–1·49	diopside	biaxial positive	1·69–1·71
calcite	uniaxial negative	1·49–1·66	idocrase	uniaxial positive	1·71–1·72
heulandite	biaxial positive	1·49–1·50	clinozoisite	biaxial positive	1·71–1·72
dolomite	uniaxial negative	1·50–1·68	spinel	isotropic	1·72
leucite	uniaxial positive	1·51	augite	biaxial positive	1·71–1·75
orthoclase	biaxial negative	1·52–1·53	kyanite	biaxial negative	1·71–1·73
strontianite	biaxial negative	1·52–1·67	pyrope	isotropic	1·74
gypsum	biaxial positive	1·52–1·53	grossular	isotropic	1·74
albite	biaxial positive	1·53–1·54	azurite	biaxial positive	1·73–1·84
chalcedony	isotropic	1·54	epidote	biaxial positive	1·73–1·77
cordierite	biaxial negative	1·53–1·54	staurolite	biaxial positive	1·74–1·75
aragonite	biaxial negative	1·53–1·69	chrysoberyl	biaxial positive	1·74–1·75
nepheline	uniaxial negative	1·53–1·54	benitoite	uniaxial positive	1·76–1·80
wavellite	biaxial positive	1·53–1·55	corundum	uniaxial negative	1·76–1·77
marialite	uniaxial negative	1·52–1·53	almandine	isotropic	1·78
halite	isotropic	1·54	spessartine	isotropic	1·80
muscovite	biaxial negative	1·56–1·60	cerussite	biaxial negative	1·80–2·08
quartz	uniaxial positive	1·55	zircon	uniaxial positive	1·93–1·99
brucite	uniaxial positive	1·56–1·58	sphene	biaxial positive	1·90–2·03
beryl	uniaxial negative	1·57–1·58	anglesite	biaxial positive	1·88–1·89
anorthite	biaxial negative	1·57–1·59	scheelite	uniaxial positive	1·92–1·94
chlorite	biaxial positive	1·61–1·64	cassiterite	uniaxial positive	2·00–2·10
rhodochrosite	uniaxial negative	1·60–1·82	pyromorphite	uniaxial negative	2·05–2·06
topaz	biaxial positive	1·63–1·64	wulfenite	uniaxial negative	2·28–2·40
hemimorphite	biaxial positive	1·61–1·64	sphalerite	isotropic	2·37
tourmaline	uniaxial negative	1·62–1·64	diamond	isotropic	2·42
smithsonite	uniaxial negative	1·62–1·85	rutile	uniaxial positive	2·62–2·90
celestine	biaxial positive	1·62–1·63	cuprite	isotropic	2·85
hornblende	biaxial negative	1·62–1·65	pyrargyrite	uniaxial negative	2·88–3·08
apatite	uniaxial negative	1·63–1·64			

Table 4·11 Elements occurring in minerals

atomic number	name	symbol	atomic weight	atomic number	name	symbol	atomic weight
1	hydrogen	H	1·008	44	ruthenium	Ru	101·07
3	lithium	Li	6·94	45	rhodium	Rh	102·91
4	beryllium	Be	9·013	46	palladium	Pd	106·4
5	boron	B	10·82	47	silver	Ag	107·87
6	carbon	C	12·01	48	cadmium	Cd	112·4
7	nitrogen	N	14·008	50	tin	Sn	118·69
8	oxygen	O	16·00	51	antimony	Sb	121·75
9	fluorine	F	19·00	52	tellurium	Te	127·6
11	sodium	Na	22·99	53	iodine	I	126·9
12	magnesium	Mg	24·31	55	caesium	Cs	132·91
13	aluminium	Al	26·98	56	barium	Ba	137·34
14	silicon	Si	28·09	57	lanthanum	La	138·91
15	phosphorus	P	30·97	58	cerium	Ce	140·12
16	sulphur	S	32·06	59	praseodymium	Pr	140·91
17	chlorine	Cl	35·45	60	neodymium	Nd	144·24
19	potassium	K	39·10	62	samarium	Sm	150·4
20	calcium	Ca	40·08	63	europium	Eu	151·96
21	scandium	Sc	44·96	64	gadolinium	Gd	157·25
22	titanium	Ti	47·90	65	terbium	Tb	158·93
23	vanadium	V	50·94	66	dysprosium	Dy	162·50
24	chromium	Cr	52·00	67	holmium	Ho	164·93
25	manganese	Mn	54·94	68	erbium	Er	167·26
26	iron	Fe	55·85	69	thulium	Tm	168·93
27	cobalt	Co	58·93	70	ytterbium	Yb	173·04
28	nickel	Ni	58·71	71	lutecium	Lu	174·99
29	copper	Cu	63·55	72	hafnium	Hf	178·49
30	zinc	Zn	65·37	73	tantalum	Ta	180·95
31	gallium	Ga	69·72	74	tungsten	W	183·85
32	germanium	Ge	72·59	76	osmium	Os	190·2
33	arsenic	As	74·92	77	iridium	Ir	192·22
34	selenium	Se	78·96	78	platinum	Pt	195·09
35	bromine	Br	79·92	79	gold	Au	196·97
37	rubidium	Rb	85·47	80	mercury	Hg	200·59
38	strontium	Sr	87·62	82	lead	Pb	207·2
39	yttrium	Y	88·91	83	bismuth	Bi	208·98
40	zirconium	Zr	91·22	90	thorium	Th	232·05
41	niobium	Nb	92·91	92	uranium	U	238·07
42	molybdenum	Mo	95·94				

Jennifer Bevan, Dr Andrew Clark, Robert Symes

The mineral kingdom

In this chapter, descriptions are given of about 300 of principally the most common and widely distributed mineral species. Some that are less common, or found in only a few localities, are included because of their interest to mineral collectors. The minerals are arranged in the now most widely used crystallochemical type of classification which groups together similar types of chemical compounds under the following broad headings:

1 native elements
2 sulphides, arsenides, tellurides, sulphosalts
3 oxides and hydroxides
4 halides
5 carbonates, nitrates, borates
6 sulphates, chromates
7 tungstates, molybdates
8 phosphates, arsenates, vanadates
9 silicates

For a complete index to known mineral species, reference can be made to *An Index of Mineral Species and Varieties Arranged Chemically* by M H Hey published by the British Museum (Natural History) in 1955 with two subsequent appendices (1963 and 1974).

1 Native elements

Gold

Composition Au. Often rich in silver (electrum) and usually with small amounts of copper.
Crystal system Cubic.
Habit Crystals (cubic, octahedral and dodecahedral) are rare; usually occurs as alluvial grains, scales or dendritic growths; occasionally as rounded masses known as nuggets.
Twinning Common on octahedral planes.
Specific gravity 19·3 when pure, lower when alloyed with silver.
Hardness 2½–3.
Fracture Hackly fracture; very malleable and ductile.
Colour and transparency Golden yellow becoming lighter in shade as the silver content increases; opaque.
Streak Golden yellow, shining.
Lustre Metallic.
Distinguishing features Colour; specific gravity; softness. Not attacked by hydrochloric acid and can be distinguished readily from pyrite, which is

brittle, and chalcopyrite which crumbles when cut with a knife blade.
Formation and occurrence Usually in hydrothermal veins, associated with quartz and sulphide minerals, mainly pyrite. On account of its resistance to chemical action, when the host rock disintegrates, it becomes concentrated as alluvial deposits in stream beds. Can be recovered from these placer deposits by washing the sand and gravel from auriferous stream beds. Principal areas now producing gold are Russia, South Africa, United States, Canada and Australia.
Uses Formerly as a currency standard; in the decorative arts of jewellery and plating; in scientific and electrical apparatus.

Silver

Composition Ag. Often contains considerable gold or mercury; mercury-rich varieties are known as amalgam.
Crystal system Cubic.
Habit Crystals are rare, sometimes cubes, octahedra, dodecahedra; usually occurs massive, scaly, or as arborescent and wiry aggregates.
Twinning Common on octahedral planes.
Specific gravity 10·5 for the pure metal.
Hardness 2½–3.
Fracture Hackly fracture; malleable and ductile.
Colour and transparency Silver-white tarnishing rapidly to grey or black; opaque.
Streak Bright silver-white.
Lustre Metallic.
Distinguishing features Colour; tarnish; density; habit. Soluble in nitric acid.
Formation and occurrence Usually in the oxidized part of silver sulphide deposits or hydrothermal veins. Sometimes in conglomerates or placer deposits.
Uses Metallic silver is widely used in the electronics and photographic industries, and in jewellery and plating. On account of its increasing value it is less used now in coinage.

Copper

Composition Cu. Often contains small amounts of gold, silver, bismuth, or iron.
Crystal system Cubic.
Habit Crystals are cubic, dodecahedral, tetrahexahedral, sometimes octahedral; commonly

Above
Fine crystalline gold nugget, the 'Latrobe nugget'; from McIvor, Victoria, Australia.

Right
Dendritic growth of silver on quartz; from Michigan.

forms distorted, branching, or wire-like crystal growths.
Twinning Common on octahedral planes.
Specific gravity 8·9.
Hardness 2½–3.
Fracture Hackly fracture; malleable and ductile.
Colour and transparency Bright copper red on fresh surfaces, quickly tarnishing to a dull brown colour; often superficially coated with black or green crusts; opaque.
Streak Metallic pale red.
Lustre Metallic.
Distinguishing features Characteristic colour; habit; softness. It is soluble in nitric acid.
Formation and occurrence Usually as hydro-thermal deposits filling cracks and cavities in certain basaltic lava-flows, or as a cement in sandstones.
Uses Metallic copper is of immense importance in the electrical industry and in the production of alloys, chiefly brass.

Platinum
Composition Pt. Always with some iron, often copper and other platinum-group metals.
Crystal system Cubic.
Habit Occasionally found as small distorted cubes; usually as grains, scales or nuggets.
Twinning On octahedral planes.
Specific gravity 14–19 (21·5 for pure metal).
Hardness 4–4½.
Fracture Hackly fracture; malleable.
Colour and transparency Steel grey to silver white; opaque.
Streak White, steel grey.
Lustre Metallic.
Distinguishing features Specific gravity; colour; infusibility. It is insoluble in all acids apart from hot aqua regia.
Formation and occurrence In ultrabasic igneous rocks, associated with chromite. Also in nickel-bearing noritic rocks and placer deposits. Major deposits in Russia, Canada, and South Africa.
Uses On account of its resistance to acids platinum is very widely used in chemical laboratories. In industry it finds application as a catalytic agent.

Iron and Nickel-iron
Composition Fe and Fe-Ni.
Crystal system Cubic.
Habit Native iron as a terrestrial mineral is very rare, occurring as grains and sometimes larger masses; in meteorites, iron and nickel-iron occur intergrown as kamacite-taenite lamellar masses.
Specific gravity 7·3–8·2.
Hardness 4–5.
Cleavage and fracture Poor cubic cleavage; hackly fracture; malleable.
Colour and transparency Steel grey to iron black; opaque.
Streak White, steel grey.
Lustre Metallic.
Distinguishing features Strongly magnetic.
Formation and occurrence Terrestrially native iron occurs in volcanic rocks, particularly where they meet coal seams or other reducing agents. The most important occurrence is on Disko Island, west Greenland.

Arsenic
Composition As. Often with some antimony.
Crystal system Hexagonal (Trigonal).
Habit Crystals are very rare, but occasionally show a pseudocubic form; usually as compact granular masses with reniform or botryoidal forms.
Twinning Rare on $\{10\bar{1}4\}$.
Specific gravity 5·6–5·8.
Hardness 3½.
Cleavage and fracture Perfect basal cleavage; uneven granular fracture; brittle.
Colour and transparency Tin white on fresh surfaces, rapidly tarnishing to dark grey; opaque.
Streak Tin white.
Lustre Submetallic.
Distinguishing features Standard chemical tests; characteristic garlic odour when heated.
Formation and occurrence As a minor consti-tuent of some hydrothermal veins, associated with

silver, cobalt, and nickel ores.

Uses Principally in industry and agriculture for its properties as a poison, in weedkillers, insecticides. Only a minor source of the metal.

Antimony

Composition Sb. Often contains some arsenic, silver or iron.

Crystal system Hexagonal (Trigonal).

Habit Rarely as pseudocubic or tabular crystals; usually massive, lamellar, botryoidal or reniform.

Twinning Common on $\{10\bar{1}4\}$.

Specific gravity 6·6–6·7.

Hardness $3–3\frac{1}{2}$.

Cleavage and fracture Perfect basal cleavage; uneven fracture; brittle.

Colour and transparency Tin white; opaque.

Streak Grey.

Lustre Metallic.

Distinguishing features Gives similar chemical tests to arsenic but the absence of characteristic fumes on heating is diagnostic.

Formation and occurrence In hydrothermal veins, associated with silver, antimony and arsenic ores, often occurring with stibnite.

Uses In the production of certain alloys. Its compounds in the manufacture of pigments, medicines. Supplies only a very small part of the raw material.

Bismuth

Composition Bi. Sometimes with small amounts of sulphur, arsenic and tellurium.

Crystal system Hexagonal (Trigonal).

Habit Rare as crystals; usually in granular masses or arborescent forms.

Twinning Frequent on $\{10\bar{1}4\}$.

Specific gravity 9·7–9·8.

Hardness $2–2\frac{1}{2}$.

Cleavage and fracture Perfect basal cleavage; brittle when cold, malleable when heated; sectile.

Colour and transparency Silver white with reddish hue; opaque.

Streak Silver white.

Lustre Metallic.

Distinguishing features Soluble in nitric acid, with white precipitate when the solution is diluted.

Formation and occurrence In hydrothermal veins, associated with ores of cobalt, nickel, silver and tin.

Uses As one of the principal sources of bismuth, used in medicine, pigments and certain low-melting point alloys.

Sulphur

Composition S (contains S_8 molecules).

Crystal system Orthorhombic.

Habit Predominant form is the bipyramid, often forming crystals several centimetres across; also as massive, stalactitic or powdery aggregates.

Twinning Rare.

Specific gravity 2·07.

Hardness $1\frac{1}{2}–2\frac{1}{2}$.

Cleavage and fracture Cleavage poor; uneven to subconchoidal fracture; brittle; slightly sectile.

Colour and transparency Yellow; transparent to translucent.

Streak White.

Yellow translucent sulphur crystals on matrix; from Agrigento, Italy.

Lustre Resinous to greasy.

Distinguishing features Colour; low specific gravity; burns with a pale blue flame and is soluble in carbon disulphide, insoluble in water and acids.

Formation and occurrence By sublimation around volcanic vents, and is also deposited from hot springs. The important commercial deposits are in sedimentary rocks where the element is associated with gypsum, and particularly in the cap rocks of salt domes.

Uses In the manufacture of sulphuric acid, matches, gunpowder and insecticides.

Diamond

Composition C. Pure carbon.

Crystal system Cubic.

Habit Usually octahedral, often with curved faces; also cubic, dodecahedral, tetrahedral.

Twinning Common on the octahedral plane.

Specific gravity 3·50–3·53.

Hardness 10.

Cleavage and fracture Perfect octahedral cleavage; conchoidal fracture; brittle.

Colour and transparency Colourless, sometimes pale yellow, blue, red, or black; transparent to translucent.

Lustre Adamantine to greasy.

Distinguishing features Hardness; habit; lustre.

Formation and occurrence In ultrabasic rocks rich in olivine and phlogopite (kimberlites) which are formed as cylindrical bodies or pipes. Also in alluvial deposits associated with other heavy and persistent minerals.

Uses Widely used in jewellery and as an industrial abrasive; for the latter purpose a greyish to black diamond aggregate known as bort is often used.

Graphite (Plumbago, Black Lead)

Composition C. Pure carbon.

Crystal system Hexagonal.

Habit Hexagonal scales or soft tabular crystals; more often massive, foliated, granular, or earthy.

Specific gravity 2·09–2·25.

Hardness 1–2.

Cleavage Perfect basal cleavage; sectile.

Colour and transparency Iron black to steel

Sulphur.

Diamond: octahedron.

Diamond: octahedron with curved faces.

grey; opaque.
Streak Dark grey (pencil lead).
Lustre Submetallic; sometimes dull or earthy.
Distinguishing features Colour; ability to mark paper. Molybdenite, the only mineral of similar appearance, has a much brighter lustre.
Formation and occurrence By the metamorphism of organic matter trapped in sediments.
Uses Electrodes and in nuclear reactors; crucibles are made from compressed flaky graphite. Also used as an industrial lubricant. The quantity consumed as pencil lead accounts for only a very small proportion of the total production.

2 Sulphides, Arsenides, Tellurides, Sulphosalts

Tetradymite

Composition Bi_2Te_2S.
Crystal system Hexagonal.
Habit Crystals rare, pyramidal; usually foliated or granular massive.
Specific gravity 7·2.
Hardness $1\frac{1}{2}$–2.
Cleavage Perfect basal cleavage; flexible.
Colour and transparency Pale steel grey, tarnishing to dull or iridescent; opaque.
Streak Pale steel grey.
Lustre Metallic.
Distinguishing features Cleavage; colour; standard chemical tests for tellurium.
Formation and occurrence In gold-quartz veins and in contact metamorphic deposits; usually associated with gold tellurides.
Uses A minor ore of bismuth and tellurium.

Dyscrasite

Composition Ag_3Sb.
Crystal system Orthorhombic.
Habit Pyramidal crystals; usually massive, granular, or foliated.
Twinning Repeatedly on {110} forming hexagonal aggregates.
Specific gravity 9·7.
Hardness $3\frac{1}{2}$–4.
Cleavage and fracture {001}, {011}, distinct cleavages; uneven fracture; sectile.
Colour and transparency Silver white, tarnishing to yellow, grey or black; opaque.
Lustre Metallic.
Distinguishing features Sectile; colour; association.
Formation and occurrence A primary mineral in some silver deposits, commonly associated with other sulphides and calcite.
Uses An ore of silver.

Argentite (Silver Glance)

Composition Ag_2S. Sometimes copper-bearing (jalpaite).
Crystal system Cubic (acanthite is the monoclinic modification stable below 179°C).
Habit Crystals occur as cubes or octahedra (paramorphs of acanthite after argentite); also as massive or arborescent aggregates.
Twinning On octahedral planes.

Specific gravity 7·2–7·4.
Hardness 2–$2\frac{1}{2}$.
Cleavage and fracture Poor cleavage; sub-conchoidal fracture; very sectile.
Colour and transparency Dark lead grey; opaque.
Streak Dark grey shining.
Lustre Metallic.
Distinguishing features Hardness; cleavage; much less brittle than galena. It is soluble in dilute nitric acid.
Formation and occurrence As a primary mineral by hydrothermal deposition in silver-bearing veins; often as inclusions in galena.
Uses One of the most important ores of silver.

Chalcosine (Chalcocite, Copper Glance)

Composition Cu_2S.
Crystal system Orthorhombic.
Habit Crystals are short prismatic or thick tabular; usually massive.
Twinning Common on {110} giving stellate grouping.
Specific gravity 5·5–5·8.
Hardness $2\frac{1}{2}$–3.
Cleavage and fracture Indistinct prismatic cleavage; conchoidal fracture; brittle.
Colour and transparency Dark grey; opaque.
Streak Dark grey, sometimes shining.
Lustre Metallic.
Distinguishing features Hardness; association. Soluble in hot nitric acid.
Formation and occurrence By the alteration of primary copper sulphides in zones of secondary enrichment.
Uses One of the most important ores of copper.

Bornite (Erubescite, Peacock Ore)

Composition Cu_5FeS_4.
Crystal system Cubic.
Habit As cubes, dodecahedra, and octahedra; often massive.
Twinning On octahedral plane.
Specific gravity 5·1.
Hardness 3.
Fracture Uneven to subconchoidal fracture; brittle.

Chalcosine.

Iridescent massive bornite ('peacock ore') with brassy chalcopyrite and quartz; from Morocco.

Colour and transparency Copper red to brown, tarnishing rapidly to purple iridescent; opaque.
Streak Pale grey to black.
Lustre Metallic.
Distinguishing features Colour and tarnish. Soluble in nitric acid.
Formation and occurrence Common in hydrothermal ore deposits, both as a primary constituent and as a product of secondary enrichment.
Uses An important ore of copper.

Galena

Composition PbS. Usually contains some silver.
Crystal system Cubic.
Habit Crystals are usually cubes, often with octahedral modification; also massive.
Twinning Penetration and contact twins on octahedral plane.
Specific gravity 7·58.
Hardness 2½.
Cleavage and fracture Perfect cubic cleavage; subconchoidal or stepped fracture; brittle.
Colour and transparency Lead grey; opaque.
Streak Lead grey.
Lustre Metallic.
Distinguishing features Specific gravity; colour; lustre; cleavage.
Formation and occurrence Very widespread in most hydrothermal sulphide bodies usually associated with sphalerite, especially in metasomatic or replacement deposits. It also occurs in sedimentary formations.
Uses Ore of lead; the metal is used in batteries, sheeting, pigments, and a number of alloys. Also, important source of silver.

Sphalerite (Blende)

Composition ZnS. Usually contains some iron.
Crystal system Cubic.
Habit Crystals occur as tetrahedra; also massive and sometimes fibrous.
Twinning Common on octahedral plane.
Specific gravity 3·9–4·1.
Hardness 3½–4.
Cleavage and fracture Perfect on {011}; conchoidal fracture; brittle.
Colour and transparency Brown to black (with increasing iron content), sometimes yellow or red; transparent to translucent.
Streak Brown-yellow, sometimes white.
Lustre Resinous to adamantine.
Distinguishing features Colour; lustre; cleavage; slowly soluble in hydrochloric acid, giving 'rotten egg' odour of hydrogen sulphide.
Formation and occurrence Widely occurring in hydrothermal ore veins, usually associated with galena and other sulphides. Also found in contact metasomatic and sedimentary environments.
Uses Major ore of zinc. The metal is used in several alloys, notably brass, in the galvanizing process for the coating of iron, and in pigments.

Chalcopyrite (Copper Pyrites)

Composition CuFeS$_2$.
Crystal system Tetragonal.
Habit Crystals often of tetrahedral appearance; generally massive, compact.

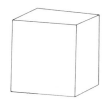

Galena: cube.

Galena: cube and octahedron.

Black, tetrahedral crystals of sphalerite with white globular calcite; from Trepča, Yugoslavia.

Twinning On {112}
Specific gravity 4·1–4·3.
Hardness 3½–4.
Cleavage and fracture Poor cleavage; uneven fracture; brittle.
Colour and transparency Brass yellow, sometimes with iridescent tarnish; opaque.
Streak Greenish black.
Lustre Metallic.
Distinguishing features From pyrite by habit, colour, greater brittleness. From gold by its lack of malleability, and its solubility in nitric acid.
Formation and occurrence Very widespread in medium- to high-temperature hydrothermal ore veins, associated with pyrite and other sulphides. Also in contact metamorphic deposits.
Uses One of the principal copper ores.

Stannite

Composition Cu$_2$FeSnS$_4$.
Crystal system Tetragonal.
Habit Crystals rare, appear tetrahedral due to twinning; usually massive or granular.
Twinning Common on {112}.
Specific gravity 4·4.
Hardness 4.
Cleavage and fracture Poor cleavage; uneven fracture.
Colour and transparency Steel grey to iron black; opaque.
Streak Black.
Lustre Metallic.
Distinguishing features Habit. Massive specimens require optical examination in reflected light for positive identification.
Formation and occurrence Rare mineral found in tin-bearing veins, especially from Cornwall, England, and Bolivia.
Uses A minor ore of tin.

Sphalerite: combination of two tetrahedra and cube.

Chalcopyrite.

Wurtzite

Composition ZnS. Usually contains some iron.
Crystal system Hexagonal.
Habit As pyramidal crystals.
Specific gravity 4·0–4·1.
Hardness $3\frac{1}{2}$–4.
Cleavage and fracture Distinct $\{11\bar{2}0\}$ cleavage, also with difficulty on the basal plane; even to conchoidal fracture; brittle.
Colour and transparency Brownish black; opaque to translucent.
Streak Brown.
Lustre Resinous.
Distinguishing features Habit.
Formation and occurrence Rarer, unstable, high-temperature modification of sphalerite with which it often occurs intergrown.

Greenockite

Composition CdS.
Crystal system Hexagonal.
Habit Usually as an earthy coating, but occasionally occurs as hemimorphic pyramidal crystals.
Specific gravity 4·9.
Hardness 3–$3\frac{1}{2}$.
Cleavage and fracture Good prismatic, basal imperfect cleavages; brittle.
Colour and transparency Yellow-orange; translucent.
Streak Orange-yellow to brick red.
Lustre Adamantine to resinous.
Distinguishing features Colour; zeolitic association. Soluble in concentrated hydrochloric acid liberating hydrogen sulphide gas.
Formation and occurrence Rare, as a coating on zinc ores.
Uses Most important cadmium mineral, although its rarity means that it is not of economic importance.

Greenockite.

Pyrrhotine (Pyrrhotite, Magnetic Pyrites)

Composition FeS. Usually deficient in iron.
Crystal system Hexagonal.
Habit Mostly massive but occasionally as rosettes of hexagonal flat plates.
Twinning On $\{10\bar{1}2\}$.

Radiating group of acicular brassy millerite crystals with dolomite; from Kladno, Czechoslovakia.

Specific gravity 4·6–4·64 (troilite to 4·79).
Hardness $3\frac{1}{2}$–$4\frac{1}{2}$.
Fracture Basal parting, uneven to subconchoidal fracture; brittle.
Colour and transparency Bronze yellow tarnishing to brown; opaque.
Streak Dark greyish black.
Lustre Metallic.
Distinguishing features Habit; colour; magnetism.
Formation and occurrence By magmatic segregation in basic igneous rocks, associated with pentlandite. Also, in high-temperature sulphide veins and replacement bodies. In meteorites, when it is known as troilite, usually with a composition close to FeS.
Uses Ores are valuable for minor nickel content.

Nickeline (Niccolite, Kupfernickel)

Composition NiAs.
Crystal system Hexagonal.
Habit Usually massive; occasionally as pyramidal crystals.
Specific gravity 7·8.
Hardness 5–$5\frac{1}{2}$.
Fracture Uneven fracture; brittle.
Colour and transparency Pale copper red, sometimes with grey tarnish; opaque.
Streak Pale brownish black.
Lustre Metallic.
Distinguishing features Colour. Soluble in nitric acid, giving a green solution.
Formation and occurrence Frequently in ore deposits associated with norites accompanied by other nickel arsenides and sulphides, pyrrhotine, and chalcopyrite. Weathers to annabergite.
Uses A source of metallic nickel.

Millerite

Composition NiS.
Crystal system Hexagonal (Trigonal).
Habit Crystals usually acicular needles in radiating groups.
Specific gravity 5·3–5·6.
Hardness 3–$3\frac{1}{2}$.
Cleavage and fracture Perfect rhombohedral cleavage; uneven fracture; brittle.
Colour and transparency Pale brass yellow, often with grey tarnish; opaque.
Streak Greenish black.
Lustre Metallic.
Distinguishing features Colour; habit.
Formation and occurrence As a late-stage product in hydrothermal deposits, associated with, and often replacing other nickel minerals.
Uses A source of metallic nickel.

Pentlandite

Composition $(Fe,Ni)_9S_8$. Often contains a little cobalt.
Crystal system Cubic.
Habit Massive, often granular.
Specific gravity 4·6–5·0.
Hardness $3\frac{1}{2}$–4.
Fracture Conchoidal fracture; brittle.
Colour and transparency Bronze yellow; opaque.

Streak Bronze brown.

Lustre Metallic.

Distinguishing features Not readily distinguished from pyrrhotine, except in a polished section.

Formation and occurrence Usually as an orientated intergrowth with pyrrhotine, due to exsolution of the two minerals at relatively low temperatures.

Uses One of the most important ores of nickel. The metal is used in steel making, plating, and coinage.

Covelline (Covellite)

Composition CuS.

Crystal system Hexagonal.

Habit Hexagonal platy crystals; usually massive.

Specific gravity 4·6–4·76.

Hardness $1\frac{1}{2}$–2.

Cleavage and fracture Perfect basal cleavage; uneven fracture; brittle to sectile; thin plates are flexible.

Colour and transparency Indigo blue; very thin plates are green translucent, otherwise opaque.

Streak Dark grey, shining.

Lustre Varies from submetallic on crystals to dull on massive specimens.

Distinguishing features Colour; cleavage.

Formation and occurrence Usually associated with chalcosine in the zones of secondary enrichment of copper sulphide deposits.

Uses An important ore of copper.

Cinnabar

Composition HgS.

Crystal system Hexagonal (Trigonal).

Habit Crystals are usually rhombohedral, or thick tabular, rarely prismatic; also granular, massive, and earthy coatings.

Twinning Common on $\{0001\}$, often interpenetrant.

Specific gravity 8·09.

Hardness 2–$2\frac{1}{2}$.

Cleavage and fracture Perfect prismatic cleavage; subconchoidal fracture; slightly sectile.

Colour and transparency Scarlet, sometimes brownish red; transparent to translucent.

Streak Vermilion.

Lustre Adamantine, dull when massive.

Distinguishing features Colour, high specific gravity; cleavage.

Formation and occurrence Low-temperature hydrothermal mineral associated with volcanic activity. Also deposited in hot springs. It usually occurs associated with pyrite, chalcopyrite, realgar, and stibnite.

Uses Cinnabar is the major mercury-bearing mineral. The metal is widely used in electrical and other scientific instruments, and in the manufacture of drugs and chemicals.

Realgar

Composition AsS.

Crystal system Monoclinic.

Habit Granular or massive aggregates; occasionally short striated prismatic crystals.

Twinning Contact twins on $\{100\}$.

Specific gravity 3·56.

Hardness $1\frac{1}{2}$–2.

Cleavage and fracture Good pinacoidal cleavage $\{010\}$; conchoidal fracture; sectile.

Colour and transparency Bright red-orange; fresh material transparent.

Streak Red-orange.

Lustre Resinous.

Distinguishing features Colour; lustre; hardness.

Formation and occurrence Low-temperature hydrothermal deposits associated with orpiment, stibnite, and other arsenic minerals. Also in deposits from hot springs and as a volcanic sublimate. Realgar alters rapidly on exposure to light and air.

Uses A minor constituent of arsenic ores.

Orpiment

Composition As_2S_3.

Crystal system Monoclinic.

Habit Rarely as short prismatic crystals, but usually as foliated or granular masses associated with realgar.

Twinning On $\{100\}$.

Vivid red crystal of cinnabar with quartz and dolomite; from Almadén, Spain.

Cinnabar: thick tabular habit.

Realgar.

Metallic, elongated, striated prismatic crystals of stibnite; from Iyo, Japan.

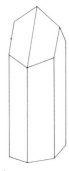

Stibnite.

Specific gravity 3·5.
Hardness 1½–2.
Cleavage and fracture Perfect {010}, flexible cleavages.
Colour and transparency Gold yellow to lemon yellow; transparent to translucent.
Streak Yellow.
Lustre Resinous to pearly.
Distinguishing features Colour; association with realgar. Both minerals are soft and have a perfect cleavage.
Formation and occurrence Low-temperature mineral formed in hydrothermal veins and hot springs, usually as an alteration product of realgar and other arsenic minerals.
Uses A constituent of arsenic ores.

Stibnite (Antimony Glance)

Composition Sb_2S_3.
Crystal system Orthorhombic.
Habit Elongated striated prisms, commonly in masses of radiating crystals; sometimes granular.

Twinning Rare.
Specific gravity 4·63.
Hardness 2.
Cleavage and fracture Perfect {010}, cleavage along the length of the crystals; subconchoidal fracture; slightly sectile.
Colour and transparency Lead grey to steel grey sometimes with blackish or iridescent tarnish; opaque.
Streak Lead grey.
Lustre Metallic.
Distinguishing features Habit; specific gravity. Soluble in hydrochloric acid.
Formation and occurrence At low-temperatures in hydrothermal and replacement deposits, and in hot spring deposits. Often associated with realgar, orpiment, galena, pyrite and cinnabar.
Uses Stibnite is the most important ore of antimony.

Bismuthinite (Bismuth Glance)

Composition Bi_2S_3.
Crystal system Orthorhombic; isostructural with stibnite.
Habit Prismatic to acicular crystals; usually massive, foliated or fibrous.
Specific gravity 6·8.
Hardness 2.
Cleavage and fracture Perfect {010} cleavage; flexible and sectile.
Colour and transparency Lead grey to tin white, sometimes with iridescent tarnish; opaque.
Streak Lead grey.
Lustre Metallic.
Distinguishing features From stibnite by higher specific gravity; less flexible but more sectile. Dissolves in nitric acid with a white precipitate on dilution.
Formation and occurrence In high-temperature hydrothermal veins associated with native bismuth, arsenopyrite, quartz and other sulphides.
Uses An ore of bismuth.

Pyrite

Composition FeS_2.
Crystal system Cubic.
Habit Commonly occurs as cubes and sometimes pyritohedra or octahedra; cubic faces often striated, with the striations at right angles to those on adjacent faces; also found in massive form, as nodules, and sometimes replacing fossil material.
Twinning Frequently twinned forming inter-penetrant crystals.
Specific gravity 5·01.
Hardness 6–6½.
Cleavage and fracture Poor cubic {100} cleavage, conchoidal to uneven fracture; brittle.
Colour and transparency Pale brass yellow, iridescent tarnish; opaque.
Streak Greenish or brownish black.
Lustre Metallic.
Distinguishing features From chalcopyrite by hardness and colour. Soluble in nitric acid, but insoluble in hydrochloric acid.
Formation and occurrence Most common sulphide mineral found in many geological environments. Common in hydrothermal veins, contact

121

metamorphic deposits and as an accessory mineral in sedimentary rocks. Often associated with chalcopyrite, sometimes with gold. Rapidly decomposes on weathering.

Uses Deposits are often worked for the copper contained in the associated minerals; and its sulphur content for the manufacture of sulphuric acid.

Marcasite

Composition FeS_2.
Crystal system Orthorhombic.
Habit Crystals are commonly tabular; often stalactitic and in radiating forms.
Twinning Often repeatedly twinned on $\{101\}$ giving stepped or fan-like crystals (cockscomb pyrites).
Specific gravity 4·9.
Hardness 6–6$\frac{1}{2}$.
Cleavage and fracture Distinct cleavage on $\{101\}$; uneven fracture; brittle.
Colour and transparency Pale bronze yellow, tin white on fresh fractures; opaque.
Streak Greyish or brownish black.
Lustre Metallic.
Distinguishing features Habit; colour; decomposes more readily than pyrite.
Formation and occurrence At lower temperatures than pyrite and from acid solutions. Found as concretions in sedimentary rocks, such as the English Chalk, and associated with low-temperature galena and sphalerite ores.

Sperrylite

Composition $PtAs_2$.
Crystal system Cubic.
Habit Cubes or cubo-octahedra, often with rounded edges and corners.
Specific gravity 10·6.
Hardness 6–7.
Fracture Conchoidal fracture; brittle.
Colour and transparency Tin white; opaque.
Streak Black.
Lustre Metallic.
Distinguishing features Colour; specific gravity; it melts at a lower temperature than native platinum.
Formation and occurrence In platinum deposits, especially at Sudbury, Ontario and the Bushveld igneous complex, South Africa.
Uses A minor ore of platinum.

Löllingite

Composition $FeAs_2$.
Crystal system Orthorhombic.
Habit Crystals rare, elongated prismatic; striated along the prism; frequently granular and massive.
Twinning On $\{011\}$ sometimes to produce trillings.
Specific gravity 7·4.
Hardness 5–5$\frac{1}{2}$.
Cleavage and fracture Sometimes crystals show distinct $\{010\}$, cleavage; uneven fracture; brittle.
Colour and transparency Silver white to steel grey; opaque.
Streak Greyish black.
Lustre Metallic.
Distinguishing features From arsenopyrite by

habit; massive specimens are difficult to distinguish.
Formation and occurrence In medium- to high-temperature hydrothermal veins, often with calcite and sulphides of iron and copper. Also found with silver and gold ores.
Uses An ore of arsenic.

Cobaltite

Composition CoAsS. Some iron and nickel are usually present.
Crystal system Cubic.
Habit Sometimes as cubes or pyritohedra; faces striated as in pyrite; commonly massive.
Twinning Rare.
Specific gravity 6·3.
Hardness 5$\frac{1}{2}$.
Cleavage and fracture Perfect cubic cleavage; uneven fracture; brittle.
Colour and transparency Tin white inclining to reddish; opaque.
Streak Grey black.
Lustre Metallic.
Distinguishing features Hardness; colour. Weathers to pink erythrite.
Formation and occurrence High-temperature mineral found in metamorphic rocks and in hydrothermal veins, associated with other cobalt and nickel sulphides.
Uses An ore of cobalt.

Arsenopyrite (Mispickel)

Composition FeAsS. Sometimes with up to 12 per cent iron replaced by cobalt.
Crystal system Monoclinic.
Habit Short prismatic crystals; faces often striated; also occurs massive or granular.
Twinning Common, forming pseudo-orthorhombic crystals, penetration twins or cross-shaped twins.
Specific gravity 5·9–6·2.
Hardness 5$\frac{1}{2}$–6.
Cleavage and fracture Poor $\{101\}$ cleavage; uneven fracture; brittle.
Colour and transparency Silver white to greyish white; often tarnished; opaque.
Streak Greyish black.
Lustre Metallic.
Distinguishing features Colour; habit. Gives chemical tests for arsenic.
Formation and occurrence The most abundant arsenic mineral, occurring in a wide variety of deposits. Being a high-temperature mineral it is one of the first sulphides to form and is found in contact metamorphic sulphide deposits, associated with other sulphides, and sometimes with gold-quartz deposits. It is also found in high-temperature tin-bearing veins and sometimes disseminated in limestones and schists.
Uses An important ore of arsenic.

Glaucodot

Composition (Co,Fe)AsS.
Crystal system Orthorhombic.
Habit Short prismatic striated crystals; also massive.
Twinning On $\{012\}$ producing cruciform twins, similar to arsenopyrite.

Pyrite: striated cube.

Pyrite: pyritohedron.

Marcasite: spear-shaped twin.

Arsenopyrite.

Wedge-shaped crystals of arsenopyrite with brassy chalcopyrite and pinkish dolomite; from Freiberg, East Germany.

Specific gravity 5·9–6·1.
Hardness 5.
Cleavage and fracture Perfect {010}, cleavage; uneven fracture; brittle.
Colour and transparency Greyish tin white to reddish silver white; opaque.
Streak Black.
Lustre Metallic.
Distinguishing features Not easy to distinguish from cobaltian arsenopyrite, with which it forms a complete composition series; often associated with cobaltite.
Formation and occurrence As for arsenopyrite, but rarer and found in cobalt-bearing deposits.
Uses An ore of cobalt and arsenic.

Molybdenite

Composition MoS_2.
Crystal system Hexagonal.
Habit Hexagonal, tabular crystals; horizontally striated on prism faces; commonly foliated, massive, or scaly.
Specific gravity 4·6–4·7.
Hardness $1–1\frac{1}{2}$.
Cleavage and fracture Perfect basal cleavage; flexible laminae; sectile.
Colour and transparency Bright lead grey; opaque.
Streak Green to bluish-grey.
Lustre Metallic.
Distinguishing features From graphite by its lustre and density; specimens often contain well-formed hexagonal crystals.
Formation and occurrence Widespread, in small amounts in granites, pegmatites, and aplites. Also occurs in deep-seated hydrothermal veins with scheelite, wolframite, topaz and fluorite.
Uses Most common molybdenum mineral and the chief ore for this metal, used in the manufacture of special steels.

Calaverite

Composition $AuTe_2$. Always contains a small amount of silver. One of a number of gold tellurides, others being orthorhombic krennerite ($AuTe_2$) and monoclinic sylvanite [$(Au,Ag)Te_2$].
Crystal system Monoclinic.

Skutterudite: octahedral habit.

Habit In small bladed or lath-like crystals; often massive or granular.
Twinning On {101}, {310} and {111}.
Specific gravity 9·3.
Hardness $2\frac{1}{2}$–3.
Fracture Subconchoidal to uneven fracture; very brittle.
Colour and transparency Brass yellow to silver white; opaque.
Streak Yellowish to greenish grey.
Lustre Metallic.
Distinguishing features Colour; specific gravity. Standard chemical tests for gold and tellurium.
Formation and occurrence In low-temperature veins, usually associated with other tellurides, pyrite and other sulphides, and sometimes gold.
Uses As a source of metallic gold.

Skutterudite, Smaltite and Chloanthite

Composition $(Co,Ni)As_3$, $(Co,Ni)As_{3-x}$ and $(Ni,Co)As_{3-x}$. X lies between 0·5 and 1. The three minerals form an isomorphous series. Iron is usually present, sometimes up to 12 per cent.
Crystal system Cubic.
Habit Crystals occur as cubes, octahedra or cubo-octahedra; usually massive.
Twinning Sixling twins occur but are rare.
Specific gravity 6·1–6·9.
Hardness $5\frac{1}{2}$–6.
Cleavage and fracture Distinct cubic and octahedral cleavages; conchoidal to uneven fracture; brittle.
Colour and transparency Tin white to silver grey, sometimes tarnished or iridescent; opaque.
Streak Black.
Lustre Metallic.
Distinguishing features Habit. Massive specimens require chemical tests to distinguish the group from arsenopyrite and to differentiate between the three members of the group.
Formation and occurrence In medium-temperature hydrothermal veins with other cobalt and nickel minerals, especially cobaltite and nickeline.
Uses As ores of cobalt and nickel.

Stephanite

Composition Ag_5SbS_4.
Crystal system Orthorhombic.
Habit Flat tabular prisms; often massive and disseminated.
Twinning Repeated twinning on {110} occurs.
Specific gravity 6·26.
Hardness $2–2\frac{1}{2}$.
Cleavage and fracture Poor cleavage; uneven fracture; brittle.
Colour and transparency Iron black; opaque.
Streak Iron black.
Lustre Metallic.
Distinguishing features Decomposed by dilute nitric acid, giving a residue of sulphur and antimony oxide.
Formation and occurrence Found associated with other primary silver minerals, particularly argentite.
Uses Minor ore of silver.

Pyrargyrite

Composition Ag_3SbS_3. Usually contains a small amount of arsenic. Pyrargyrite and proustite are the two most common members of the ruby silver group.

Crystal system Hexagonal (Trigonal).

Habit Usually in hexagonal prisms with blunt pyramidal terminations; also massive and compact.

Twinning Common on $\{10\bar{1}4\}$ forming 'swallow-tail' crystal groups.

Specific gravity 5·8.

Hardness $2\frac{1}{2}$.

Cleavage and fracture Distinct $\{10\bar{1}1\}$ cleavage; conchoidal to uneven fracture; brittle.

Colour and transparency Deep red; translucent.

Streak Dark red.

Lustre Adamantine.

Distinguishing features Habit; colour.

Formation and occurrence Low-temperature mineral and one of the last primary silver minerals to crystallize. It is usually associated with proustite, argentite, tetrahedrite and native silver.

Uses An ore of silver.

Proustite

Composition Ag_3AsS_3. Usually contains a small amount of antimony.

Crystal system Hexagonal (Trigonal).

Habit Short prismatic crystals, sometimes rhombohedral and scalenohedral; also massive, compact.

Twinning Common on $\{10\bar{1}4\}$ and $\{10\bar{1}1\}$.

Specific gravity 5·6.

Hardness $2-2\frac{1}{2}$.

Cleavage and fracture Distinct rhombohedral $\{10\bar{1}1\}$, cleavage; conchoidal to uneven fracture; brittle.

Colour and transparency Bright red, blackens on exposure to light; transparent to translucent.

Streak Bright red.

Lustre Adamantine.

Distinguishing features Colour. On prolonged heating gives a malleable globule of silver, whereas pyrargyrite gives a brittle globule.

Formation and occurrence Similar to pyrargyrite, with which it is usually associated, but in rather smaller quantities.

Uses An ore of silver.

Tetrahedrite and Tennantite

Composition $Cu_{12}Sb_4S_{13}$ and $Cu_{12}As_4S_{13}$. The minerals form a continuous solid solution series through antimony-arsenic replacement. Iron and zinc always substitute for some copper. Other elements are often present, notably silver (freibergite) and mercury.

Crystal system Cubic.

Habit Crystals are often tetrahedral; also massive, granular, compact.

Twinning [111] twin axis.

Specific gravity 4·6–5·1; tetrahedrite higher than tennantite.

Hardness $3-4\frac{1}{2}$.

Fracture Subconchoidal to uneven fracture; brittle.

Colour and transparency Grey to iron black; opaque.

Streak Brown to black.

Lustre Metallic.

Distinguishing features Habit. Chemical tests are often required to distinguish massive specimens from other sulphides, although the lack of cleavage is a useful guide.

Formation and occurrence Mainly in low- to medium-temperature hydrothermal veins associated with copper, lead, silver and zinc minerals.

Uses An ore of copper and, sometimes, silver.

Germanite

Composition $Cu_3Ge(As,S)_4$. Gallium, iron and zinc always replace some germanium.

Crystal system Cubic.

Habit Always massive.

Specific gravity 4·4–4·6.

Hardness 4.

Fracture Brittle.

Colour and transparency Dark reddish grey; opaque.

Streak Dark grey to black.

Lustre Metallic.

Distinguishing features Colour; positive identification only from a polished section.

Top
Brilliant red translucent crystals of proustite; from Dolores mine, Chile.

Above
Black tetrahedrite crystals dusted with pyrite; from Hesse, Germany.

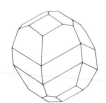

Proustite: combination of prism, scalenohedron and two rhombohedra.

Tetrahedrite: tetrahedron.

Formation and occurrence Rare, found at Tsumeb, south-west Africa; associated with pyrite, tennantite, enargite, galena and sphalerite.
Uses An ore of germanium.

Enargite

Composition Cu_3AsS_4. Often contains some iron and antimony.
Crystal system Orthorhombic.
Habit Crystals are commonly striated prisms with flat basal terminations; also tabular in form; often massive; granular.
Twinning Occasionally forms star-shaped trillings.
Specific gravity 4·4.
Hardness 3.
Cleavage and fracture Perfect {110} cleavage; {100} and {010} uneven fracture; brittle.
Colour and transparency Grey-black to sooty black; opaque.
Streak Greyish black.
Lustre Metallic, tarnishing to dull.
Distinguishing features Habit; cleavage; low melting point.
Formation and occurrence A low- to medium-temperature mineral from hydrothermal veins usually associated with other sulphides and quartz.
Uses An ore of copper.

Bournonite

Bournonite: cog-wheel shaped twins.

Composition $PbCuSbS_3$. Small amounts of arsenic often substitute for antimony.
Crystal system Orthorhombic.
Habit Simple crystals are usually short prismatic or tabular; also massive, granular.
Twinning Repeated twinning on {110} is common producing wheel-like crystal aggregates (cog-wheel ore).
Specific gravity 5·7–5·9.
Hardness $2\frac{1}{2}$–3.
Cleavage and fracture Imperfect {010}, cleavage; subconchoidal to uneven fracture; brittle.
Colour and transparency Steel grey to black; opaque.
Streak Grey to black.
Lustre Metallic.
Distinguishing features Habit; twinning; lustrous on the edges of the cog-wheels and dull on broad tabular faces.
Formation and occurrence One of the commonest sulphosalts. It is found in medium-temperature hydrothermal veins, commonly associated with galena, tetrahedrite, sphalerite, chalcopyrite and pyrite.
Uses As an ore of copper.

Cuprite: octahedron.

Right
Dark red modified octahedron of cuprite; from Onganja mine, south-west Africa.

Boulangerite

Composition $Pb_5Sb_4S_{11}$.
Crystal system Monoclinic.
Habit Elongated prismatic to acicular crystals; striated along elongation direction; often in fibrous masses.
Specific gravity 6·0–6·2.
Hardness $2\frac{1}{2}$–3.
Cleavage and fracture Good {100}, cleavage; generally brittle but thin fibres are flexible.
Colour and transparency Bluish lead grey; opaque.

Streak Brownish grey to brown.
Lustre Metallic.
Distinguishing features Higher specific gravity than jamesonite.
Formation and occurrence In low- to medium-temperature hydrothermal veins, often associated with jamesonite and other lead sulphosalts.
Uses An ore of lead.

Jamesonite

Composition $Pb_4FeSb_6S_{14}$.
Crystal system Monoclinic.
Habit Acicular to fibrous crystals; striated along the elongation direction; often in felted masses; also massive, columnar.
Specific gravity 5·6.
Hardness 2–3.
Cleavage and fracture Good basal {001}, cleavage; brittle.
Colour and transparency Grey-black, sometimes with iridescent tarnish; opaque.
Streak Grey-black.
Lustre Metallic.
Distinguishing features Lower specific gravity than boulangerite; inflexible, unlike stibnite.
Formation and occurrence In medium-temperature hydrothermal veins, associated with other sulphides and sulphosalts.
Uses An ore of lead.

3 Oxides and Hydroxides

Cuprite

Composition Cu_2O.
Crystal system Cubic.
Habit Crystals are usually small modified octahedra, dodecahedra and cubes; also as fine acicular crystals (chalcotrichite) and often massive, granular.
Specific gravity 6·14.

Hardness $3\frac{1}{2}$–4.
Cleavage and fracture Poor octahedral cleavage; conchoidal to uneven fracture; brittle.
Colour and transparency Dark red; translucent.
Streak Brownish red.
Lustre Adamantine to submetallic.
Distinguishing features Association; hardness. Cuprite is soluble in hydrochloric acid staining the solution blue on dilution.
Formation and occurrence In the oxidized zones of copper sulphide veins, often associated with native copper, malachite, azurite, and chalcosine.
Uses An important ore of copper.

Periclase

Composition MgO. Often contains some iron, zinc or manganese.
Crystal system Cubic.
Habit Octahedral crystals, sometimes cubic; most often in irregular or rounded grains.
Twinning Not seen in natural material; synthetic magnesia sometimes forms spinel twins on octahedral plane.
Specific gravity 3·6.
Hardness 5½.
Cleavage and fracture Perfect cubic cleavage; imperfect octahedral.
Colour and transparency Colourless to greyish white, often coloured yellow green or black by impurities; transparent or cloudy.
Streak White.
Distinguishing features Association; cleavage. Soluble in dilute hydrochloric and nitric acids.
Formation and occurrence In metamorphosed limestones where it is formed when dolomite in the limestone is heated, dissociating into periclase and carbon dioxide.

Zincite

Composition ZnO.
Crystal system Hexagonal.
Habit Pyramidal crystals occur but are rare; usually massive, foliated or granular.
Twinning On basal plane.
Specific gravity 5·7.
Hardness 4.
Cleavage and fracture Perfect prismatic cleavage; conchoidal fracture; brittle.
Colour and transparency Orange-yellow to red; translucent.
Streak Orange-yellow.
Lustre Subadamantine.
Distinguishing features Rare, the foremost examples coming from Franklin and Sterling Hill, New Jersey, United States, where the association with willemite, franklinite and calcite is characteristic; fluorescent.
Formation and occurrence Associated with calcite in a few zinc deposits.
Uses A minor ore of zinc.

Corundum (Ruby, Sapphire)

Composition Al_2O_3.
Crystal system Hexagonal (Trigonal).
Habit Crystals are commonly tabular, steep pyramidal, or barrel shaped; water worn, in alluvial deposits; also massive, granular (emery).
Twinning Lamellar twinning on $\{10\bar{1}1\}$.
Specific gravity 4·0–4·1.
Hardness 9.
Fracture Parting on basal plane; uneven to conchoidal fracture; brittle.
Colour and transparency Common corundum is bluish grey to brown; transparent in gem varieties to translucent. Ruby is coloured red due to the presence of a little chromium. Sapphire's blue colour is thought to be caused by iron or titanium.
Lustre Adamantine to vitreous.
Distinguishing features Habit; specific gravity; hardness.

Formation and occurrence Most abundant in metamorphic rocks, in pegmatites and in some nepheline syenites. Gemstones are found mainly in placer deposits.
Uses As abrasive (emery) and as a precious stone.

Hematite (Haematite)

Composition Fe_2O_3.
Crystal system Hexagonal (Trigonal).
Habit Crystals often thin tabular, sometimes in rosettes (iron rose); also rhombohedral; usually massive, compact aggregates; sometimes in mamillated or botryoidal forms (kidney ore).
Twinning Penetration twins on basal plane.
Specific gravity 5·26.
Hardness 5–6.
Fracture Parting on twin planes; subconchoidal to uneven fracture; brittle.
Colour and transparency Steel grey to black; earthy and compact material reddish; opaque.
Streak Dark red.
Lustre Metallic to dull (earthy material).
Distinguishing features Colour; hardness; streak. Not generally magnetic.
Formation and occurrence Widespread, chiefly found in thick beds of sedimentary origin. Also as an accessory mineral in igneous rocks and hydrothermal veins, as a volcanic sublimate, in metamorphosed sediments and contact metamorphic deposits.
Uses One of chief iron ores. Pigment (red ochre).

Ilmenite

Composition $FeTiO_3$. Magnesium and manganese can substitute for iron.
Crystal system Hexagonal (Trigonal).
Habit Crystals are thick tabular, or acute rhombohedral; often massive or as sand grains.
Twinning On basal pinacoid $\{0001\}$.
Specific gravity 4·8.
Hardness 5–6.
Fracture Parting on basal plane; conchoidal to subconchoidal fracture; brittle.
Colour and transparency Iron black; opaque.
Streak Black.
Lustre Metallic to dull.
Distinguishing features Streak; generally non-magnetic.
Formation and occurrence Common accessory mineral in igneous rocks such as gabbros, diorites, and anorthosites. Also in ore veins and pegmatites and in heavy black beach sands.
Uses Important as a source of titanium.

Braunite

Composition $(Mn,Si)_2O_3$. Manganese is always greatly in excess of silicon.
Crystal system Tetragonal.
Habit Pyramidal crystals, appear octahedral due to their departing only slightly from cubic symmetry; also massive, granular.
Twinning On $\{112\}$, contact twins.
Specific gravity 4·7–4·8.
Hardness 6–6½.
Cleavage and fracture Perfect pyramidal cleavage; uneven fracture.
Colour and transparency Brownish black to

Zincite.

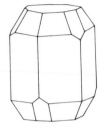

Corundum.

Hematite.

Ilmenite.

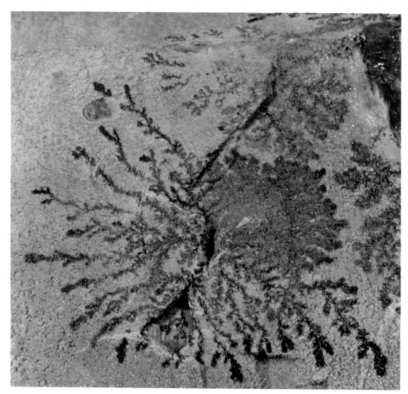

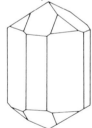

Rutile.

Pyrolusite:
dendritic form.

Cassiterite.

steel grey; opaque.
Streak Dark brownish black to steel grey.
Lustre Submetallic.
Distinguishing features Colour; habit. Soluble
in hydrochloric acid, leaving a residue of silica.
Formation and occurrence In hydrothermal
veins with other manganese oxides; also as a
secondary mineral with pyrolusite and psilomelane.
Uses Sometimes as an ore of manganese.

Rutile

Composition TiO_2. Sometimes contains considerable iron and niobium (ilmenorutile) or
tantalum (strüverite).
Crystal system Tetragonal.
Habit Striated prismatic crystals, sometimes
acicular, rarely pyramidal; also massive, granular.
Twinning Common on bipyramid $\{011\}$ forming
knee-shaped twins; also repeated, cyclic; reticulated
(sagenite), common.
Specific gravity 4·2–5·6 for niobium and tantalum varieties.
Hardness 6–6½.
Cleavage and fracture Distinct prismatic $\{110\}$
cleavage; parting on twin plane; conchoidal to
uneven fracture; brittle.
Colour and transparency Usually reddish
brown to red, sometimes black (iron-, tantalum-
and niobium-bearing); transparent to translucent,
black specimens opaque.
Streak Pale brown.
Lustre Adamantine to metallic.
Distinguishing features Habit; specific gravity;
twinning.
Formation and occurrence Widespread accessory mineral to many kinds of igneous rocks, and
in metamorphic rocks such as quartzites, schists,
and gneisses. Also in alluvial deposits and sometimes as acicular inclusions in quartz and feldspar.
Uses A source of titanium.

Pyrolusite

Composition MnO_2.
Crystal system Tetragonal.
Habit Rarely as prismatic crystals elongated
parallel to the c-axis; usually as reniform sooty
masses, which soil the hands when touched, and in
earthy form; also dendritic encrustations.
Specific gravity 5·06 for crystals, lower when
massive, earthy.
Hardness Crystals 6–6½; massive 2–6.
Cleavage and fracture Perfect prismatic $\{110\}$,
cleavage; uneven fracture; brittle.
Colour and transparency Dark grey black;
opaque.
Streak Black or bluish black.
Lustre Metallic to dull.
Distinguishing features Habit. Dissolves in conc.
hydrochloric acid liberating chlorine gas. Not easy
to distinguish from psilomelane except when
crystals are visible.
Formation and occurrence Secondary mineral
formed by the alteration of manganite or other
primary manganese minerals. Commonly associated with hausmannite, braunite, goethite and
limonite.
Uses An ore of manganese.

Wad

Wad is not a true mineral but a mixture of several
hydrous manganese oxides (*see* pyrolusite and
psilomelane). It is found in earthy or concretionary
masses with a black or brownish-black colour, and
the material is usually very soft, soiling the fingers
when touched. It occurs in bog, lake and shallow-
marine deposits and is formed under highly oxidizing conditions.

Cassiterite (Tinstone)

Composition SnO_2. Often contains a little iron
and sometimes tantalum.
Crystal system Tetragonal.
Habit Crystals are usually short prismatic, sometimes slender prismatic or bipyramidal; also in
botryoidal fibrous crusts (wood tin) or waterworn
pebbles (stream tin).
Twinning Common on $\{011\}$ forming contact
and penetration twins; knee-shaped as rutile.
Specific gravity 7·0.
Hardness 6–7.
Cleavage and fracture Imperfect prismatic $\{100\}$
cleavage; subconchoidal to uneven fracture;
brittle.

Colour and transparency Generally brown to black, rarely colourless; nearly transparent to opaque.
Streak White or light grey to brown.
Lustre Adamantine.
Distinguishing features Colour; specific gravity; streak, hardness; translucent outer layers of crystals.
Formation and occurrence High-temperature mineral occurring in hydrothermal veins associated with granites and altered rocks; also very abundant in some alluvial deposits (for example, Malaya).
Uses Cassiterite is one of the very few tin minerals and its most important ore.

Anatase (Octahedrite)

Composition TiO_2.
Crystal system Tetragonal.
Habit Sharp bipyramids; often striated at right angles to the c-axis; also tabular.
Twinning Rare.
Specific gravity 3·8–3·9.
Hardness $5\frac{1}{2}$–6.
Cleavage and fracture Perfect basal {001}, and pyramidal {011}, cleavages; subconchoidal fracture; brittle.
Colour and transparency Brown, deep blue or black; transparent to nearly opaque.
Streak White to pale yellow.
Lustre Adamantine.
Distinguishing features Habit; colour.
Formation and occurrence Most frequently in vein or fissure deposits of Alpine type, in gneisses or schists. Also widespread as a minor constituent of igneous and metamorphic rocks and as a detrital mineral.

Brookite

Composition TiO_2.
Crystal system Orthorhombic.
Habit Only found as crystals, which are usually thin, tabular, or prismatic.
Specific gravity 4·1.
Hardness $5\frac{1}{2}$–6.
Cleavage and fracture Poor cleavage; subconchoidal to uneven fracture; brittle.
Colour and transparency Various shades of brown, often variegated, also black; transparent to opaque.
Streak White to yellowish.
Lustre Adamantine.
Distinguishing features Habit; colour.
Formation and occurrence Formed at relatively low temperatures and is found chiefly in veins in gneisses and schists, as an accessory mineral in igneous and metamorphic rocks and as a detrital mineral.

Tungstite (Tungsten Ochre)

Composition $WO_3.H_2O$.
Crystal system Orthorhombic.
Habit Sometimes as microscopic platy crystals; generally massive, earthy.
Specific gravity 5·5.
Hardness $2\frac{1}{2}$.
Cleavage Perfect basal {001}, cleavage.
Colour and transparency Yellow to yellowish green; translucent.

Streak Yellow.
Lustre Resinous.
Distinguishing features Colour; association.
Formation and occurrence By the oxidation of wolframite.
Uses An ore of tungsten.

Uraninite (Pitchblende)

Composition UO_2. Some uranium is always present in the oxidized form U_3O_8; lead and thorium are also usually present.
Crystal system Cubic.
Habit Cubes, sometimes with octahedral modification; usually massive (pitchblende), earthy.
Twinning Rare.
Specific gravity Crystals 7·5–9·7; massive 6·5–9·0.
Hardness 5–6.
Cleavage and fracture Octahedral cleavage; conchoidal to uneven fracture; brittle.
Colour and transparency Brownish black or greyish black; opaque.
Streak Black, often brownish or greenish black.
Lustre Submetallic, pitchy or greasy.
Distinguishing features Habit; specific gravity. Bright yellow or green alteration patches. Highly radioactive.
Formation and occurrence Mainly in high- and medium-temperature hydrothermal veins and in granite and syenite pegmatites. Also in low-grade deposits of sedimentary origin.
Uses Primary ore of uranium.

Brucite

Composition $Mg(OH)_2$. Some manganese, iron or zinc can substitute for some magnesium.
Crystal system Hexagonal (Trigonal).
Habit Usually as broad tabular crystals; commonly foliated massive, fibrous, or fine granular.
Specific gravity 2·4.
Hardness $2\frac{1}{2}$.
Cleavage and fracture Perfect basal cleavage; separable, flexible plates; sectile.
Colour and transparency White, pale green, pale blue; manganoan varieties yellow to brown; transparent.
Streak White.
Lustre Pearly on cleavage faces, waxy to vitreous elsewhere.
Distinguishing features Lustre; hardness. From gypsum by its solubility in hot, dilute hydrochloric acid.
Formation and occurrence Low-temperature hydrothermal mineral, often found in veins in serpentine, in chloritic or dolomitic schists, and in marble. Usually associated with calcite, aragonite, talc and magnesite.
Uses Source of magnesium and refractory materials.

Lepidocrocite

Composition $FeO(OH)$.
Crystal system Orthorhombic.
Habit Tabular or scaly crystals in micaceous, fibrous or massive aggregates.
Specific gravity 3·9–4·1.
Hardness 5.
Cleavage and fracture Perfect {010}, cleavage; brittle.

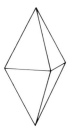

Anatase: bipyramidal habit.

Colour and transparency Red to brown; transparent.
Streak Dull orange.
Lustre Submetallic.
Distinguishing features Colour; streak.
Formation and occurrence Secondary mineral, usually occurring with goethite.
Uses A constituent of some iron ores.

Boehmite

Composition AlO(OH).
Crystal system Orthorhombic.
Habit Microscopic tabular crystals; commonly disseminated or in pisolitic aggregates.
Specific gravity 3·0–3·1.
Hardness 3.
Cleavage and fracture Good {010}, cleavage.
Colour White.
Streak White.
Lustre Dull, earthy.
Distinguishing features Identified microscopically. Chemical tests for aluminium.
Formation and occurrence A constituent of bauxite, formed by the weathering of aluminium silicate rocks low in quartz.
Uses Aluminium is obtained from bauxite.

Manganite

Composition MnO(OH).
Crystal system Monoclinic (pseudo-orthorhombic).
Habit Crystals prismatic; striated with blunt wedge-shaped or flat terminations, often grouped in bundles; also columnar to coarse fibrous.
Twinning Contact or penetration twins on prism face {011}.
Specific gravity 4·3.
Hardness 4.

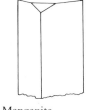

Manganite.

Brilliant black prismatic crystals of manganite; from Ilfeld, East Germany.

Cleavage and fracture Perfect pinacoidal {010}, cleavage; uneven fracture; brittle.
Colour and transparency Black to dark grey; opaque.
Streak Reddish brown to black.
Lustre Submetallic to dull.
Distinguishing features Habit; streak. Dissolves in conc. hydrochloric acid liberating chlorine.
Formation and occurrence A low-temperature mineral found in hydrothermal veins associated with granitic rocks. Also deposited in bogs, lakes, and shallow marine environments; often found with pyrolusite, goethite, and baryte.
Uses An ore of manganese.

Gibbsite (Hydrargillite)

Composition Al(OH)$_3$.
Crystal system Monoclinic.
Habit Tabular crystals, often with hexagonal appearance; usually massive as botryoidal or stalactitic incrustations with fibrous structure.
Twinning Common on {001}.
Specific gravity 2·4.
Hardness $2\frac{1}{2}$–$3\frac{1}{2}$.
Cleavage and fracture Perfect basal {001}, cleavage; tough.
Colour and transparency White, often with grey, green or red hue; transparent.
Streak White.
Lustre Vitreous; pearly on cleavages.
Distinguishing features Hardness. Not easy to distinguish from boehmite when fine grained.
Formation and occurrence Secondary mineral resulting from the alteration of aluminium minerals. It is present in bauxite and laterite deposits. Also occurs as a low-temperature hydrothermal mineral.
Uses With boehmite and diaspore, gibbsite is a constituent of bauxite deposits.

Bauxite

Bauxite is the name given to material rich in hydrous aluminium oxides. It is not a single mineral but a mixture of several (*see* gibbsite, boehmite and diaspore). Bauxite is a massive or earthy yellow to brown material, sometimes occurring as concretionary masses. It is a secondary product formed by the leaching away of silica from rocks containing aluminium silicates.

Psilomelane

Composition $BaMn^{2+}Mn_8^{4+}O_{16}(OH)_4$.
Crystal system Monoclinic.
Habit Massive, botryoidal, and stalactitic; sometimes earthy.
Specific gravity 4·4–4·7.
Hardness 5–6; lower for earthy material.
Colour and transparency Iron black to steel grey; opaque.
Streak Brownish black, shining.
Lustre Submetallic to dull.
Distinguishing features General appearance; streak.
Formation and occurrence A secondary mineral formed by the weathering of manganese carbonates or silicates and is usually found in concretionary form in lake and swamp deposits and clay.
Uses An ore of manganese.

Diaspore

Composition AlO(OH). Sometimes contains manganese or iron.
Crystal system Orthorhombic.
Habit Crystals often thin platy, also prismatic; often in lamellar, foliated, and scaly aggregates.
Specific gravity 3·3–3·5.
Hardness $6\frac{1}{2}$–7.
Cleavage and fracture Perfect {010} cleavage; conchoidal fracture; brittle.
Colour and transparency Colourless to white, sometimes green or brownish, manganese varieties red; transparent to translucent.
Streak White.
Lustre Vitreous, pearly on cleavages.
Distinguishing features Chemical tests for Al.
Formation and occurrence With corundum in emery deposits and in limestones. Massive material occurs in bauxite deposits.
Uses A source of aluminium.

Goethite

Composition FeO(OH).
Crystal system Orthorhombic.
Habit Prismatic crystals elongated along *c*-axis; also thin tabular; usually massive, stalactitic, or radiating fibrous; sometimes earthy.
Specific gravity 3·3–4·3.
Hardness 5–$5\frac{1}{2}$.
Cleavage and fracture Perfect {010} cleavage; uneven fracture; brittle.
Colour and transparency Dark brown to yellowish brown in earthy varieties; opaque except in thin splinters.
Streak Brownish yellow.
Lustre Adamantine to dull.
Distinguishing features From hematite by

streak and colour. Magnetic after strong heating.
Formation and occurrence Secondary mineral found in the oxidation zone (iron hat) of veins containing iron minerals, usually formed by the oxidation at surface temperatures and pressures of minerals such as siderite, pyrite and magnetite. It is also formed as a direct precipitate in bogs and springs. Goethite is the major constituent of the impure hydrated iron oxide known as limonite.
Uses An important iron ore.

Limonite

Limonite is an impure hydrated iron oxide, often composed largely of goethite. It is brown coloured, in various shades with a yellow-brown streak and is found in botryoidal forms or as weathering crusts. It is a secondary material formed under oxidizing conditions at ordinary temperatures and pressures, either by the oxidation of iron minerals or as a precipitate in bog, lake or marine deposits. Used as a pigment (yellow ochre).

Spinel

Composition $MgAl_2O_4$ (spinel), $FeAl_2O_4$ (hercynite), $ZnAl_2O_4$ (gahnite), $MnAl_2O_4$ (galaxite). Chemical substitution between the four end-member compositions is common.
Crystal system Cubic.
Habit Octahedral crystals are common.
Twinning On octahedral planes (spinel twins).
Specific gravity Spinel 3·58; hercynite 4·32; gahnite 4·61; galaxite 4·08.
Hardness $7\frac{1}{2}$–8.
Fracture Octahedral parting; uneven to conchoidal fracture; brittle.
Colour and transparency Variable: red, blue, green, brown or black (spinel); dark green (gahnite); black (hercynite and galaxite); transparent (magnesium-rich) to opaque.
Streak White (spinel); grey (gahnite); green (hercynite); red-brown (galaxite).
Lustre Vitreous.
Distinguishing features Habit; hardness; infusible.
Formation and occurrence High-temperature minerals found as accessories in igneous rocks, in metamorphosed aluminous schists, in contact metamorphosed limestones and sometimes in high-temperature ore veins. Gem quality stones come from Sri Lanka, Burma and India in alluvial deposits.
Uses Sometimes as a gemstone.

Magnetite

Composition Fe_3O_4 or $Fe^{2+}Fe_2^{3+}O_4$. A member of the spinel group, often containing some magnesium, zinc, manganese, aluminium or chromium.
Crystal system Cubic.
Habit Commonly as octahedral crystals, also dodecahedra; often massive, granular.
Twinning On octahedral planes.
Specific gravity 5·18.
Hardness $5\frac{1}{2}$–$6\frac{1}{2}$.
Fracture Octahedral parting; subconchoidal to uneven fracture; brittle.
Colour and transparency Iron black; opaque.
Streak Black.

Psilomelane: botryoidal stalactitic form.

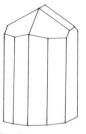

Goethite.

Spinel: octahedron.

Chrysoberyl: twinned crystal.

Lustre Metallic to dull.
Distinguishing features Strongly magnetic; habit; colour; streak.
Formation and occurrence Abundant and widespread high-temperature mineral. As an accessory product in igneous rocks, in contact and regionally metamorphosed rocks, and in high-temperature sulphide veins. Specimens known as lodestones were used as compasses in early times.
Uses An important iron ore.

Franklinite

Composition $(Zn,Mn^{2+},Fe^{2+})(Fe^{3+},Mn^{3+})_2O_4$. A member of the spinel group.
Crystal system Cubic.
Habit As octahedral crystals; also massive, granular.
Twinning On octahedral planes.
Specific gravity 5·1–5·3.
Hardness $5\frac{1}{2}$–$6\frac{1}{2}$.
Fracture Parting on octahedral plane; uneven fracture; brittle.
Colour and transparency Brownish black; opaque.
Streak Reddish brown.
Lustre Metallic to submetallic.
Distinguishing features Less magnetic than magnetite; habit; streak.
Formation and occurrence In the zinc deposits at Franklin and Sterling Hill, New Jersey, United States associated with calcite.
Uses An ore of zinc.

Chromite

Composition $FeCr_2O_4$. A member of the spinel group; it often contains some magnesium (magnesiochromite) or aluminium.
Crystal system Cubic.
Habit Octahedral crystals occur but are rare; usually massive, granular.
Specific gravity 4·5–5·1 (4·1 for magnesiochromite).
Hardness $5\frac{1}{2}$.
Fracture Uneven fracture; brittle.
Colour and transparency Black; opaque.
Streak Brown.
Lustre Metallic.
Distinguishing features Weakly magnetic; streak.
Formation and occurrence As an accessory mineral in igneous rocks, principally peridotites and serpentinites derived from them. Often in alluvial deposits.
Uses Valuable refractory material used for lining furnaces. Also the sole source of chromium metal which is widely used in stainless steels.

Hausmannite

Composition $Mn_2^{2+}Mn^{4+}O_4$.
Crystal system Tetragonal.
Habit Pseudo-octahedral crystals; also granular, massive.
Twinning On {112}, often repeated as fivelings.
Specific gravity 4·84.
Hardness $5\frac{1}{2}$.
Cleavage and fracture Perfect {001}, cleavage; uneven fracture; brittle.

Colour and transparency Brown black; opaque.
Streak Chestnut brown.
Lustre Submetallic.
Distinguishing features Cleavage. Dissolves in hydrochloric acid liberating chlorine, whereas braunite gelatinizes in hot hydrochloric acid.
Formation and occurrence Primary manganese mineral occurring in high-temperature hydrothermal veins and contact metamorphic deposits. Usually associated with hematite, baryte and other manganese oxides.
Uses A constituent of manganese ores.

Chrysoberyl (Alexandrite)

Composition $BeAl_2O_4$. Small amounts of iron and chromium are often present.
Crystal system Orthorhombic.
Habit Crystals usually tabular or short prismatic; also granular, massive.
Twinning Common on {130}, forming cyclic groups.
Specific gravity 3·7–3·8.
Hardness $8\frac{1}{2}$.
Cleavage and fracture Distinct {110} cleavage; conchoidal to uneven fracture; brittle.
Colour and transparency Green, yellow, sometimes brownish; variety alexandrite is reddish in artificial light; transparent to translucent.
Streak White.
Lustre Vitreous.
Distinguishing features Habit; chatoyancy; hardness. Alexandrite by colour in artificial light.
Formation and occurrence In granite pegmatites, mica schists, and dolomitic marble. Also in alluvial deposits.
Uses Fine specimens are cut as gemstones.

Perovskite

Composition $CaTiO_3$. Some specimens are rich in niobium (dysanalyte), others in cerium (knopite).
Crystal system Orthorhombic.
Habit Crystals are commonly cubic or octahedral in appearance.
Twinning Lamellar or penetration twinning on {111} faces.
Specific gravity 4·0 (higher for niobium, cerium varieties).
Hardness $5\frac{1}{2}$.
Cleavage and fracture Poor {001} cleavage; uneven to subconchoidal fracture; brittle.
Colour and transparency Yellow, brown, or black; transparent to opaque.
Streak White to grey.
Lustre Adamantine, metallic or dull.
Distinguishing features Habit.
Formation and occurrence Rare mineral occurring in basic igneous rocks and their associated pegmatites. Also in metamorphosed limestones and chlorite or talc schists.

Pyrochlore and Microlite

Composition $(Ca,Na)_2(Nb,Ta)_2O_6(O,OH,F)$ and $(Ca,Na)_2(Ta,Nb)_2O_6(O,OH,F)$. Pyrochlore: cerium, other rare earth elements and uranium frequently substitute for calcium and sodium; titanium is invariably present in substitution for niobium and tantalum. Microlite: all the above elements may be present in minor amounts; fluorine content is generally less than that of pyrochlore.

Crystal system Cubic.

Habit Crystals commonly octahedral; also in irregular masses and embedded grains.

Twinning Rare.

Specific gravity 4·2 (pyrochlore) – 6·4 (microlite).

Hardness 5–5½.

Cleavage and fracture Octahedral cleavage; subconchoidal to splintery fracture.

Colour and transparency Pyrochlore – various shades of brown to black; microlite – pale yellow to brown, sometimes shades of red or green; subtranslucent to opaque.

Streak Paler than colour.

Lustre Vitreous to resinous.

Distinguishing features Habit; colour; specific gravity. Some specimens are radioactive due to substitutions by radioactive elements such as uranium and thorium.

Formation and occurrence Usually in pegmatites in, or near to, alkaline igneous rocks, associated with zircon and apatite, tantalite and columbite. Also found within some alkaline rocks and in some contact metamorphic rocks associated with them.

Uses A source of niobium and tantalum.

Fergusonite

Composition $(Yt,Er)(Nb,Ta)O_4$. Other rare earth elements, and uranium, iron or calcium, can substitute for yttrium and erbium; titanium frequently (and tin or tungsten less often) substitute for niobium and tantalum. Most specimens are partially altered to a water-bearing material of similar composition.

Crystal system Tetragonal.

Habit Crystals prismatic to pyramidal; also in irregular masses and grains.

Specific gravity 5·6–5·8 (decreasing with alteration but increasing with tantalum content).

Hardness 5½–6½.

Fracture Subconchoidal.

Colour and transparency Grey, yellow or brown, altering to dark brown and black; translucent to opaque.

Streak Brown to grey.

Lustre Vitreous to submetallic.

Distinguishing features Lustre; specific gravity; colour; association; some varieties contain substituted radioactive elements and so are radioactive.

Formation and occurrence As a primary mineral of chemically complex granite pegmatites, associated with other rare earth minerals. Also found in 'placer' deposits derived from these rocks by weathering and erosion.

Stibiotantalite and Stibiocolumbite

Composition $SbTaO_4$ and $SbNbO_4$. A complete series extends between these end-members, with mutual substitution of tantalum and niobium; some bismuth may substitute for antimony.

Crystal system Orthorhombic.

Habit Prismatic, striated crystals.

Twinning Common; individuals are often irregularly intergrown.

Specific gravity 5·9–7·4 (increasing with increasing tantalum content).

Hardness 5½.

Cleavage and fracture Pinacoidal distinct cleavage; subconchoidal to granular fracture.

Colour and transparency Dark brown to light yellowish or reddish brown, also greenish yellow; transparent.

Streak Pale yellow to yellow brown.

Lustre Resinous to adamantine.

Distinguishing features Colour; transparency; specific gravity.

Formation and occurrence In granite pegmatites of complex mineralogy, as a primary, but often as a secondary mineral. Also occurs in 'placer' deposits derived from such rocks by weathering and erosion. The tantalum end-member is by far the more common of the two.

Tapiolite

Composition $Fe(Ta,Nb)_2O_6$. Manganese can substitute for iron.

Crystal system Tetragonal.

Habit Short or equant prismatic crystals.

Twinning Very common, giving rise to forms with pseudo-orthorhombic symmetry.

Specific gravity 7·9 (variable).

Hardness 6–6½.

Fracture Uneven to subconchoidal.

Colour and transparency Black, sometimes brownish on surface; subtranslucent to opaque.

Streak Brown to black.

Lustre Subadamantine to submetallic.

Distinguishing features Colour; habit; specific gravity.

Formation and occurrence In granite pegmatites, and as a detrital mineral in areas of granite pegmatites. Relatively rare.

Columbite and Tantalite

Composition $(Fe,Mn)(Nb,Ta)_2O_6$ and $(Fe,Mn)(Ta,Nb)_2O_6$. A complete series exists between these minerals, with mutual substitution of niobium and tantalum. Small amounts of tin substitute for the first pair of elements in the formula, and minor amounts of tungsten for the second. The variety 'manganotantalite' is tantalite which contains more manganese than iron.

Crystal system Orthorhombic.

Habit Short or equant prismatic crystals; sometimes thin or thick tabular; may form large groups of subparallel crystals; also massive.

Twinning Common, usually simple contact twins forming heart-shaped, striated crystals; also repeated twins which may have a pseudohexagonal form.

Specific gravity 5·2 (columbite) to 8·0 (tantalite).

Hardness 6 (columbite) to 6½ (tantalite).

Cleavage and fracture Pinacoidal, distinct {010} cleavage; subconchoidal to uneven fracture.

Colour and transparency Iron black to brownish black; reddish brown in thin splinters; frequently iridescent surface tarnish; subtranslucent to opaque.

Streak Dark red to black.

Lustre Submetallic to subresinous.

Distinguishing features Colour; specific gravity; crystal form.

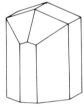

Columbite: tantalite.

Formation and occurrence Usually as a primary mineral in granite pegmatites of complex mineralogy. Also found in detrital 'placer' deposits derived from such pegmatites by weathering and erosion. Columbite-tantalite series minerals have worldwide distribution and are relatively abundant.
Uses As ores of niobium and tantalum. Columbium was the American name for the element niobium.

Euxenite and Polycrase

Composition (Yt,Er,Ce,La,U)(Nb,Ti,Ta)$_2$ (O,OH)$_6$ and (Yt,Er,Ce,La,U)(Ti,Nb,Ta)(O,OH)$_6$. The end-members of this series are defined according to whether titanium exceeds niobium plus tantalum. Thorium and calcium may also be present. Commonly metamict.
Crystal system Orthorhombic.
Habit Stout prismatic, sometimes flattened with striated faces; often in parallel or slightly radial aggregates of crystals; also massive.
Twinning Common.
Specific gravity 5·0–5·9.
Hardness 5½–6½.
Fracture Subconchoidal to conchoidal.
Colour and transparency Black, sometimes with a greenish or brownish tint; transparent in thin splinters.
Streak Yellowish, greyish or reddish brown.
Lustre Brilliant to submetallic or greasy.
Distinguishing features Lustre; colour; specific gravity; habit.
Formation and occurrence In granite pegmatites of complex chemistry, and in detrital deposits associated with them.

4 Halides

Halite (Rock Salt)

Composition NaCl.
Crystal system Cubic.
Habit Cubic crystals often with hollow faces (hopper crystals); also massive and granular.
Specific gravity 2·16.
Hardness 2½.
Cleavage and fracture Perfect cubic cleavage; conchoidal fracture; very brittle.
Colour and transparency Colourless, sometimes stained by impurities; transparent.
Streak White.
Lustre Vitreous.
Distinguishing features Habit; cleavage; transparency; hardness; taste; soluble in water.
Formation and occurrence Very common, occurring in extensive beds formed by the evaporation of enclosed bodies of sea-water. In some places the sedimentary beds are over 300 metres (1 000 feet) thick. Also occurs as a sublimate in volcanic regions and as an efflorescence in arid areas.
Uses In many manufacturing processes, such as the production of sodium carbonate and hydrochloric acid, and for culinary purposes.

Sylvine

Composition KCl.
Crystal system Cubic.
Habit Cubic crystals, sometimes octahedral; also massive and as crusts.
Twinning On octahedral faces, only seen in artificial crystals.
Specific gravity 1·99.
Hardness 2.
Cleavage and fracture Perfect cubic cleavage; uneven fracture; less brittle than halite.
Colour and transparency Colourless or white, sometimes slightly grey, blue, or red; transparent to translucent.
Streak White.
Lustre Vitreous.
Distinguishing features From halite by more bitter taste; slightly sectile.
Formation and occurrence In bedded salt deposits associated with, but much less common than, halite. Also in volcanic fumaroles.
Uses As a source of potassium salts, widely used in fertilizers.

Carnallite

Composition KMgCl$_3$.6H$_2$O.
Crystal system Orthorhombic.
Habit Crystals rare, sometimes pseudohexagonal or tabular; usually massive granular.
Specific gravity 1·60.
Hardness 2½.
Fracture Conchoidal fracture; brittle.
Colour and transparency Colourless to white, often reddish due to minute hematite inclusions; transparent to translucent.
Streak White.
Lustre Greasy, dull to shining.
Distinguishing features From other salts by lack of cleavage; deliquescent; soluble in water; taste.
Formation and occurrence In the upper layers of bedded salt deposits, associated with sylvine and halite.
Uses As a source of potassium salts for fertilizers.

Fluorite (Fluorspar)

Composition CaF$_2$. Sometimes contains some of the rare earth elements.
Crystal system Cubic.
Habit Cubic crystals very common, less frequently octahedral; also granular, compact.
Twinning On octahedron, usually as cubic penetration twins.
Specific gravity 3·18; higher for rare earth-bearing varieties.
Hardness 4.
Cleavage and fracture Perfect octahedral cleavage; subconchoidal to splintery fracture; brittle.
Colour and transparency Colourless when pure but often purple, blue, green, yellow, or brown, commonly with colour zones parallel to cubic or octahedral faces; generally fluorescent; transparent to translucent.
Streak White.
Lustre Vitreous.
Distinguishing features Habit; cleavage; hardness. Does not dissolve readily in hydrochloric acid.
Formation and occurrence Very widespread occurring most commonly in hydrothermal veins, particularly those associated with galena and

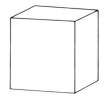

Halite: cube.

Fluorite: modified cube.

Fluorite: modified cube.

Fluorite: interpenetration twin.

sphalerite. Also usually associated with quartz and baryte. Also in pneumatolytic tin veins and in some granites. The banded variety from Derbyshire, England is known as Blue John.

Uses Pure transparent fluorite is used for optical purposes. Lower-grade material in large quantities as a flux for steel making. Also used in the manufacture of hydrofluoric acid.

Cryolite

Composition Na_3AlF_6.
Crystal system Monoclinic.
Habit Crystals rare, sometimes pseudocubic; usually massive, coarsely granular.
Twinning Very common.
Specific gravity 2·96.
Hardness $2\frac{1}{2}$.
Fracture Parting gives pseudocubic forms; uneven fracture; brittle.
Colour and transparency Colourless to white, also brownish, reddish; transparent to translucent.
Streak White.
Lustre Vitreous to greasy.
Distinguishing features Habit; colour; lustre. Becomes invisible when immersed in water because it has the same refractive index.
Formation and occurrence Found in major quantities at only a few localities. At Ivigtut in Greenland it occurs in a pegmatite body associated with microcline, quartz, siderite and fluorite.
Uses In the molten form as an electrolyte in the reduction of aluminium ores to metallic aluminium; for this purpose, however, it has now largely been superseded by an artificial compound.

Chlorargyrite (Cerargyrite, Horn Silver) and Bromyrite

Composition AgCl and AgBr. A complete solid solution series occurs between these two end-members in the chlorargyrite group.
Crystal system Cubic.
Habit Crystals cubic, but rare; usually massive or in crusts resembling wax or horn.
Twinning On octahedral planes.
Specific gravity 5·55 (chlorargyrite) – 6·50 (bromyrite).
Hardness $2\frac{1}{2}$.
Fracture Uneven to subconchoidal fracture; sectile and ductile, very plastic.
Colour and transparency Colourless when fresh, sometimes greyish; chlorargyrite turns brown or purple on exposure to light; transparent to translucent.
Streak White.
Lustre Resinous to adamantine.
Distinguishing features Colour; sectile. Gives a silver globule when heated on a charcoal block.
Formation and occurrence Secondary minerals occurring in the oxidized zones of silver deposits, particularly in arid regions.
Uses As an ore of silver.

Atacamite

Composition $Cu_2(OH)_3Cl$.
Crystal system Orthorhombic.
Habit Slender prismatic crystals; also tabular; in confused crystalline aggregates; also massive, fibrous, granular.
Twinning Sometimes as doublets, trillings, and complex groups.
Specific gravity 3·76.
Hardness $3–3\frac{1}{2}$.
Cleavage and fracture Perfect pinacoidal $\{010\}$, cleavage; conchoidal fracture; brittle.
Colour and transparency Various shades of green; transparent to translucent.
Streak Apple green.
Lustre Adamantine to vitreous.
Distinguishing features Colour; habit. Does not effervesce in hydrochloric acid while malachite does.
Formation and occurrence Secondary mineral in the oxidized zones of copper deposits, especially in arid, saline conditions. Often associated with malachite, cuprite, brochantite, chrysocolla, gypsum and limonite.
Uses A constituent of some copper ores.

Botallackite

Composition $Cu_2Cl(OH)_3$.
Crystal system Monoclinic.
Habit In thin crystalline or powdery crusts.
Specific gravity 3·6.
Cleavage One good cleavage.
Colour and transparency Green, sometimes bluish green.
Distinguishing features Colour; habit. Specimens are only known from one locality.
Formation and occurrence Rare secondary mineral, found at the Botallack mine, Cornwall, England, associated with atacamite and paratacamite.

Diaboleite

Composition $Pb_2CuCl_2(OH)_4$.
Crystal system Tetragonal.
Habit Square tabular crystals; also as aggregates of thin plates.
Specific gravity 5·4.
Hardness $2\frac{1}{2}$.
Cleavage and fracture Perfect basal cleavage; conchoidal fracture; brittle.
Colour and transparency Deep blue; transparent.
Streak Pale blue.
Lustre Vitreous.
Distinguishing features Colour; association.
Formation and occurrence Found as small crystals associated with linarite and other secondary minerals from Tiger, Arizona and with cerussite and mendipite in a small number of oxidized iron and manganese deposits.

Chloroxiphite

Composition $Pb_3CuCl_2O_2(OH)_2$.
Crystal system Monoclinic.
Habit Bladed crystals in subparallel groups.
Specific gravity 6·9 – 7·0.
Hardness $2\frac{1}{2}$.
Cleavage and fracture $\{\bar{1}01\}$, perfect, $\{100\}$, distinct cleavages, very brittle.
Colour and transparency Olive green; translucent.
Streak Pale yellowish green.
Lustre Resinous to adamantine.

Cryolite.

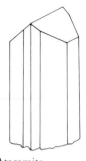

Atacamite.

Diaboleite.

Creamy-white radiating scalenohedra of calcite; from Bound Brook, New Jersey.

Distinguishing features Colour; habit; association.

Formation and occurrence Rare secondary mineral, embedded in mendipite at Higher Pitts Farm, Priddy, Somerset, England and at Merehead Quarry, Somerset.

Mendipite

Composition $Pb_3O_2Cl_2$.
Crystal system Orthorhombic.
Habit Crystals bladed or fibrous, sometimes massive.
Specific gravity 7·2.
Hardness $2\frac{1}{2}$.
Cleavage and fracture Perfect $\{110\}$, and two other good cleavages; fracture conchoidal to uneven.
Colour and transparency Colourless to white or grey; transparent to translucent.
Streak White.
Lustre Pearly on cleavages, resinous elsewhere.
Distinguishing features Habit; association.
Formation and occurrence Associated with cerussite and other lead oxychlorides, in the manganese deposits of the Mendip Hills, Somerset, England.

5 Carbonates, Nitrates, Borates

Calcite

Composition $CaCO_3$. Manganese and ferrous iron, as well as lead and zinc, may substitute for calcium: a complete series extends to rhodochrosite and partial series towards smithsonite, siderite and cerussite. Other elements, such as magnesium, may also substitute for calcium but in smaller amounts.
Crystal system Hexagonal (Trigonal).
Habit Four habits are common: tabular; prismatic; rhombohedral; and scalenohedral; but all possible combinations and variations of these types may be found, as well as fibrous, granular, stalactitic or massive aggregates.
Twinning Common; usually the twin plane is either the basal pinacoid or a rhombohedron; lamellar twinning may also be produced by pressure.
Specific gravity 2·72.
Hardness 3 on cleavage; $2\frac{1}{2}$ on base.
Cleavage and fracture Perfect cleavage, parallel to the unit rhombohedron; conchoidal fracture rarely seen because of this.

Colour and transparency Colourless (transparent) or white (opaque) when pure; often yellow or brownish when iron-containing, pinkish when manganese- or cobalt-containing, and various other tints.

Streak White.

Lustre Vitreous to earthy.

Distinguishing features Cleavage; colour; hardness. Dissolves readily, effervescing freely, in cold, dilute hydrochloric acid. Many specimens also fluoresce in ultraviolet light according to impurities present.

Formation and occurrence Very common, and in many different environments: origin may be igneous as in intrusive carbonatites; sedimentary, either precipitated directly from sea-water or, having formed the shells of many living organisms, accumulating on their death to give limestone; and metamorphic, when such rocks are subjected to high temperatures and/or pressures to produce marble. Often a major constituent of metallic ore veins, as well as forming many other different kinds of vein in a wide range of rocks. May also replace other minerals, especially in igneous rocks. Stalactites and stalagmites can be formed of calcite deposited from solutions within caves in limestone; travertine and tufa are similarly deposited in areas of hot (and cold) calcareous springs.

Uses Most important in the manufacture of cement; also as flux in the smelting of metallic ores, and as a fertilizer. Frequently in the form of limestone or marble, as a decorative or building stone.

Magnesite

Composition $MgCO_3$. Although there appears to be complete solid solution between magnesite and siderite ($FeCO_3$), calcium and manganese substitute in only small amounts.

Crystal system Hexagonal (Trigonal).

Habit Crystals rare but rhombohedral or prismatic when found; usually massive or fibrous; may be compact or granular.

Specific gravity 3·0–3·2.

Hardness $3\frac{1}{2}$–$4\frac{1}{2}$.

Cleavage and fracture Perfect rhombohedral cleavage; conchoidal fracture.

Colour and transparency White or colourless when pure, grey, brown, or yellowish when iron-bearing; transparent to translucent.

Streak White.

Lustre Vitreous to earthy.

Distinguishing features From dolomite and calcite by specific gravity and lack of twinning; from chert by hardness. Will dissolve with effervescence in hydrochloric and nitric acids only on heating.

Formation and occurrence Much less common than calcite, and does not directly form sedimentary rocks. 'Sedimentary' magnesite results from the replacement of existing limestones by magnesium-bearing solutions (often forming dolomite in the process) and from the action of carbonate-rich solutions on rocks containing magnesium minerals. It occurs also as veins in magnesium-rich metamorphic rocks such as talc schists and serpentinites.

Uses For making cements, and refractory bricks and crucibles.

Siderite (Chalybite)

Composition $FeCO_3$. Rarely found pure; complete solid solutions exist between it and both magnesite and rhodochrosite, and calcium can substitute in small amounts.

Crystal system Hexagonal (Trigonal).

Habit Usually as rhombohedra (same as cleavage form); crystals frequently have curved faces due to their composite nature; also massive, granular, fibrous, compact, botryoidal and earthy.

Twinning On rhombohedron; often lamellar.

Specific gravity 3·96 when pure.

Hardness $3\frac{1}{2}$–$4\frac{1}{2}$.

Cleavage and fracture Perfect rhombohedral cleavage; uneven fracture.

Colour and transparency Grey-brown to yellowish brown, brown or dark brown.

Streak White.

Lustre Vitreous.

Distinguishing features Habit; cleavage; from other carbonates by colour and specific gravity. Soluble in hot hydrochloric acid with effervescence.

Formation and occurrence Large sedimentary deposits of massive siderite have been formed by direct precipitation either in lakes or in the sea; these are often impure, containing clays (clay ironstone), carbonaceous matter (blackband iron ore), or calcium carbonate, and nodules of siderite are common in clay and shale rocks. Equally important are the replacement deposits caused by the action of ferrous solutions on limestones. It is also a common mineral in hydrothermal metallic ore veins, some of which are also of economic importance.

Uses As an ore of iron.

Rhodochrosite

Composition $MnCO_3$. Calcium and ferrous iron may substitute for manganese resulting in complete solid solution with siderite and incomplete with calcite; limited amounts of magnesium and zinc may also substitute for manganese.

Crystal system Hexagonal (Trigonal).

Habit Crystals rhombohedral, rarely scaleno-hedral, usually with curved faces; commonly massive, compact, cleavable or granular.

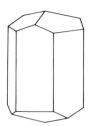

Calcite: prism and flat rhombohedron.

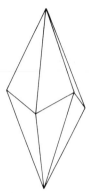

Calcite: scalenohedron.

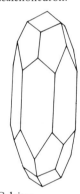

Calcite: combination of prism, scalenohedron and rhombohedron.

Calcite: rhombohedron.

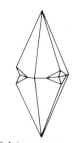

Calcite: scalenohedron twinned on basal pinacoid.

Rose-pink rhombohedra of rhodochrosite; from Silverton, Colorado.

Calcite:
scalenohedron
twinned on
rhombohedron.

Calcite:
scalenohedron
twinned on
rhombohedron.

Curved composite
creamy-white
rhombohedra of
dolomite on
quartz; from
Germany.

Twinning Rare, lamellar on rhombohedron.
Specific gravity 3·5–3·7.
Hardness 3½–4.
Cleavage and fracture Perfect, rhombohedral cleavage; uneven fracture.
Colour and transparency Pale to deep rose pink, yellowish grey, brownish; transparent to translucent.
Streak White.
Lustre Vitreous.
Distinguishing features From other rhombohedral carbonates by colour; cleavage; from rhodonite by inferior hardness and alteration to black or brown manganese oxides. Dissolves with effervescence in hot, dilute hydrochloric acid.
Formation and occurrence In high-temperature metasomatic deposits, as well as a primary mineral in hydrothermal metallic ore veins. Also in pegmatites and some metamorphic rocks.
Uses A minor ore of manganese.

Smithsonite

Composition $ZnCO_3$. Ferrous iron and manganese may substitute for some zinc, as may calcium, magnesium, cadmium, copper, cobalt and, less commonly, lead.
Crystal system Hexagonal (Trigonal).
Habit Crystals, rhombohedra or scalenohedra with curved faces, rare; usually reniform, botryoidal, stalactitic, as crystalline encrustations or massive; honeycombed masses known as 'dry-bone ore'.
Specific gravity 4·3–4·5.
Hardness 4–4½.
Cleavage and fracture Perfect rhombohedral cleavage (when seen); uneven fracture.
Colour and transparency Grey, brown or greyish white, also green, yellow and other shades; yellow variety ('turkey-fat ore') contains cadmium; translucent.
Streak White.
Lustre Vitreous.
Distinguishing features Cleavage; specific gravity. Soluble in warm hydrochloric acid with effervescence.
Formation and occurrence In most lead and zinc mining centres, occurring in the oxidized zone of these ore deposits, and may be secondary after primary sphalerite. Replaces limestone: often found as pseudomorphs after calcite. Also stalactitic.
Uses An ore of zinc. As an ornamental stone, especially the yellow stalactitic (Sardinia) and green (Greece, New Mexico) varieties.

Dolomite

Composition $CaMg(CO_3)_2$. Ferrous iron may substitute for magnesium in large amounts, as may manganese, zinc and lead in much smaller amounts. Some substitution of magnesium by calcium may also occur.
Crystal system Hexagonal (Trigonal).
Habit Crystals common, with curved composite rhombohedral faces; also coarse and granular to fine grained and compact.
Twinning Common, especially on basal plane and rhombohedron.
Specific gravity 2·8–2·9.
Hardness 3½–4.
Cleavage and fracture Perfect rhombohedral

Dolomite:
showing curved
composite faces.

cleavage; subconchoidal fracture.
Colour and transparency White, sometimes colourless, often yellow, brown or pinkish; transparent to translucent.
Streak White.
Lustre Vitreous to pearly.
Distinguishing features Habit; colour. Other properties similar to calcite but reacts less readily with dilute hydrochloric acid: large fragments react only when hot.
Formation and occurrence Much 'sedimentary' dolomite has resulted from the action of magnesium-bearing solutions on pre-existing sediments, either soon after their deposition or much later, when they had become rock. Some less extensive deposits, however, are primary evaporites. Metamorphic events may recrystallize both types to dolomite marble. Hydrothermal veins can contain good crystals of dolomite in association with metallic ores, and, similarly to calcite, it has been found in igneous rocks to a greater (in the rock beforsite) or lesser extent.
Uses As a building and ornamental stone, and in the manufacture of refractory bricks for furnace linings.

Ankerite

Composition Ca(Mg,Fe)(CO$_3$)$_2$. Manganese may also substitute for magnesium.
Crystal system Hexagonal (Trigonal).
Habit Crystals rhombohedral; also massive.
Twinning Common, similar to dolomite.
Specific gravity 2·93–3·10.
Hardness 3½–4.
Cleavage Perfect, rhombohedral cleavage.
Colour and transparency White, yellow, yellowish brown, brown, rarely grey; becomes dark brown on weathering; translucent.
Streak White.
Lustre Vitreous.
Distinguishing features From dolomite by colour which becomes darker on heating; from siderite by specific gravity.
Formation and occurrence In sedimentary rocks as the result of both hydrothermal and low-temperature metasomatism, like dolomite; frequently in coal seams. Also in metamorphic rocks.

Aragonite

Composition CaCO$_3$. Strontium, lead and, more rarely, barium, may substitute for calcium. Atomic arrangement differs from calcite.
Crystal system Orthorhombic.
Habit Prismatic, terminated by very steep pyramids, crystals usually found in radiating groups; tabular; stalactitic; pseudohexagonal, repeatedly twinned crystals; also encrusting.
Twinning Common on {110}; an intergrowth of three individuals twinned in this way and with basal planes in common produces the distinctive pseudohexagonal twin crystal.
Specific gravity 2·94.
Hardness 3½–4.
Cleavage and fracture Distinct cleavage {010}; subconchoidal fracture.
Colour and transparency Colourless, white, grey, yellowish, sometimes green or violet; transparent to translucent.
Streak White.
Lustre Vitreous.
Distinguishing features From calcite by habit, cleavage and specific gravity. Dissolves with effervescence in cold dilute hydrochloric acid.
Formation and occurrence Main component of the shells of many organisms (such as corals and oysters) both fossil and recent, and is the primary calcium carbonate precipitate from sea-water, but it is metastable and its atoms may rearrange to give the calcite structure, given time, under normal temperatures and pressures. Much less common than calcite. As well as occurring in sedimentary deposits, it can be deposited from hot springs, occur as a secondary mineral in cavities in volcanic rocks, and form in the oxidized zone of ore deposits. Some stalactites in limestone caves are aragonite, as are 'cave pearls' found in underground pools (and true pearls formed in oyster shells). Also widespread in some metamorphic rocks.

Witherite

Composition BaCO$_3$. Some limited replacement of barium by calcium, strontium or magnesium.
Crystal system Orthorhombic.
Habit Crystals are always twinned, on {110}, giving rise to pseudohexagonal prisms and pyramids which resemble quartz; also massive, granular or columnar, or botryoidal.
Twinning Always twinned, on {110}.
Specific gravity 4·29–4·30.
Hardness 3½.
Cleavage and fracture Distinct, {010}; fracture uneven.
Colour and transparency Colourless, white, greyish or light yellowish brown; transparent to translucent.
Distinguishing features High specific gravity. Dissolves with effervescence in dilute hydrochloric acid; sulphuric acid added to this solution gives a white precipitate. Powdered witherite moistened with hydrochloric acid will colour the flame apple green.
Formation and occurrence In low-temperature hydrothermal veins, nearly always in sedimentary rocks. Although not widespread, it is the second most common barium mineral after baryte.
Uses As an ore of barium.

Strontianite

Composition SrCO$_3$. Significant amounts of calcium and barium may substitute for strontium.
Crystal system Orthorhombic.
Habit Prismatic crystals, often acicular and radiating; like aragonite, frequently repeatedly twinned to give pseudohexagonal forms; also massive, fibrous or granular.
Twinning Common, on {110}, single, repeated or lamellar.
Specific gravity 3·72.
Hardness 3½.
Cleavage and fracture Nearly perfect {110} cleavage; uneven fracture.
Colour and transparency Colourless, white, yellow, greenish or brownish; transparent to translucent.
Streak White.
Lustre Vitreous.
Distinguishing features Specific gravity. Dissolves with effervescence in hydrochloric acid; powder moistened with this acid will colour the flame crimson.
Formation and occurrence Usually in sedimentary rocks; also in low-temperature hydrothermal veins. Has been recorded in association with the igneous rocks, carbonatites. May replace celestine.
Uses In fireworks and red flares; as a source of strontium.

Cerussite

Composition PbCO$_3$.
Crystal system Orthorhombic.
Habit Crystals common, showing many forms: prismatic; tabular parallel to {010}; bipyramidal; or pseudohexagonal multiple twins; also massive, granular, compact and sometimes stalactitic.
Twinning Frequent; may be multiple, forming reticulated groups or star-shaped pseudohexagonal crystals; or single, giving arrowhead twins.
Specific gravity 6·57.

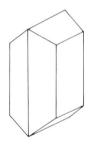

Aragonite.

Witherite: pseudohexagonal twin.

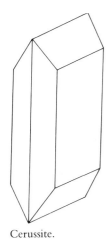
Cerussite.

Twinned mass of cerussite on copper; from Tsumeb, south-west Africa.

Hardness 3–3½.
Cleavage and fracture Distinct {110} and {021} cleavages; conchoidal fracture.
Colour and transparency White or greyish, sometimes tinted due to impurities; transparent to translucent.
Streak White.
Lustre Adamantine.
Distinguishing features High specific gravity; lustre. Dissolves with effervescence in warm hydrochloric and nitric acids.
Formation and occurrence In oxidation zones of lead veins, usually secondary after other lead minerals.
Uses A valuable ore of lead.

Malachite

Composition $Cu_2CO_3(OH)_2$.
Crystal system Monoclinic.
Habit Almost always botryoidal, stalactitic or encrusting, often with a smooth surface and internal structure of compact radiating fibres, banded in various shades of green; also granular, earthy, rare crystals.
Specific gravity 3·9–4·05.
Hardness 3½–4.
Cleavage and fracture Perfect {$\bar{2}01$} cleavage, but rarely seen; subconchoidal, uneven fracture.
Colour and transparency Bright green, various shades; crystals translucent to transparent.
Streak Pale green.
Lustre Vitreous to adamantine in crystals; otherwise silky to dull.
Distinguishing features Colour; habit. Dissolves with effervescence in dilute hydrochloric acid.
Formation and occurrence Widely distributed secondary copper mineral, occurring in the oxidized zones of copper deposits (especially copper veins in limestone), associated with other, primary copper minerals from which it formed and which it may replace. A very large deposit occurs in southern Africa, in the 'copper belt' of Zaire and Zambia.
Uses A valuable ore of copper: also used as an ornamental, semiprecious stone.

Azurite (Chessylite)

Composition $Cu_3(CO_3)_2(OH)_2$.
Crystal system Monoclinic.
Habit Crystals frequently complex in habit and malformed; generally modified prismatic or tabular; also in radiating groups, massive or earthy.
Specific gravity 3·80.
Hardness $3\frac{1}{2}$–4.
Cleavage and fracture Perfect prismatic cleavage; also pinacoidal; fracture conchoidal.
Colour and transparency Intense azure blue various shades; crystals transparent to translucent.
Streak Pale blue.
Lustre Vitreous to adamantine.
Distinguishing features Colour. Dissolves with effervescence in dilute hydrochloric acid. Frequently alters to malachite.
Formation and occurrence As a secondary copper mineral, occurring in the oxidized zone of copper deposits, often associated with, though not as common as, malachite.
Uses A minor ore of copper.

Sphaerocobaltite (Cobaltocalcite)

Composition $CoCO_3$. Small amounts of nickel, ferrous iron and calcium substitute for cobalt.
Crystal system Hexagonal (Trigonal).
Habit Crystals rare; as small spherical masses, with a crystalline surface and concentric and radiated structure; as crusts.
Specific gravity 4·11–4·13.
Hardness 4.
Cleavage and fracture Rhombohedral cleavage.
Colour and transparency Rose red, altering superficially to brown or black; transparent to subtranslucent.
Streak Pale rose-red.
Lustre Vitreous.
Distinguishing features Colour; habit. Dissolves with effervescence with difficulty in cold but easily in hot hydrochloric and nitric acids.
Formation and occurrence Rare, associated with cobalt mineralization as a secondary mineral.

Alstonite (Bromlite)

Composition $CaBa(CO_3)_2$.
Distinguishing features From witherite by specific gravity (3·67 to 3·71) and sharpness of the bipyramids.

Bismutite

Composition $(BiO)_2CO_3$.
Crystal system Tetragonal.
Habit Usually as dense, earthy masses, opaline crusts; radially fibrous crusts or spheroidal aggregates; may have pseudomorphous prismatic structure.
Specific gravity 6·7–7·4 (natural), 8·2 (artificial).
Hardness $3\frac{1}{2}$, but variable.
Cleavage and fracture Basal cleavage where observable.
Colour and transparency Pale yellow, brown, grey or greenish tints; more intense colours due to chemical impurities.
Streak Grey.
Lustre Vitreous to pearly.

Distinguishing features Habit; association. Dissolves with effervescence in hydrochloric and nitric acids.
Formation and occurrence In the oxidized zones of bismuth-bearing metallic ore deposits, as a common secondary bismuth mineral. May pseudomorph bismuthinite.

Azurite.

Leadhillite

Composition $Pb_4(SO_4)(CO_3)_2(OH)_2$. Copper may substitute for lead in small amounts.
Crystal system Monoclinic.
Habit Usually pseudohexagonal, sometimes due to twinning, thick to thin tabular; often with striated or curved faces; also massive, granular.
Twinning Very common, several different modes giving rise to pseudohexagonal, lamellar, contact and interpenetration twins.
Specific gravity 6·55–6·57.
Hardness $2\frac{1}{2}$–3.
Cleavage and fracture Perfect basal cleavage; splits into flexible laminae; conchoidal fracture; sectile.
Colour and transparency Colourless, white, grey, pale green, blue and yellow; transparent to translucent.
Streak White.
Lustre Resinous to adamantine.
Distinguishing features High specific gravity. Dissolves with effervescence in nitric acid leaving a residue (lead sulphate). Breaks up in hot water.
Formation and occurrence A secondary mineral in the oxidized zone of lead deposits. May pseudomorph galena and calcite.

Phosgenite

Composition $Pb_2(CO_3)Cl_2$.
Crystal system Tetragonal.
Habit Short prismatic to prismatic crystals; rarely thick tabular; also massive, granular.
Specific gravity 6·13.
Hardness 3.
Cleavage $\{001\}$ and $\{110\}$, distinct cleavages, $\{100\}$, less distinct cleavage.
Colour and transparency Colourless, white, various pale tints especially brown and yellowish brown; transparent to translucent.
Streak White.
Lustre Adamantine.
Distinguishing features Specific gravity. Dissolves with effervescence in dilute nitric acid; the resulting solution will give a white precipitate on addition of silver nitrate. Fragment will fuse in a candle flame.
Formation and occurrence A secondary mineral in the oxidized zones of lead deposits, formed by alteration of galena and other lead minerals. Has been formed at Laurium, Greece, due to the action of sea-water on old lead slags. Readily alters to cerussite.

Hydromagnesite

Composition $Mg_4(OH)_2(CO_3)_3.3H_2O$.
Crystal system Monoclinic.
Habit Crystals small, as tufts, rosettes or crusts of acicular or bladed crystals; also massive, chalky.
Twinning Common, on $\{100\}$.

Specific gravity 2·2–2·3.
Hardness 3½.
Cleavage Perfect {010} cleavage.
Colour and transparency Colourless to white; transparent.
Streak White.
Lustre Vitreous.
Distinguishing features Habit. Dissolves with effervescence in hydrochloric and nitric acids.
Formation and occurrence As a low-temperature hydrothermal mineral; may be formed by weathering processes. Found in altered magnesium-rich igneous and metamorphic rocks.

Turquoise-blue radiating needles of aurichalcite on limonite; from Stockton, Utah.

Aurichalcite

Composition $(Zn,Cu)_5(OH)_6(CO_3)_2$. Zinc and copper substitute mutually over a considerable range.
Crystal system Orthorhombic.
Habit Delicate acicular or lath-like crystals, often striated or furrowed with wedge-shaped terminations, forming tufted or feathery incrustations; also (rarely) columnar, laminated, granular.
Specific gravity 3·6–3·9.
Hardness 1–2.
Cleavage Perfect, {010} cleavage.
Colour and transparency Pale green to greenish blue and sky blue; transparent.
Streak Pale green or blue.
Lustre Silky to pearly.
Distinguishing features Colour; habit; dissolves in hydrochloric and nitric acids and ammonia.
Formation and occurrence Secondary mineral in the oxidized zone of copper and zinc deposits.

Hydrozincite

Composition $Zn_5(OH)_6(CO_3)_2$.
Crystal system Monoclinic.
Habit Usually massive, earthy, porous to compact; as encrustations, sometimes concentrically banded with fibrous radial structure, stalactitic; rare minute crystals are tapering, lath- or blade-like.
Specific gravity 3·6–4.
Hardness 2–2½.
Cleavage Perfect {100} cleavage.
Colour and transparency White to grey, yellowish, brownish, pinkish, pale lilac; transparent to translucent.
Streak Dull to shining.

Lustre Pearly (in crystals).
Distinguishing features Dissolves easily in hydrochloric and nitric acids. Fluoresces pale blue or lilac in ultraviolet light.
Formation and occurrence A secondary mineral, formed in the oxidized zone of ore deposits by the alteration of sphalerite (blende) and other zinc minerals; may pseudomorph dolomite.

Nitratine (Soda Nitre, Chile Saltpetre)

Composition $NaNO_3$
Crystal system Hexagonal (Trigonal).
Habit Crystals rare, rhombohedral; usually massive; as encrustation or in beds.
Twinning Common, similar to that of calcite.
Specific gravity 2·29.
Hardness 1–2.
Cleavage and fracture Perfect rhombohedral cleavage; conchoidal fracture but rarely seen.
Colour and transparency Colourless or white, also reddish brown, greyish or yellowish; transparent to translucent.
Streak White.
Lustre Vitreous.
Distinguishing features From calcite by specific gravity and hardness. Dissolves easily and completely in water: deliquescent; taste cooling. Will fuse in candle flame with bright yellow emission (due to sodium).
Formation and occurrence Only in regions of very low rainfall, such as the Atacama Desert of northern Chile where it is an economic deposit occurring interbedded with sand, halite, and gypsum. The elements forming these minerals have probably been dissolved out of surrounding volcanic rocks, to be redeposited as sedimentary rocks.
Uses In fertilizers and explosives.

Nitre

Composition KNO_3.
Crystal system Orthorhombic.
Habit Usually as thin encrustations or as silky acicular crystals; also as multiple twins with pseudohexagonal forms.
Twinning {110}, often multiple giving rise to pseudohexagonal twins like those of aragonite.
Specific gravity 2·1.
Hardness 2.
Cleavage Perfect, {011} cleavage.
Colour and transparency White; translucent.
Streak White.
Lustre Vitreous.
Distinguishing features Dissolves easily in water; not deliquescent; saline and cooling taste. Fragment will fuse in candle flame with violet emission (due to potassium).
Formation and occurrence As delicate crusts on surfaces of earth, walls, rocks; also as a constituent of certain soils, including some of those in limestone caves. Also found in the Chile nitratine deposits. Manufactured artificially from some soils for commercial purposes.
Uses As a source of nitrates for explosives, fertilizer, and metallurgical and chemical processes.

Borax

Composition $Na_2B_4O_7.10H_2O$.
Crystal system Monoclinic.
Habit Prismatic crystals, sometimes very large; also massive, cellular or as encrustations.
Specific gravity 1·7.
Hardness 2–2½.
Cleavage and fracture Perfect {100} cleavage, also prismatic; conchoidal fracture; soft and brittle.
Colour and transparency Colourless or white, sometimes with tinges of blue, green or grey; translucent.
Streak White.
Lustre Vitreous.
Distinguishing features Low specific gravity; hardness. Readily soluble in water; sweetish-alkaline taste. Fuses in candle flame with swelling and bright yellow emission (due to sodium).
Formation and occurrence In evaporite deposits, as a deposit from the evaporation of salt lakes and as an efflorescence on the surface of the ground in regions of low rainfall.
Uses In cleansing, as an antiseptic and preservative, and as a solvent and flux in industrial processes. It is an ore of boron, which itself has many important scientific and industrial applications.

Colemanite

Composition $Ca_2B_6O_{11}.5H_2O$.
Crystal system Monoclinic.
Habit Short, prismatic crystals usually projecting into cavities, often highly modified; also massive, compact and granular.
Specific gravity 2.42.
Hardness 4–4½.
Cleavage and fracture Perfect parallel to {010}, also parallel to {001} (less good); uneven fracture.
Colour and transparency Colourless, white or greyish; transparent to translucent.
Streak White.
Lustre Vitreous to adamantine.
Distinguishing features Cleavage; hardness. Soluble in hot hydrochloric acid, with separation of white precipitate (boric acid) on cooling. Fuses easily in a gas flame, crumbling and breaking up.
Formation and occurrence As nodules in clays, and in beds, especially of Tertiary age, as a secondary mineral derived from primary borates such as ulexite.
Uses As an ore of boron, once very important economically but since superseded by the discovery of large deposits of kernite.

Ulexite

Composition $NaCaB_5O_9.8H_2O$.
Crystal system Triclinic.
Habit Usually as rounded masses of fine fibrous crystals ('cotton balls') or as parallel fibrous aggregates; rare elongate crystals.
Twinning Common, possibly due in part to mechanical deformation.
Specific gravity 1·96
Hardness 2½ (of crystals).
Cleavage Perfect {010} cleavage (in crystals).
Colour and transparency White (crystals colourless); transparent to translucent.

Streak White.
Lustre Vitreous (crystals); silky (aggregates).
Distinguishing features Habit. Tasteless. Dissolves slightly in cold water, more so in hot. Fuses with swelling in candle flame.
Formation and occurrence An evaporite mineral, crystallizing from saline lakes in areas of low rainfall with other borate minerals. Also found in bedded gypsum deposits.
Uses A source of borax.

Boracite

Composition $Mg_3B_7O_{13}Cl$. Ferrous iron may substitute for magnesium.
Crystal system Orthorhombic at ordinary temperatures, but crystals may show cubic (high-temperature) forms.
Habit As isolated, embedded crystals, showing a range of cubic forms; faces often pitted due to etching; also massive, as fine grained to fibrous.
Twinning Penetration twins, rare.
Specific gravity 2·9–3·0.
Hardness 7–7½.
Fracture Conchoidal fracture.
Colour and transparency Colourless to white or grey, also shades of green in iron-bearing varieties; transparent to translucent.
Streak White.
Lustre Vitreous to adamantine.
Distinguishing features Hardness; pseudo-habit. Insoluble in water but dissolves in hot hydrochloric acid. Fuses in a gas flame with green emission (due to boron).
Formation and occurrence In evaporite deposits, associated with beds of halite, anhydrite and gypsum, and in potash deposits, where it may occur as small crystals or concretions.

Kernite

Composition $Na_2B_4O_7.4H_2O$.
Crystal system Monoclinic.
Habit Usually in coarse, cleavable aggregates; crystals are stubby, sometimes wedge-shaped or rounded; markedly striated; irregularly developed.
Specific gravity 1·95.
Hardness 2½–3.

Crystals of colemanite; from Death Valley, California.

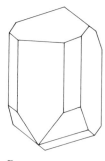

Borax.

Cleavage Perfect {100} and {001} cleavages giving long splintery fragments.
Colour and transparency Colourless, but usually white due to surface alteration to tincalconite ($Na_2B_4O_7.5H_2O$); transparent.
Streak White.
Lustre Vitreous to pearly.
Distinguishing features Cleavage. Slowly dissolves in cold water; readily dissolves in hot water and acids. Fuses in gas flame.
Formation and occurrence Found only in the Mojave Desert of California, United States, in a large deposit containing crystals 0·6 to 0·9 metre (2 to 3 feet) thick. It is believed to have formed from an existing borax evaporite deposit by recrystallization caused by increased temperature and pressure.
Uses A major source of borax and boron compounds.

6 Sulphates, Chromates

Baryte (Barytes, Barite)

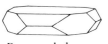

Baryte: tabular habit.

Celestine: prismatic habit.

Geode of blue celestine crystals; from Madagascar.

Composition $BaSO_4$. Strontium substitutes for barium in a solid solution series from baryte to celestine; lead and calcium may also substitute for barium but in minor amounts.
Crystal system Orthorhombic.
Habit Crystals usually tabular, parallel to base, often diamond shaped due to a vertical prism, a shape which may be further modified, especially by other prisms; prismatic crystals common, elongated parallel to the *a*-axis; also as globular concretions and as fibrous, lamellar, granular and earthy aggregates; some 'desert roses' are rosettes of baryte crystals enclosing sand grains.
Twinning Glide twinning of secondary origin produced by pressure in massive material.
Specific gravity 4·5.
Hardness 3–3½.
Cleavage and fracture Perfect {001} cleavage, less perfect cleavage parallel to {210}; uneven fracture.
Colour and transparency Colourless, white, yellowish, grey, pale green, pale blue, red, brown; yellow, red and brown varieties result from impurities of iron minerals; blue varieties may be due to irradiation from small amounts of radium in the crystal structure; transparent to opaque.
Streak White.
Lustre Vitreous, approaching resinous, sometimes pearly on {001} cleavage.
Distinguishing features High specific gravity; cleavage; habit. Insoluble in dilute acids. Powdered, will colour gas flame apple green (barium). Some specimens are fluorescent (and phosphorescent) in ultraviolet light.
Formation and occurrence Most common barium mineral, mainly as a gangue mineral, in metalliferous hydrothermal veins. Also as veins, cement, or concretions in sedimentary rocks. Also in residual surface deposits, in hot spring deposits, and in cavities in some igneous rocks. May be a primary or a secondary (replacement) mineral.
Uses Over 80 per cent is used as a drilling mud for oil and gas wells. Also used in the manufacture of paints, some floor coverings and textiles, and as a

filler for paper. As 'barium meal' in medical X-ray technology. Chief source of barium for chemicals.

Celestine

Composition $SrSO_4$. Complete substitution of strontium by barium can occur; calcium may also substitute but to a limited extent.
Crystal system Orthorhombic.
Habit As well-formed crystals with a tabular or prismatic habit, closely resembling those of baryte (combinations of prism and basal pinacoid with prominent development of other prisms); also radiating fibrous, granular.
Specific gravity 3·96.
Hardness 3–3½.
Cleavage and fracture Perfect basal cleavage; good {210} cleavage; imperfect conchoidal fracture.

Colour and transparency Colourless, white, pale blue, reddish, greenish, brownish; most colours caused by impurities; blue tints may be due to irradiation; transparent to translucent.
Streak White.
Lustre Vitreous, inclined to pearly.
Distinguishing features Specific gravity; habit; cleavage. Insoluble in dilute acids. Powdered, will colour gas flame crimson (strontium).
Formation and occurrence Chiefly in sedimentary rocks, either disseminated throughout the rock or as veins or crystal-lined nodules and cavity infills, having formed sometimes as a direct deposit from sea-water, sometimes as a secondary, replacement deposit. Also as a primary mineral in some hydrothermal ore veins and as a cavity- or vein-filling mineral in some igneous rocks.
Uses The principal ore of strontium. Strontium compounds are used in the manufacture of (red) fireworks and flares; also in the production of paints, ceramics, plastics and in some electrolytic refining processes.

Anglesite

Composition $PbSO_4$. Some substitution of lead by barium may occur.
Crystal system Orthorhombic.
Habit Crystals prismatic, or tabular parallel to the basal pinacoid, or occasionally as pyramids variously modified; many complex and varied forms; some faces may be striated; also granular, compact,

stalactitic, and as concentrically banded massive varieties which may enclose an unaltered core of galena.

Specific gravity 6·3–6·4.
Hardness 3.
Cleavage and fracture Good {001}, distinct {210}, cleavages; conchoidal fracture.
Colour and transparency Colourless, white, grey, pale shades of yellow, blue or green; transparent to opaque.
Streak White.
Lustre Adamantine when pure.
Distinguishing features High specific gravity. Does not effervesce in acids. Small fragments will fuse in a candle flame.
Formation and occurrence Common secondary mineral, usually formed from galena and accompanied and followed by cerussite in the oxidized zone of lead deposits.
Uses As an ore of lead.

Anhydrite

Composition $CaSO_4$. Small amounts of strontium and barium may substitute for calcium.
Crystal system Orthorhombic.
Habit Crystals not common, may be equant or thick tabular; usually massive, fibrous or granular, or as contorted concretionary forms ('tripe-stone'). sometimes pseudomorphous after halite cubes.
Twinning Simple or repeated on {011} or {120}.
Specific gravity 2·9–3·0.
Hardness 3–3½.
Cleavage and fracture Three pinacoidal cleavages, one perfect, one very good, one good, giving rise to rectangular (apparently cubic) cleavage fragments; uneven fracture.
Colour and transparency White or colourless when pure, often grey, more rarely bluish mauve due to natural radiation, or red or brown due to iron oxide impurities; transparent to sub-translucent.
Streak White.
Lustre Vitreous to pearly on cleavage.
Distinguishing features Cleavage, hardness. Does not dissolve readily or effervesce in dilute acids.
Formation and occurrence Important rock-forming mineral, found in bedded evaporite deposits, associated with other evaporite minerals. Some may have been a primary precipitate, but most deposits are thought to be secondary after gypsum. Also as a minor component of some sedimentary rocks, and as a product of hydrothermal alteration of limestones and dolomites. May occur as a gangue mineral in some metal ore veins, and filling cavities and veins in some igneous rocks.
Uses As a fertilizer; in the manufacture of plaster and cement, and as a source of sulphur for chemicals, especially in the manufacture of sulphuric acid.

Gypsum

Composition $CaSO_4.2H_2O$.
Crystal system Monoclinic.
Habit Crystals may be of prismatic habit; tabular parallel to side pinacoid, or diamond shaped with edges bevelled; other forms are uncommon; also granular massive or rock-forming; foliated; fibrous.

Twinning Very common on {100} ('swallow-tail') and on {Ī01} ('arrow-head'); as well as simple twins, crystals are often in radiating interpenetrating groups.
Specific gravity 2·30–2·37.
Hardness 2.
Cleavage and fracture Perfect cleavage parallel to {010} giving very thin, flexible, non-elastic plates; also distinct cleavage parallel to {100} and {011}, giving rise to lozenge-shaped cleavage fragments.
Colour and transparency Colourless, white, grey, sometimes yellow or brownish; transparent to translucent.
Streak White.
Lustre Subvitreous; pearly parallel to cleavage.
Distinguishing features Hardness; habit; cleavage. Dissolves in hot dilute hydrochloric acid.
Formation and occurrence As an evaporite deposit, being one of the first primary precipitates from an evaporating brine; as a secondary mineral deposited from percolating ground waters ('desert roses' of gypsum crystal rosettes enclosing sand grains may be formed in this way) or replacing other sedimentary rocks, even anhydrite beds which are themselves secondary; and in calcareous muds and clays in association with the decomposition of pyrite, often resulting in particularly well-formed crystals. Any well-formed crystals of gypsum are known as selenite; silky, fibrous varieties produced by the crystallization of secondary gypsum in veins are known as satin spar; and the pure, finely compact massive variety is known as alabaster. Also found where limestones have been reacted upon by volcanic vapours or decomposing sulphide ore deposits, and as a gangue material in metallic veins.
Uses In the production of plaster of Paris for which it is heated to drive off 75 per cent of the water combined in the structure, and ground into powder. When water is added it recombines and the mass sets hard through the formation of interlocking crystals of gypsum. Also used as a filler in various materials, and as a fertilizer. Decorative varieties have ornamental uses.

Chalcanthite

Composition $CuSO_4.5H_2O$. Iron may substitute for copper.
Crystal system Triclinic.
Habit Crystals commonly short prismatic, less commonly thick tabular; also stalactitic, reniform; as cross-fibre veinlets; massive, granular, encrusting.
Twinning Rarely, as cruciform intergrowths.
Specific gravity 2·28.
Hardness 2½.
Cleavage and fracture Imperfect {1Ī0} cleavage; conchoidal fracture.
Colour and transparency Deep azure blue, sometimes greenish; subtransparent to translucent.
Streak White.
Lustre Vitreous.
Distinguishing features Colour. Metallic taste. Dissolves in water (a piece of iron dipped in this solution will become coated with metallic copper).
Formation and occurrence In the zone of weathering of copper ore deposits, from the

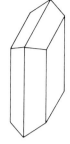

Gypsum.

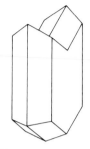

Gypsum: twinned crystal.

Chalcanthite.

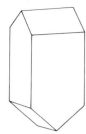

Epsomite.

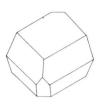

Glauberite: tabular habit.

alteration of copper sulphides.

Uses A minor ore of copper and as a fungicide.

Epsomite

Composition $MgSO_4.7H_2O$. Nickel, manganese and zinc can substitute for magnesium in varying amounts.

Crystal system Orthorhombic.

Habit Rarely in crystals; usually in botryoidal masses and delicately fibrous crusts.

Twinning Rarely seen.

Specific gravity 1·68.

Hardness $2-2\frac{1}{2}$.

Cleavage Perfect {010} cleavage.

Colour and transparency Colourless to white; transparent to translucent.

Streak White.

Lustre Vitreous; fibrous types silky to earthy.

Distinguishing features Occurrence; habit. Bitter saline taste. Dissolves in water.

Formation and occurrence Usually encrusting walls of caves and mine workings. Also in mineral waters and in evaporite deposits and fumaroles.

Uses Medicinal.

Alunite

Composition $KAl_3(OH)_6(SO_4)_2$.

Crystal system Hexagonal (Trigonal).

Habit Crystals uncommon, small rhombohedra with basal plane; usually massive, granular, fibrous, earthy.

Specific gravity 2·6–2·8.

Hardness 4.

Cleavage and fracture Basal cleavage; conchoidal fracture (of crystals), splintery fracture (of masses).

Colour and transparency White when pure, also grey, yellow, red; transparent to subtranslucent.

Streak White.

Lustre Vitreous to pearly.

Distinguishing features Insoluble in water and practically insoluble in acids (slowly dissolves in sulphuric acid). Difficult to identify positively without chemical tests.

Formation and occurrence As an alteration or replacement of volcanic rocks, sometimes in large masses. Also around fumaroles.

Uses In the production of alum and other potassium and aluminium salts.

Jarosite

Composition $KFe_3(OH)_6(SO_4)_2$.

Crystal system Hexagonal (Trigonal).

Habit As crusts or coatings of microscopic (pseudocubic or tabular) crystals; granular massive, fibrous, concretionary.

Specific gravity 2·91–3·25.

Hardness $2\frac{1}{2}-3\frac{1}{2}$.

Cleavage Basal, distinct.

Colour and transparency Ochrous, amber yellow to dark brown; translucent.

Streak Pale yellow, sometimes glistening.

Lustre Subadamantine to vitreous.

Distinguishing features Habit; association. Insoluble in water, soluble in hydrochloric acid.

Formation and occurrence A secondary mineral widespread as crusts and coatings on and associated with iron ores.

Plumbojarosite

Composition $PbFe_6(OH)_{12}(SO_4)_4$.

Crystal system Hexagonal (Trigonal).

Habit As tiny tabular crystals, hexagonal plates, making up crusts or compact masses; also earthy.

Specific gravity 3·67.

Hardness Soft, talc-like feel.

Colour Golden brown to dark brown.

Streak Brown, glistening.

Lustre Dull to glistening or silky.

Distinguishing features Habit; colour; occurrence. Only slightly soluble in acids.

Formation and occurrence A secondary mineral in the oxidized zones of lead deposits.

Thenardite

Composition Na_2SO_4.

Crystal system Orthorhombic.

Habit Bipyramidal or tabular crystals, often twinned; some faces striated; also as crusts or powdery.

Twinning Cross-shaped and 'butterfly' twins common.

Specific gravity 2·67.

Hardness $2\frac{1}{2}-3$.

Cleavage and fracture {010} perfect, {101} fair, cleavages; uneven fracture.

Colour and transparency Colourless when pure, also white and various pale tints; transparent to translucent.

Streak White.

Lustre Vitreous.

Distinguishing features Taste faintly salty. Very soluble in water. Transparent varieties take up water from the air and become cloudy.

Formation and occurrence In saline lakes and as a surface deposit on the soils of arid regions, and around fumaroles and so on.

Glauberite

Composition $Na_2SO_4.CaSO_4$.

Crystal system Monoclinic.

Habit Crystals tabular parallel to the basal plane; or bipyramidal with striated faces and rounded-off apices.

Specific gravity 2·8.

Hardness $2\frac{1}{2}-3$.

Cleavage and fracture Perfect basal cleavage; conchoidal fracture.

Colour and transparency Usually grey or yellowish, also colourless or reddish due to included iron oxide; transparent to translucent.

Streak White.

Lustre Vitreous to pearly.

Distinguishing features Habit. Slightly salty taste. Whitens in water (due to partial solution, leaving a surface deposit of gypsum).

Formation and occurrence An evaporite mineral occurring in bedded salt deposits.

Uses In chemical and industrial processes.

Polyhalite

Composition $K_2Ca_2Mg(SO_4)_4.2H_2O$.

Crystal system Triclinic.

Habit Crystals very rare, usually twinned; characteristically in compact granular, fibrous or

lamellar masses.
Twinning On {010} and {100}, often multiple.
Specific gravity 2·78.
Hardness 2½–3.
Cleavage and fracture {10$\bar{1}$}, perfect, cleavage.
Colour and transparency Grey, flesh red or brick red due to minute inclusions of iron oxides; translucent.
Lustre Resinous.
Distinguishing features Colour. Bitter taste. Readily soluble in hydrochloric acid. Decomposes in water with the formation of gypsum.
Formation and occurrence An evaporite mineral, one of the last to be precipitated from saline solutions, and occurring in bedded salt deposits.
Uses A source of potassium.

Linarite

Composition $(Pb,Cu)_2SO_4(OH)_2$.
Crystal system Monoclinic.
Habit Usually prismatic, also tabular parallel to base, either as single crystals or as groups or crusts of radiating or confused aggregates of prismatic crystals.
Twinning On {100}, common.
Specific gravity 5·35.
Hardness 2½.
Cleavage and fracture {100}, perfect cleavage, {001}, imperfect cleavage; conchoidal fracture.
Colour and transparency Deep azure blue; translucent.
Streak Pale blue.
Lustre Vitreous to subadamantine.
Distinguishing features Colour; association and occurrence; cleavage. Soluble in dilute nitric acid, and in dilute sulphuric acid giving a white coating of lead sulphate.
Formation and occurrence Rare, but with world-wide distribution, found in the oxidized zone of some lead-copper ores.

Brochantite

Composition $Cu_4(SO_4)(OH)_6$.
Crystal system Monoclinic.
Habit Stout prismatic to acicular, also tabular parallel to base; commonly as loose aggregates of acicular crystals, crystalline crusts; also massive, granular.
Twinning Common, on {100}, often giving pseudo-orthorhombic, symmetrical twinned crystals.
Specific gravity 4.
Hardness 3½–4.
Cleavage and fracture Perfect, {100} cleavage; uneven to conchoidal fracture.
Colour and transparency Emerald green to dark green, also light green; transparent to translucent.
Streak Pale green.
Lustre Vitreous.
Distinguishing features Colour; habit; association. Soluble in hydrochloric and nitric acids.
Formation and occurrence Secondary mineral found in the oxidized zone of copper deposits, especially in arid regions, but of worldwide occurrence.

Antlerite

Composition $Cu_3(SO_4)(OH)_4$.
Crystal system Orthorhombic.
Habit Usually thick tabular, also short prismatic crystals; also fibrous or as felty aggregates of acicular crystals; granular.
Specific gravity 3·88.
Hardness 3½.
Cleavage Perfect {010} cleavage.
Colour and transparency Emerald green to dark green, also light green; translucent.
Streak Light green.
Lustre Vitreous.
Distinguishing features Habit; colour; association. Soluble in dilute sulphuric acid.
Formation and occurrence Secondary mineral found in the oxidized zone of copper deposits mainly in arid regions.

Caledonite

Composition $Cu_2Pb_5(SO_4)_3(CO_3)(OH)_6$.
Crystal system Orthorhombic.
Habit Elongated crystals, usually small and in divergent groups; often striated; also as coatings.
Specific gravity 5·6–5·7.
Hardness 2½–3.
Cleavage and fracture Perfect {010} cleavage.
Colour and transparency Deep bluish green; translucent.
Streak Greenish white.
Lustre Resinous.
Distinguishing features Soluble with effervescence in nitric acid, leaving a precipitate; colour; association; crystal form.
Formation and occurrence A secondary mineral of the oxidized zone of copper-lead deposits.

Connellite

Composition $Cu_{19}(SO_4)Cl_4(OH)_{32}.2$ (or 3) H_2O. The extremely rare mineral, buttgenbachite, similar in properties and composition, has the (SO_4) group replaced by $(NO_3)_2$; there is a complete series between the two end members.
Crystal system Hexagonal.
Habit Acicular and striated; in radiated groups of needles or as felted aggregates.
Specific gravity 3·36.
Hardness 3.
Colour and transparency Azure blue.
Streak Pale greenish blue.
Lustre Vitreous.
Distinguishing features Soluble in hydrochloric and nitric acids, insoluble in water; colour; form; association.
Formation and occurrence A secondary mineral from the oxidized zone of copper ores; first described from a Cornish locality.

Crocoite

Composition $PbCrO_4$.
Crystal system Monoclinic.
Habit Usually prismatic or acicular crystals; often striated and with smooth, brilliant faces, sometimes hollow; as columnar aggregates or granular, massive.
Specific gravity 5·9–6·1.
Hardness 2½–3.

Crocoite.

Wolframite.

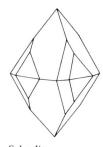

Scheelite.

Sheaves of bladed, reddish-brown crystals of hübnerite; from Adams mine, Silverton, Colorado.

Cleavage and fracture {110}, distinct cleavage; fracture conchoidal or uneven.
Colour and transparency Bright orange-red, various shades; translucent.
Streak Orange-yellow.
Lustre Adamantine.
Distinguishing features Colour; high lustre; high specific gravity; crystal form.
Formation and occurrence A rare secondary mineral from the oxidized portions of lead-chromium (and sometimes vanadium) veins.
Uses Of little commercial importance, but historically important in that the element, chromium, was first discovered in this mineral. Used as a pigment (chrome yellow).

7 Tungstates, Molybdates

Wolframite

Composition (Fe,Mn)WO$_4$. Iron and manganese substitute mutually; the iron end member is known as ferberite; the manganese end member as hübnerite.
Crystal system Monoclinic.
Habit Hübnerite forms prismatic to long prismatic (less frequently short or tabular) crystals. Wolframite crystals commonly short prismatic (less often long prismatic or tabular). Ferberite commonly as elongated, flattened, often wedge-shaped, bladed (also short prismatic). All forms may have striated faces; may form subparallel groups; may be found massive, granular.
Twinning Twinning on {100} and {023} faces occurs in all members; usually contact but occasionally interpenetrating.
Specific gravity Increases with iron content: 7·0 (hübnerite) to 7·5 (ferberite).
Hardness Increases with iron content: 4–4½.
Cleavage and fracture Perfect {010} cleavage; uneven fracture.
Colour and transparency Hübnerite is yellowish brown to reddish brown, rarely brownish black; wolframite is dark greyish to brownish black; ferberite is black; members of the series go from transparent, through translucent, to opaque, with increasing iron content.
Streak Same as mineral colour.

Lustre Resinous (hübnerite) through metallic-adamantine to submetallic (ferberite).
Distinguishing features Colour; high specific gravity; one good cleavage; iron-rich members weakly magnetic; crystals often colour zoned due to variations in iron content; lustre.
Formation and occurrence In quartz veins and pegmatites associated with metallic ore mineralization in granites; more rarely, in high-temperature hydrothermal veins. Also in some alluvial deposits where it has been concentrated by water action often forming economic deposits.
Uses The chief ore of tungsten which has many metallurgical and industrial applications.

Scheelite

Composition CaWO$_4$. Molybdenum can substitute for tungsten, and a partial series extends towards powellite. Growth zones of varying colour and molybdenum content are often found.
Crystal system Tetragonal.
Habit Typically bipyramidal, often with striated faces, sometimes tabular; also massive, granular, columnar.
Twinning Penetration twins (on {110}) are more common than the contact type.
Specific gravity 5·9–6·1.
Hardness 4½–5.
Cleavage and fracture Good cleavage parallel to {101}; fracture uneven.
Colour and transparency Colourless to white, usually pale yellow, also greenish, grey, brownish, reddish; coloured tints usually due to molybdenum content; transparent to translucent.
Streak White.
Lustre Vitreous to adamantine.
Distinguishing features Pyramidal habit; high specific gravity; colour; most scheelite will fluoresce under ultraviolet light.
Formation and occurrence Widespread, in many granite pegmatites, (limestone) contact metamorphic deposits, and high-temperature ore veins associated with granitic rocks.
Uses An ore of tungsten.

Stolzite

Composition PbWO$_4$. Molybdenum may substitute for tungsten.
Crystal system Tetragonal.
Habit As crystals; usually bipyramidal; also thick tabular or (infrequently) prismatic; all with some striated faces.
Specific gravity 7·9–8·3.
Hardness 2½–3.
Cleavage and fracture Imperfect basal cleavage; fracture conchoidal to uneven.
Colour and transparency Reddish brown, pale shades of brown, grey and yellow, also green, red; transparent in thin pieces.
Streak White.
Lustre Resinous, subadamantine.
Distinguishing features Form; specific gravity (very high); colour.
Formation and occurrence A secondary mineral, found in the oxidized zone of ore deposits containing primary lead and tungsten minerals. Much less common than wulfenite.

Wulfenite

Composition $PbMoO_4$. Tungsten can substitute for molybdenum, so forming a series towards stolzite; calcium can also substitute for lead, with at least a partial series towards powellite.
Crystal system Tetragonal.
Habit Crystals usually square tabular with prominent base, sometimes extremely thin; more rarely bipyramidal or short prismatic; also massive, coarse to fine granular.
Specific gravity 6·5–7·0.
Hardness 3.
Cleavage and fracture {011}, distinct cleavage; fracture subconchoidal to uneven.
Colour and transparency Yellow, orange, red, grey, white, olive green to brown: transparent to subtranslucent.
Streak White.
Lustre Vitreous to adamantine.
Distinguishing features Tabular crystals; high lustre; colour; association.
Formation and occurrence A secondary mineral, formed in the oxidized zone of deposits of lead- and molybdenum-bearing minerals, and associated with other secondary lead minerals such as vanadinite and pyromorphite.
Uses A minor source of molybdenum.

Powellite

Composition $CaMoO_4$. Tungsten substitutes for molybdenum.
Crystal system Tetragonal.
Habit Crystals rare, usually pyramidal; also thin tabular, sometimes striated; also massive, foliated or earthy; or as crusts of merged crystals. May pseudomorph molybdenite.
Specific gravity 4·2 (when pure) increasing to 4·5 with increasing tungsten content.
Hardness $3\frac{1}{2}$–4.
Fracture Fracture uneven.
Colour and transparency Straw yellow, brown, shades of pale green, grey and blue to dark blue (almost black); transparent.
Lustre Subadamantine to greasy.
Distinguishing features Association; crystal form; fluoresces yellow in ultraviolet light.
Formation and occurrence A secondary mineral, often formed by alteration of molybdenite and usually found in contact metamorphic deposits.

8 Phosphates, Arsenates, Vanadates

Xenotime

Composition YPO_4. Erbium and other rare earth elements may substitute for yttrium.
Crystal system Tetragonal.
Habit Short to long prismatic; also equant, pyramidal crystals, often found in parallel growth with zircon, which it resembles; sometimes as radial aggregates or rosettes of coarse crystals.
Twinning Rare, on {111}.
Specific gravity 4·4–5·1.
Hardness 4–5.
Cleavage and fracture {100}, perfect cleavage; fracture uneven.
Colour and transparency Yellowish to reddish brown; also paler and greyish tints; translucent to opaque.
Streak Pale variant of mineral colour.
Lustre Vitreous to resinous.
Distinguishing features Crystal form; occurrence; cleavage; distinguished from zircon by inferior hardness.
Formation and occurrence A primary, minor accessory mineral in granitic and alkaline igneous rocks, and, in larger crystals, in their associated pegmatites. Also found in some mica- and quartz-rich metamorphic rocks and associated pegmatites, and as a detrital mineral in some alluvial deposits.

Monazite

Composition $(Ce,La,Th)PO_4$. Yttrium and other rare earths can also be present.
Crystal system Monoclinic.
Habit Crystals rare and usually small, flattened parallel to {100} or short prismatic; faces often striated; usually found as granular masses, frequently as sand.
Twinning Common, both contact and penetration twins on {100}.
Specific gravity 4·6–5·4, mostly around 5.
Hardness 5–$5\frac{1}{2}$.
Cleavage and fracture {100}, distinct cleavage; also basal parting; fracture uneven.
Colour and transparency Yellowish or reddish brown to brown, also greenish, whitish; sub-transparent to subtranslucent.
Streak White or pale shades of above.
Lustre Variable, resinous to adamantine.
Distinguishing features Inferior in hardness to zircon.
Formation and occurrence Widespread as an accessory in granitic igneous and gneissic metamorphic rocks, from which it is extracted by weathering and water action to form often large detrital deposits of 'monazite sands' which are of commercial importance. Found in larger crystals in pegmatites associated with these rocks.
Uses An important source of thorium, cerium and other rare earths; monazite sands are relatively easy to exploit commercially.

Vivianite

Composition $Fe_3(PO_4)_2.8H_2O$. Small amounts of manganese, magnesium and calcium can

Brilliant orange platy crystals of wulfenite; from Sonora, Mexico.

Wulfenite: tabular crystal.

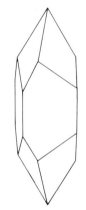

Wulfenite: bipyramidal habit.

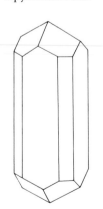

Xenotime: prismatic habit.

Monazite.

Dark green translucent prismatic crystal of vivianite on matrix; from Chihuahua, Mexico.

Amblygonite

Composition $(Li,Na)Al(PO_4)(F,OH)$. Lithium is almost always greatly in excess of sodium, but varieties in which hydroxyl is in excess of fluorine are more common and are called montebrasite.

Crystal system Triclinic.

Habit Small crystals are equant to short prismatic; large crystals may be ill-formed and have rough faces; usually in large, cleavable masses; columnar; compact.

Twinning Contact twins on $\{\bar{1}\bar{1}1\}$ common; usually tabular, with twinned individuals equal or unequal in size.

Specific gravity 3·1 (amblygonite) to 2·98 (montebrasite).

Hardness $5\frac{1}{2}$–6.

Cleavage and fracture Perfect $\{100\}$ cleavage; one other good cleavage; fracture uneven to subconchoidal.

Colour and transparency White to milky or creamy white; various other pale shades; rarely colourless; transparent to translucent.

Streak White.

Lustre Vitreous to greasy.

Distinguishing features Two cleavages; specific gravity; not easily soluble in acids; occurrence and association; fragments fuse easily and colour a gas flame red.

Formation and occurrence As constituents of granitic pegmatites of the lithium- and phosphate-rich type; often in crystals of enormous size, associated with other lithium and phosphate minerals.

Uses A source of lithium.

Apatite

Composition $Ca_5(PO_4)_3(F,Cl,OH)$. From the formula, it is apparent that fluorine, chlorine and hydroxyl can substitute for one another, to give fluor-, chlor- and hydroxy-apatite according to whichever is the greater. In addition, manganese, strontium and other metals can substitute for calcium, and carbonate (CO_3) can replace phosphate to a lesser extent.

Crystal system Hexagonal.

Habit Crystals short to long prismatic, some tabular, usually terminated by a bipyramid with or without the basal pinacoid; also massive, coarse granular to compact; also globular or reniform with poor fibrous internal structure; stalactitic; earthy; or as nodules.

Specific gravity 3–3·2.

Hardness 5 (can just be scratched by a knife).

Cleavage and fracture Imperfect basal cleavage; fracture conchoidal and uneven.

Colour and transparency Usually pale sea green and bluish green (darker blue-green varieties due to manganese substitution); shades of blue, green, violet, red, grey, brown, also colourless; transparent to opaque.

Streak White.

Lustre Vitreous.

Distinguishing features Habit; hardness. Dissolves in hydrochloric acid (addition of sulphuric acid to the solution gives a white precipitate). Most fluoresce in ultraviolet light.

substitute for iron. Ideally all the iron should be in the ferrous state, but some may become oxidized to the ferric state.

Crystal system Monoclinic.

Habit Usually prismatic, sometimes flattened, and may become blade-like in radiating groups, or as reniform or encrusting masses with an internal, radiating, fibrous structure. Also powdery.

Twinning Rare, lamellar.

Specific gravity 2·68–2·7.

Hardness $1\frac{1}{2}$–2.

Cleavage One perfect $(\{010\})$ cleavage; thin cleavage plates are flexible; sectile.

Colour and transparency Colourless, or green when fresh, becoming blue, almost black due to rapid oxidation of iron (see above); most specimens are deep blue or greenish blue; transparent to translucent.

Streak White or pale blue, rapidly changing to dark blue or brown.

Lustre Vitreous.

Distinguishing features Colour and streak; low hardness; easily soluble in acids.

Formation and occurrence A secondary mineral in the oxidized zone of metallic ore deposits containing iron sulphides; in weathered, phosphate-bearing pegmatites; and widespread in clays and other sediments and sedimentary rocks, especially those containing bone and other organic matter. An alteration mineral of fossilized bones and teeth.

Apatite.

149

Formation and occurrence One of the commonest and most widely distributed minerals, occurring in almost all igneous rocks, including pegmatites and hydrothermal veins; in contact and regional metamorphic rocks; and in sedimentary deposits, as sometimes extensive bedded marine sequences, as rocks derived from such deposits, or as replacements of limestones by phosphate-rich solutions. The main inorganic part of bone and teeth is apatite; fossil bones and teeth and sediments derived from such organic remains are composed largely of a cryptocrystalline form of apatite known as collophane.

Uses As fertilizers; generally the bedded phosphate deposits are most economically important. The mineral is usually treated with sulphuric acid to give 'superphosphate' which is more soluble in the weak acids of soils.

Pyromorphite

Composition $Pb_5(PO_4)_3Cl$. Arsenic substitutes for phosphorus and a complete series extends to mimetite; calcium may also replace lead, to a lesser extent, and vanadium may replace phosphorus.

Crystal system Hexagonal.

Habit Prismatic crystals made up of hexagonal prism, bipyramid, and basal pinacoid, like apatite; often in rounded, barrel-shaped forms (then known as campylite); also hollow: often globular, reniform, with poor columnar internal structure; as crusts; granular.

Specific gravity 6·5–7·1.

Hardness 3½–4.

Cleavage and fracture Poor, prismatic cleavage; subconchoidal to uneven fracture.

Colour and transparency Various shades of green and brown; also orange-yellow and red; rarely colourless; subtransparent to translucent.

Streak White or yellowish white.

Lustre Resinous to subadamantine.

Distinguishing features Colour; habit; high specific gravity; lustre; association. Dissolves in hydrochloric and nitric acids.

Formation and occurrence A secondary mineral of the oxidized zone of lead deposits, associated with other lead minerals and sometimes in zoned crystals with the outer parts tending towards mimetite.

Uses A minor ore of lead.

Mimetite

Composition $Pb_5(AsO_4)_3Cl$. Phosphorus substitutes for arsenic, with a complete series to pyromorphite; vanadium can also replace arsenic to a considerable extent: calcium may substitute for lead, but when more calcium than lead is present the mineral is known as hedyphane.

Crystal system Hexagonal.

Habit Like pyromorphite, crystals are commonly simple, hexagonal forms, from tabular, short to long prismatic, to sometimes acicular; also as the curved, barrel-shaped crystals known as campylite; sometimes as botryoidal forms or crusts.

Specific gravity 7–7·25.

Hardness 3½.

Fracture Subconchoidal.

Colour and transparency Pale yellow to yellowish brown, orange-yellow, white or colourless; transparent to translucent.

Streak White.

Lustre Resinous to adamantine.

Distinguishing features Colour; lustre; high specific gravity; association; crystal form. Dissolves in hydrochloric acid. Difficult to distinguish from pyromorphite without chemical tests.

Formation and occurrence A rare secondary

Pyromorphite: mimetite, variety campylite, showing barrel-shaped form.

Right
Dark orange,
stubby hexagonal
prisms of
vanadinite on
sandstone; from
Mibladen,
Morocco.

Vanadinite.

mineral of the oxidized zone of arsenic-bearing
lead deposits, often associated with pyromorphite
and other lead minerals.
Uses A minor ore of lead.

Vanadinite

Composition $Pb_5(VO_4)_3Cl$. Both phosphorus
and arsenic substitute for vanadium, in significant
amounts; small amounts of calcium, zinc and
copper can substitute for lead.
Crystal system Hexagonal.
Habit Crystals short to long prismatic, usually
with smooth faces and sharp edges; also acicular;
sometimes hollow; in globular forms or crusts.
Specific gravity 6·7–7·1.
Hardness 3.
Fracture Uneven to conchoidal.
Colour and transparency Orange-red, ruby red,
brownish red to shades of brown and yellow;
subtransparent to nearly opaque.
Streak White to yellowish.
Lustre Subresinous to subadamantine.
Distinguishing features Habit; high specific
gravity; lustre; colour. Dissolves in hydrochloric
acid to give a green solution and a whitish
precipitate.
Formation and occurrence A rare secondary
mineral of the oxidized zone of lead deposits,
associated with other lead minerals; may accompany
other vanadium minerals in mineralized sediments.
Uses A source of vanadium, which is used in
steel; vanadium salts are mainly used in dyeing
and in ceramics.

Turquoise

Composition $CuAl_6(PO_4)_4(OH)_8.5H_2O$. Ferric
iron can substitute for aluminium, with a series
towards chalcosiderite, and ferrous iron for copper,
to a lesser extent.

Crystal system Triclinic.
Habit Crystals, minute, short prismatic, are rare;
usually cryptocrystalline to fine granular, as veinlets,
crusts, stalactitic or concretionary shapes.
Specific gravity 2·6–2·8.
Hardness 5–6.
Cleavage and fracture Two good cleavages in
crystals; fracture (of massive) conchoidal.
Colour and transparency Sky blue, bluish green
to apple green and greenish grey, and pale shades
of these; transparent (crystals) to opaque.
Streak White or pale green.
Lustre Vitreous (crystals) to waxy (massive).
Distinguishing features Colour; habit; hardness
(to distinguish from chrysocolla). Imitation turquoise usually fuses in a gas flame and colours it
green, whereas natural material turns brown but is
otherwise barely affected.
Formation and occurrence A secondary mineral
formed by the action of surface waters, usually in
arid regions, on aluminous igneous and sedimentary
rocks, and found in veins and irregular patches in
the rock. May pseudomorph orthoclase, apatite; also
bone and teeth, when it may be confused with
odontolite (organic apatite stained blue by vivianite).
Uses A highly prized, semiprecious stone.

Chalcosiderite

Composition $CuFe_6(PO_4)_4(OH)_8.5H_2O$. Alumi-
nium may replace iron in a complete series to
turquoise.
Crystal system Triclinic.
Habit As crusts or sheaf-like groups of distinct,
short prismatic crystals.
Specific gravity 3·25.
Hardness $4\frac{1}{2}$.
Cleavage $\{001\}$ (perfect) and $\{010\}$ (good)
cleavages.
Colour and transparency Pale green; transparent.

151

Streak White.

Lustre Vitreous.

Distinguishing features Colour; habit; association.

Formation and occurrence A very rare mineral, formed under essentially atmospheric conditions in the iron-rich gossan of ore deposits.

Erythrite (cobalt bloom) and Annabergite (nickel bloom)

Composition $Co_3(AsO_4)_2.8H_2O$ and $Ni_3(AsO_4)_2.8H_2O$. Cobalt and nickel substitute for one another to form a complete series; significant amounts of calcium, zinc, ferrous iron and magnesium (cabrerite) can also substitute for these elements.

Crystal system Monoclinic.

Habit Crystals prismatic to acicular and flattened, often deeply striated, but are usually rare and poorly formed; usually in radiating groups or crusts; often in globular or reniform shapes with coarse fibrous interior; also earthy or powdery.

Specific gravity 3·0–3·1.

Hardness $1\frac{1}{2}$–$2\frac{1}{2}$.

Cleavage One perfect cleavage.

Colour and transparency Erythrite – purplish-, crimson-red to pale pink; annabergite – palest pink, through white or grey to pale and apple green; crystals transparent to translucent.

Streak Paler than colour.

Lustre (Of crystals) weakly adamantine.

Distinguishing features Colour; association. Dissolves in acids. Powdered erythrite will turn lavender blue on heating; annabergite will become yellow and then dark green.

Formation and occurrence Rare, secondary minerals, usually formed by the alteration by weathering of cobalt and nickel arsenides in the oxidized zone of some ore deposits.

Uses As indicators in prospecting for cobalt and nickel.

Scorodite

Scorodite.

Composition $FeAsO_4.2H_2O$. Aluminium substitutes for ferric iron; a complete series probably extends to mansfieldite ($AlAsO_4.2H_2O$) which has similar properties. Phosphate may also substitute for arsenate to some extent.

Crystal system Orthorhombic.

Habit Usually in pyramidal crystals resembling octahedra; also tabular or prismatic; crystals often form irregular groups or crusts; also massive, crystalline, or porous and sinter-like.

Specific gravity 3·1 to 3·3 (iron end member).

Hardness $3\frac{1}{2}$–4.

Cleavage and fracture Prismatic, imperfect cleavage; fracture subconchoidal.

Colour and transparency Pale green to liver-brown; sometimes colourless or shades of violet, blue or yellow; transparent to translucent.

Lustre Vitreous to subadamantine.

Distinguishing features Habit; association with arsenic minerals. Dissolves in hydrochloric and nitric acids.

Formation and occurrence A secondary mineral of gossans, formed by the alteration of (usually) arsenopyrite in mineral veins. Has also been found as a rare primary mineral in a hydrothermal deposit.

Torbernite and Metatorbernite

Composition $Cu(UO_2)_2(PO_4)_2.8\text{-}12H_2O$. Small amounts of arsenic may substitute for phosphorus, and lead for copper. The water content varies with temperature and humidity; metatorbernite has none of the loosely bound water and has the formula $Cu(UO_2)_2(PO_4)_2.8H_2O$.

Crystal system Tetragonal.

Habit Square, thin to thick tabular crystals; also as foliated or scaly aggregates.

Twinning Rare.

Specific gravity 3·2 (torbernite) increasing to 3·7 (metatorbernite).

Hardness 2–$2\frac{1}{2}$.

Cleavage Perfect basal cleavage, producing thin, brittle cleavage plates.

Colour and transparency Bright emerald green to grass green; transparent to translucent.

Streak Paler than colour.

Lustre Vitreous to subadamantine, sometimes pearly.

Distinguishing features Colour; cleavage. Dissolves in hydrochloric and nitric acids. Does not fluoresce in ultraviolet light.

Formation and occurrence A secondary mineral found in association with autunite and other secondary uranium minerals as an alteration product of uraninite, especially in the gossan of veins carrying both uraninite and copper sulphides.

Group of purplish bladed crystals of erythrite on matrix; from Bou Azzer, Morocco.

Autunite and Meta-autunite

Composition $Ca(UO_2)_2(PO_4)_2.10\text{-}12H_2O$. Small amounts of barium and magnesium can substitute for calcium; surprisingly, there appears to be no natural series to torbernite, although the minerals occur in close association and have similar properties. The water content varies according to humidity and temperature – meta-autunite has about a half or less of the water in autunite.

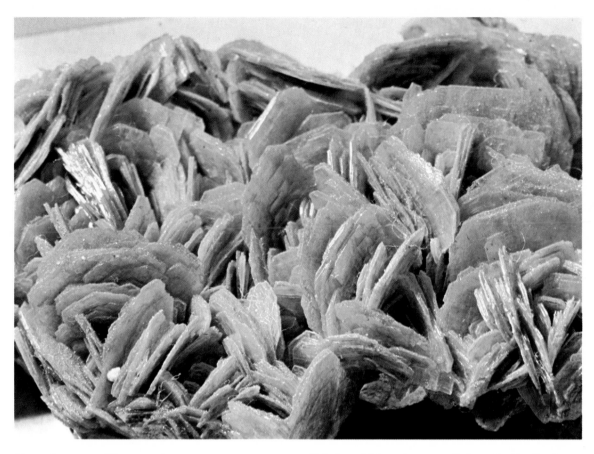

Bright yellow intersecting plates of autunite.

Crystal system Tetragonal.
Habit Square, thin to thick tabular crystals; also as foliated or scaly aggregates; or as thick crusts of subparallel crystals.
Specific gravity 3·1–3·2, varying with water content.
Hardness 2–2$\frac{1}{2}$.
Cleavage Perfect basal cleavage, producing thin cleavage plates slightly less brittle than those of torbernite.
Colour and transparency Bright lemon yellow to greenish yellow; transparent to translucent.
Streak Yellowish.
Lustre Vitreous.
Distinguishing features Colour (to distinguish from torbernite); cleavage; habit. Dissolves in hydrochloric and nitric acids. Fluoresces strongly (yellow-green) in ultraviolet light (meta-autunite less strongly).
Formation and occurrence A secondary mineral found chiefly in the zone of oxidation and weathering of hydrothermal veins and pegmatites, formed from alteration of uraninite and other uranium minerals.

Carnotite

Composition $K_2(UO_2)_2(VO_4)_2.3H_2O$. Water content may vary with humidity and temperature.
Crystal system Monoclinic.
Habit As a powder or loose microcrystalline aggregates; sometimes compact; disseminated through the matrix; rarely as crusts of poor platy crystals; microscopic crystals platy or lath-like.
Specific gravity 5 (when fully hydrated).
Hardness About 2.
Cleavage Perfect {001} cleavage.
Colour Bright yellow to greenish yellow.
Lustre Dull, earthy.

Distinguishing features Colour. Easily dissolves in hydrochloric and nitric acids. Does not fluoresce in ultraviolet light (to distinguish from autunite).
Formation and occurrence A secondary mineral, deposited from circulating ground waters which have passed through deposits of uranium minerals; generally found in sandstones, either disseminated through the rock (colouring it bright yellow) or as small masses usually associated with fossil vegetable matter. Also found as an alteration crust on some uranium ores.
Uses An ore of uranium.

Tyuyamunite

Composition $Ca(UO_2)_2(VO_4)_2.5-10H_2O$. Water content varies with humidity and temperature.
Crystal system Orthorhombic.
Habit As scales and laths flattened parallel to basal plane; sometimes as radial aggregates; crystal faces may be dull and sometimes curved; commonly massive, fine-grained or powdery.
Specific gravity 3·3–4·3 (depending on water content).
Hardness About 2.
Cleavage and fracture Perfect basal cleavage; two other cleavages; not brittle.
Colour and transparency Bright canary yellow to greenish yellow; translucent to opaque.
Lustre Adamantine (crystals); waxy (massive).
Distinguishing features Colour (more greenish than carnotite). Fluoresces greenish yellow in ultraviolet light, unlike carnotite. Dissolves in hydrochloric, sulphuric and nitric acids but not in dilute acetic acid.
Formation and occurrence A secondary mineral, deposited from uranium-bearing, circulating

153

ground waters, either in sandstones in a similar way to, and associated with, carnotite, or as veins and infill deposits in some limestones.

Uses An ore of uranium.

Francevillite

Composition $(Ba,Pb)(UO_2)_2(VO_4)_2.5H_2O$. Some varieties may be lead-free.
Crystal system Orthorhombic.
Colour Yellow.
Formation and occurrence A secondary mineral, deposited from mineral-rich, circulating ground waters, often impregnating sandstone.

Descloizite

Composition $Pb(Zn,Cu)VO_4(OH)$. A complete series exists, due to mutual substitution of zinc and copper, between descloizite and mottramite $[Pb(Cu,Zn)VO_4(OH)]$. Manganese and ferrous iron can substitute for zinc and copper; arsenic and phosphorus may substitute for vanadium.
Crystal system Orthorhombic.
Habit Crystals variable, often pyramidal or prismatic, rarely tabular; faces are usually uneven or rough, subparallel growth is frequent; commonly as crystalline crusts; also stalactitic or massive, frequently with a botryoidal surface and coarse, fibrous internal structure.
Specific gravity 5·9 (mottramite) to 6·2 (descloizite).
Hardness 3–4.
Fracture Uneven fracture.
Colour and transparency Brownish red to blackish brown; also shades of brownish green and orange-red to nearly black; transparent to opaque.
Streak Orange to brownish-red.
Lustre Greasy.
Distinguishing features Colour; streak; habit. Easily dissolves in hydrochloric and nitric acids.
Formation and occurrence A secondary mineral found in the oxidized zone of ore deposits, generally associated with vanadinite and other lead minerals. Also as a deposit in sandstones from mineral-rich, circulating ground water.

Olivenite

Composition $Cu_2AsO_4(OH)$. Phosphorus substitutes for arsenic; a partial series thus extends towards libethenite. Small amounts of iron can also substitute for copper.
Crystal system Orthorhombic.
Habit Crystals variable; often elongated, but also short prismatic to acicular; less often tabular: frequently in globular and reniform shapes with a very regular, radiating fibrous internal structure; also as curved lamellar forms; massive, granular to earthy.
Specific gravity 4·1–4·5.
Hardness 3.
Cleavage and fracture Poor cleavage; fracture conchoidal to irregular.
Colour and transparency Shades of olive green to brown; rarely paler shades through to greyish white; translucent to opaque.
Streak As colour.
Lustre Adamantine to vitreous.
Distinguishing features Colour; association.

Dissolves in hydrochloric and nitric acids.
Formation and occurrence A secondary mineral found in the oxidized zone of ore deposits, associated with other copper minerals and adamite.

Libethenite

Composition $Cu_2PO_4(OH)$. Small amounts of arsenic may substitute for phosphorus.
Crystal system Orthorhombic.
Habit Commonly as composite, poorly formed, striated, short prismatic or equant crystals.
Specific gravity 3·9.
Hardness 4.
Cleavage and fracture Very poor cleavages; fracture conchoidal to uneven.
Colour and transparency Light to dark olive green, deep to dark green; translucent.
Lustre Vitreous.
Distinguishing features Colour; association. Easily dissolves in hydrochloric and nitric acids.
Formation and occurrence A secondary mineral, found in the oxidized zone of ore deposits associated with other copper minerals.

Adamite

Composition $Zn_2AsO_4(OH)$. Copper may substitute for zinc to a considerable extent (cupro-adamite); cobalt may do so, to a lesser extent.
Crystal system Orthorhombic.
Habit Crystals often elongated, rarely tabular or equant, and usually merged together in crusts or as roughly radial aggregates.
Specific gravity 4·3–4·4.
Hardness $3\frac{1}{2}$.
Cleavage and fracture Good $\{101\}$ cleavage; uneven to subconchoidal fracture.
Colour and transparency Usually yellow, brownish yellow to green; copper-bearing varieties shades of green; cobalt varieties violet to rose; transparent to translucent.
Lustre Vitreous.
Distinguishing features Colour; association. Easily dissolves in dilute acids. Some specimens may fluoresce lemon yellow in ultraviolet light.
Formation and occurrence A rare secondary mineral, found in the oxidized zone of ore deposits containing primary zinc- and arsenic-rich minerals. May crystallize in cavities in the ore deposit or encrust adjacent sedimentary rocks.

Group of orange-brown tabular crystals of descloizite; from Grootfontein, south-west Africa.

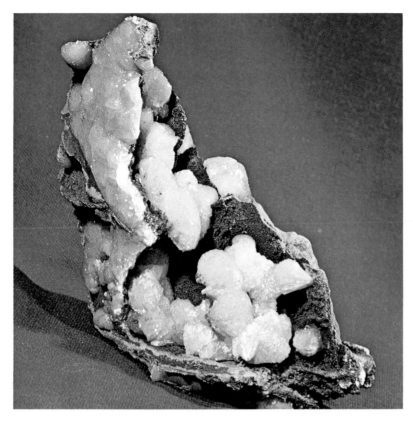

Greenish-yellow globular masses of adamite on limonite; from Mapimi, Mexico.

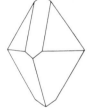

Lazulite.

Liroconite

Composition $Cu_2Al(AsO_4)(OH)_4.4H_2O$. Phosphorus may substitute for some of the arsenic.
Crystal system Monoclinic.
Habit Crystals are thin or lenticular, like flattened octahedra, with some faces striated.
Specific gravity 2·9–3·0.
Hardness 2–2$\frac{1}{2}$.
Cleavage and fracture Poor cleavage; fracture uneven to conchoidal.
Colour and transparency Sky blue to verdigris green; transparent to translucent.
Streak As colour, but paler.
Lustre Vitreous to resinous.
Distinguishing features Colour; association. Easily dissolves in hydrochloric and nitric acids.
Formation and occurrence A rare secondary mineral found associated with other copper minerals, such as azurite, in the oxidized zone of copper deposits.

Chalcophyllite

Composition $Cu_{18}Al_2(AsO_4)_3(SO_4)_3(OH)_{27}.33H_2O$. Small amounts of phosphorus may substitute for arsenic. The water content may vary according to temperature and humidity.
Crystal system Hexagonal (Trigonal).
Habit Tabular, in six-sided crystals sometimes triangularly striated; also as foliated masses or as rosettes.
Specific gravity 2·6–2·7.
Hardness 2.
Cleavage Perfect basal cleavage, producing flexible cleavage plates.
Colour and transparency Emerald green; also grass green to bluish green; transparent to translucent.

Streak Pale green.
Lustre Vitreous to subadamantine.
Distinguishing features Colour; cleavage; habit. Dissolves in hydrochloric and nitric acids; alters easily to chrysocolla.
Formation and occurrence A rare secondary mineral found with other copper minerals in the oxidized zone of copper deposits.

Lazulite

Composition $(Mg,Fe)Al_2(PO_4)_2(OH)_2$. Magnesium and ferrous iron substitute mutually; varieties in which iron exceeds magnesium are known as scorzalite, a mineral with similar properties, although much rarer than, lazulite.
Crystal system Monoclinic.
Habit Crystals are commonly steep pyramidal forms; also tabular; may be massive, compact to granular.
Twinning Common, often on base, sometimes lamellar.
Specific gravity 3·1 (iron-rich members up to 3·4).
Hardness 5$\frac{1}{2}$–6.
Cleavage and fracture Prismatic, poor to good; fracture uneven to splintery.
Colour and transparency Deep azure blue, also paler shades of blue; subtranslucent to opaque (some gem varieties are transparent).
Streak White.
Lustre Vitreous.
Distinguishing features Colour; association with (usually) silicate minerals; hardness. Not readily soluble even in hot acids.
Formation and occurrence In metamorphic rocks, as grains or masses, especially in quartz-rich types; in quartz-rich veins in metamorphic rocks; and in granite pegmatites. Found associated with high-grade metamorphic silicate minerals such as the kyanite-andalusite-sillimanite group, quartz, muscovite and garnet.
Uses A semiprecious stone.

Wavellite

Composition $Al_3(PO_4)_2(OH)_3.5H_2O$.
Crystal system Orthorhombic.
Habit Crystals rare, stout to long prismatic with striated prism faces; usually as hemispherical or globular aggregates with a fibrous, radiating internal structure; also as crusts or stalactitic; rarely opaline.
Specific gravity 2·36.
Hardness 3$\frac{1}{2}$–4.
Cleavage and fracture Perfect $\{110\}$ and good $\{101\}$ cleavages; fracture uneven to subconchoidal.
Colour and transparency Colourless, white, yellow, green and brown, and shades of these; rarely bluish.
Streak White.
Lustre Vitreous.
Distinguishing features Habit. Easily dissolves in most acids.
Formation and occurrence A secondary mineral, widespread in small amounts as a deposit on joint surfaces and in cavities in low-grade metamorphic rocks and in some sedimentary (such as phosphate-rock) deposits; rarely as a late-formed mineral in some hydrothermal veins.

Childrenite

Composition $(Fe,Mn)Al(PO_4)(OH)_2.H_2O$. Ferrous iron and manganese can substitute for each other: varieties in which manganese exceeds iron are known as eosphorite which has similar properties but is much rarer.

Crystal system Orthorhombic.

Habit Equant or pyramidal to short prismatic (eosphorite, long prismatic) and thick tabular, sometimes platy, crystals, often as radiating groups grading into botryoidal or crusty masses with internal radiating fibrous structures.

Specific gravity 3·25 (childrenite) to 3·05 (eosphorite).

Hardness 5.

Cleavage and fracture Poor cleavage; fracture subconchoidal to uneven.

Colour and transparency Brown and yellowish brown to pink or rose red (eosphorite); transparent to translucent.

Streak White.

Lustre Vitreous to resinous.

Distinguishing features Habit; colour; association. Dissolves in hydrochloric and nitric acids.

Formation and occurrence A rare mineral found in some hydrothermal veins and pegmatites.

Pseudomalachite

Composition $Cu_5(PO_4)_2(OH)_4.H_2O$.

Crystal system Monoclinic.

Habit Isolated crystals are rare; usually the small, uneven prismatic crystals are found in subparallel aggregates or radiating hemispherical forms, grading towards botryoidal or massive with a radial internal fibrous structure and concentric colour banding; also massive.

Twinning On $\{100\}$.

Specific gravity 4·0–4·3.

Hardness $4\frac{1}{2}$–5.

Cleavage and fracture Distinct, $\{100\}$ cleavage; fracture splintery.

Colour and transparency Dark emerald green to blackish green (crystals); fibrous material paler shades, also bluish green; translucent.

Streak Paler than colour.

Lustre Vitreous.

Distinguishing features Colour; habit. Dissolves in hydrochloric and nitric acids (but without effervescence, c.f. malachite).

Formation and occurrence A secondary mineral found in association with malachite, in the oxidized zone of copper deposits.

Variscite and Strengite

Composition $Al(PO_4).2H_2O$ and $Fe(PO_4).2H_2O$. A complete series extends between the two, with mutual substitution of aluminium and ferric iron, although most varieties are close to one or other end member.

Crystal system Orthorhombic.

Habit Variscite – crystals rare; slightly modified, like octahedra; normally as fine-grained masses, nodules and crusts: also opaline. Strengite – crystals variable; pyramidal, thick to thin tabular, short prismatic; generally as spherical and botryoidal aggregates with a radial fibrous internal structure

and crystalline surface; as crusts.

Specific gravity 2·6 (variscite) to 2·9 (strengite).

Hardness $3\frac{1}{2}$ (strengite) to $4\frac{1}{2}$ (variscite).

Cleavage and fracture Good, $\{010\}$ cleavage: fracture uneven to splintery (massive varieties).

Colour and transparency Variscite – pale to emerald green; also bluish green and colourless. Strengite – pinkish red, crimson and violet; also colourless; transparent to translucent.

Streak White.

Distinguishing features Colour; habit. Crystalline variscite insoluble in hydrochloric acid; strengite soluble in hydrochloric but not in nitric acid.

Formation and occurrence Secondary minerals, deposited from mineral-bearing, circulating ground waters in cavities or veins on or near the surface, or as alteration products of other phosphate minerals.

Lithiophilite and Triphylite

Composition $Li(Mn,Fe)PO_4$ and $Li(Fe,Mn)PO_4$. Ferrous iron and manganese substitute for one another; small amounts of calcium may also substitute for either of these.

Crystal system Orthorhombic.

Habit Crystals rare, and then usually coarse, short prismatic with uneven surfaces; commonly as cleavable to compact masses.

Specific gravity 3·34 (lithiophilite) to 3·58 (triphylite).

Hardness 4–5.

Cleavage and fracture Very good $\{100\}$ cleavage; fracture uneven to subconchoidal.

Colour and transparency Bluish or greenish grey (triphylite) to clove or yellowish brown (lithiophilite); may suffer surface alteration to dark brown or black: transparent to translucent.

Streak White to greyish white

Lustre Vitreous to subresinous.

Distinguishing features Colour; association. Dissolves in hydrochloric and nitric acids.

Formation and occurrence As primary minerals in some granite pegmatites, commonly associated with other lithium and phosphate minerals. They readily alter to other, secondary minerals, either while the pegmatite is crystallizing or later, under the action of circulating ground waters.

Ludlamite

Composition $(Fe,Mg,Mn)_3(PO_4)_2.4H_2O$. Most varieties are ferrous iron rich.

Crystal system Monoclinic.

Habit Tabular, sometimes in parallel aggregates; also massive, granular.

Specific gravity 3·1–3·2.

Hardness $3\frac{1}{2}$.

Cleavage Perfect $\{001\}$ cleavage.

Colour and transparency Bright green to apple green, brown when oxidized; translucent.

Streak Greenish white.

Lustre Vitreous.

Distinguishing features Colour; association. Dissolves in hydrochloric and nitric acids.

Formation and occurrence Very rare; usually as a secondary mineral after triphylite in phosphate-bearing granite pegmatites.

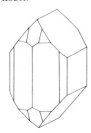

Phenakite:
rhombohedral
habit.

Olivine.

Roméite

Composition $(Ca,Na)_2Sb_2O_6(O,OH,F)$. Manganese, ferrous iron and lead may substitute for calcium and sodium; titanium may substitute for antimony.
Crystal system Cubic.
Habit Small octahedral crystals, sometimes embedded or in groups; also massive.
Twinning Rare, on $\{111\}$.
Specific gravity 4·7–5·4, varying with composition.
Hardness $5\frac{1}{2}$–$6\frac{1}{2}$.
Cleavage and fracture Imperfect octahedral cleavage: fracture splintery or uneven.
Colour and transparency Usually pale yellow to yellowish brown; also reddish to dark brown: transparent to subtranslucent.
Streak Nearly white to pale yellow.
Lustre Vitreous to greasy or subadamantine.
Distinguishing features Habit; colour; specific gravity.
Formation and occurrence A very rare mineral, found usually in manganese ore.

9 Silicates

Willemite

Composition Zn_2SiO_4. Part of the zinc may be replaced by manganese. Manganese-free willemite is not fluorescent.
Crystal system Hexagonal (Trigonal).
Habit Small prismatic or rhombohedral crystals, usually massive, granular.
Specific gravity 3·9–4·2.
Hardness $5\frac{1}{2}$.
Cleavage Basal $\{0001\}$, good cleavage.
Colour and transparency Pale greenish yellow is typical but varies from near white, brown to black; transparent to nearly opaque.
Streak White.
Lustre Vitreous to resinous.
Distinguishing features Colour; association. Willemite often shows strong fluorescence in ultraviolet light, especially specimens from Franklin, New Jersey, United States.
Formation and occurrence In the oxidized zone of zinc ore deposits. Also abundantly in the unique metamorphic ore deposit at Franklin, New Jersey where it is so common as to form a major ore. Franklin specimens occur with zincite, franklinite, often as large crystals in crystalline limestone.
Uses Minor ore of zinc.

Phenakite

Composition Be_2SiO_4.
Crystal system Hexagonal (Trigonal).
Habit Crystals often rhombohedral, sometimes prismatic; also granular and as acicular columnar aggregates.
Twinning Common on $\{10\bar{1}0\}$.
Specific gravity 3·0.
Hardness $7\frac{1}{2}$–8.
Cleavage and fracture Prismatic, distinct cleavage. Conchoidal fracture.
Colour Colourless; also white, yellow, pink and brown.

Lustre Vitreous.
Distinguishing features Habit; hardness.
Formation and occurrence Rare beryllium mineral that occurs in cavities in granite and granite pegmatites in association with beryl, topaz and apatite; also in some hydrothermal veins.
Uses Sometimes used as a gemstone.

Monticellite

Composition $CaMgSiO_4$.
Crystal system Orthorhombic.
Habit Crystals small and prismatic; usually massive, or granular.
Twinning On $\{031\}$.
Specific gravity 3·1–3·3.
Hardness $5\frac{1}{2}$.
Cleavage and fracture Indistinct $\{010\}$; brittle.
Colour and transparency Colourless to grey; transparent to translucent.
Lustre Vitreous.
Distinguishing features Occurrence.
Formation and occurrence Mainly as a metamorphic or metasomatic mineral; usually in metamorphosed calcareous, usually impure, dolomites.

Olivine

Composition $(Mg,Fe)_2SiO_4$. The olivine series is an example of continuous solid solution of two components, Mg_2SiO_4 (forsterite) and Fe_2SiO_4 (fayalite).
Crystal system Orthorhombic.
Habit Rarely found as good crystals, usually in granular masses or as isolated grains in igneous rocks.
Twinning Not common.
Specific gravity 3·2 (forsterite)–4·4 (fayalite); common olivine about 3·3–3·4.
Hardness $6\frac{1}{2}$–7.
Cleavage and fracture Pinacoidal $\{010\}$, indistinct cleavage; conchoidal fracture.
Colour and transparency Clear olive green; also white (forsterite) and brown to black (fayalite); transparent to translucent.
Streak White or grey.
Lustre Vitreous.
Distinguishing features Colour; fracture; general association. Physical and optical properties vary with increasing content of iron from forsterite to fayalite.
Alteration Alters readily; hydrothermal alteration generally results in the formation of serpentine; surface weathering leads to the oxidation of the iron content, removal of magnesium and silica, leaving red-brown pseudomorphs of iron oxides.
Formation and occurrence Important rock-forming mineral, typical of silica-poor rocks such as basalt, gabbro, troctolite and peridotite. Dunite is a rock exclusively composed of olivine, and some basalts occasionally contain nodules of granular olivine and pyroxene, possibly derived from the upper mantle. Olivine, usually forsteritic in composition, is also produced as the result of the metamorphism of magnesium-rich sediments, particularly siliceous dolomites. Fayalite occurs in rapidly cooled siliceous igneous rocks, such as pitchstone and some slags. Olivine is common in stony and stony-iron meteorites (pallasites).

Uses Magnesium-rich olivine has very high melting point, and is used in the manufacture of refractory bricks. Transparent olivines of good colour are cut as gemstones (peridot).

Tephroite

Composition Mn_2SiO_4.
Crystal system Orthorhombic.
Habit Crystals generally short prismatic; highly modified.
Twinning Not common, on $\{011\}$.
Specific gravity 4·1.
Hardness 6.
Cleavage and fracture Pinacoidal $\{010\}$, distinct cleavage; uneven to conchoidal fracture.
Colour and transparency Grey, olive green, flesh red, reddish brown; transparent to translucent.
Lustre Vitreous to greasy.
Distinguishing features Colour; occurrence.
Formation and occurrence Mainly in iron-manganese ore deposits and skarns. Sometimes in metamorphic rocks derived from manganese-rich sediments.

Humite series

Composition $Mg(OH,F)_2.1-4Mg_2SiO_4$. The group comprises four minerals, norbergite, chondrodite, humite and clinohumite, which differ only in the amount of magnesia and silica they contain.
Crystal system Humite and norbergite – orthorhombic; chondrodite and clinohumite – monoclinic (but depart only slightly from orthorhombic symmetry).
Habit Crystals usually stubby; also massive.
Specific gravity 3·1–3·3.
Hardness 6–6½.
Cleavage and fracture One poor cleavage; uneven fracture.
Colour and transparency White, pale yellow, brown; translucent.
Lustre Vitreous to resinous.
Distinguishing features Colour; association with metamorphosed limestones.
Formation and occurrence Typically in metamorphosed dolomitic limestones in association with spinel, phlogopite, garnet, idocrase and diopside.

Zircon

Composition $ZrSiO_4$. Some of the zirconium is always replaced by hafnium; part of the zirconium can also be replaced by rare earths. Zircon is often radioactive through thorium and uranium replacing zirconium in the structure.
Crystal system Tetragonal.
Habit Usually prismatic, with bipyramidal terminations.
Twinning Common, giving knee-shaped twins; twin plane $\{111\}$.
Specific gravity 4·6–4·7.
Hardness 7½.
Cleavage and fracture Prismatic $\{110\}$, indistinct cleavage; uneven, sometimes conchoidal fracture.
Colour and transparency Usually brown or reddish brown, but can be colourless, grey, green or violet; transparent to translucent.

Streak White.
Lustre Vitreous to adamantine.
Distinguishing features Habit; hardness; colour; specific gravity.
Formation and occurrence Common accessory mineral of igneous rocks such as granite, syenite and nepheline syenite. In pegmatites, the crystals sometimes reach a considerable size. Also in metamorphic rocks such as schists and gneisses. Because of its hardness, it appears as a detrital mineral in rivers and beach sands.
Uses Principal source of zirconium and hafnium for industry. Often extracted from detrital sands in beach and river deposits. Transparent blue variety produced by heat on natural stone is cut as gemstone.

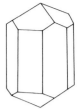

Zircon.

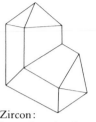

Zircon: knee-shaped twin.

Thorite

Composition $ThSiO_4$. Some of the thorium may be replaced by uranium (uranothorite), and some $-SiO_4$ by $-(OH)_4$ (thorogummite).
Crystal system Tetragonal.
Habit Crystals usually short prismatic; also massive or embedded grains.
Specific gravity Very variable with water content 4·1–6·7.
Hardness About 4½.
Cleavage and fracture Prismatic $\{100\}$ distinct cleavage, not apparent in metamict crystals; conchoidal to subconchoidal, or splintery fracture.
Colour and transparency Brownish yellow, yellow to orange, brown to black, less commonly greenish black-green; often mottled crystals; transparent in small grains, translucent to opaque in large masses.
Streak Light orange to brown.
Lustre Vitreous to greasy when fresh. Dull when partly altered.
Distinguishing features Colour; lustre; habit. Nearly always metamict. Strongly radioactive.
Formation and occurrence Widespread as a primary mineral chiefly in pegmatites; also in metasomatized zones in impure limestones; hydrothermal veins; detrital deposits.
Uses May become an important raw material if thorium comes to be used in production of atomic energy.

Andalusite.

Andalusite

Andalusite, sillimanite and kyanite provide an example of polymorphism. There is a relationship between their structure and mode of formation, the least dense andalusite forming under low-pressure metamorphism and kyanite, the densest, with a closely packed structure, forming under high-pressure conditions.
Composition Al_2SiO_5.
Crystal system Orthorhombic.
Habit Crystals prismatic, nearly square in cross-section; also massive; the variety, chiastolite, exhibits cruciform pattern of carbonaceous impurities when viewed in cross-section.
Twinning On $\{101\}$, rare.
Specific gravity 3·1–3·2.
Hardness 6½–7½.
Cleavage and fracture Prismatic, distinct cleavage; uneven to subconchoidal fracture.
Colour and transparency Commonly pink or

Kyanite: bladed crystal.

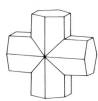

Staurolite: cruciform twin.

red, also grey, yellow, brown, green; transparent to nearly opaque.

Streak White.

Lustre Vitreous.

Distinguishing features Habit; hardness, occurrence.

Formation and occurrence Typically in thermally metamorphosed, argillaceous schists and in regionally metamorphosed rocks formed under low-pressure conditions. Rarely found in some granite pegmatites. Commonly associated minerals include sillimanite, kyanite, cordierite and corundum.

Uses Manufacture of mullite refractories, especially spark plugs; transparent green form cut as a gemstone.

Sillimanite (Fibrolite)

Composition Al_2SiO_5.

Crystal system Orthorhombic.

Habit Commonly as elongated prismatic crystals, nearly square in cross-section; striated along their length; often as fibrous or interwoven masses.

Specific gravity 3·2–3·3.

Hardness $6\frac{1}{2}$–$7\frac{1}{2}$.

Cleavage Perfect pinacoidal {010} cleavage.

Colour and transparency Colourless, white, yellowish or brownish; transparent to translucent.

Streak White.

Lustre Vitreous, often silky in fibrous material.

Distinguishing features Habit, resembling other fibrous silicates such as wollastonite and tremolite; infusible; insoluble in acids.

Formation and occurrence Typically in schists and gneisses by high-grade regional metamorphism.

Uses Sometimes as a gemstone (fibrolite).

Kyanite (Disthene)

Composition Al_2SiO_5.

Crystal system Triclinic.

Habit Crystals usually flat, bladed; also as radiating, bladed aggregates; distinctly flexible and often bent or twisted.

Specific gravity 3·5–3·7.

Hardness Variable $5\frac{1}{2}$–7 along the length of the crystals and 6–7 across them.

Blue bladed crystals of kyanite in schist; from Hurricane Mountains, North Carolina.

Cleavage and fracture {100} perfect, {010} good cleavages; parting on {001}.

Colour and transparency Blue to white but may be grey or green; often a characteristic patchy blue; transparent to translucent.

Streak White.

Lustre Vitreous, sometimes pearly on cleavage surfaces.

Distinguishing features Colour; habit; cleavage; variable hardness.

Formation and occurrence Typically in medium-grade regional metamorphic schists and gneisses, together with garnet, staurolite, mica and quartz. Also in pegmatites and quartz veins associated with schists and gneisses.

Uses The manufacture of mullite refractories. On heating to 1300 °C kyanite, sillimanite, andalusite and topaz decompose into mullite, $Al_6Si_2O_{13}$ and silica-rich glass.

Staurolite

Composition $(Fe,Mg)_2(Al,Fe)_9Si_4O_{22}(O,\dot{O}H)_2$. Some of the ferrous iron may be replaced by magnesium and manganese, some of the aluminium by ferric iron.

Crystal system Monoclinic – pseudo-orthorhombic.

Habit Usually prismatic, often with rough surfaces. Rarely massive.

Twinning Common, as cruciform twins on {031} giving a cross near 90° and {231} giving an oblique cross at about 60°.

Specific gravity 3·7–3·8.

Hardness 7–$7\frac{1}{2}$.

Cleavage and fracture Distinct {010} cleavage; uneven to subconchoidal fracture.

Colour and transparency Reddish brown to brown-black. Translucent to nearly opaque.

Streak Grey.

Lustre Vitreous to resinous.

Distinguishing features Colour; habit (particularly if twinned).

Formation and occurrence Typically as porphyroblasts in medium-grade schists and gneisses often in association with garnet, kyanite and mica.

Ilvaite

Composition $CaFe^{2+}_2Fe^{3+}Si_2O_8(OH)$.

Crystal system Orthorhombic.

Habit Crystals prismatic, often diamond-shaped in cross-section; striated along their length; also columnar or massive.

Specific gravity 3·8–4·1.

Hardness $5\frac{1}{2}$–6.

Cleavage and fracture Basal {001} distinct cleavage; uneven fracture.

Colour and transparency Black; opaque.

Streak Black.

Lustre Dull submetallic.

Distinguishing features Habit; cleavage; streak.

Formation and occurrence Chiefly as a contact metasomatic mineral; in iron, zinc and copper ore deposits.

Topaz

Composition $Al_2SiO_4(OH,F)_2$.

Crystal system Orthorhombic.

Pale brown,
transparent
crystals of sphene
with chlorite; from
Graubünden,
Switzerland.

Sphene.

Habit Well-developed short to long prismatic crystals, often with well-developed terminations; also massive granular.

Specific gravity 3·5–3·6.

Hardness 8.

Cleavage and fracture Basal {001} perfect cleavage; subconchoidal to uneven fracture.

Colour and transparency Colourless, blue, yellow, yellow-brown, rarely pink; some topaz tends to fade in sunlight; transparent to translucent.

Streak White.

Lustre Vitreous.

Distinguishing features Habit; hardness; cleavage; specific gravity.

Formation and occurrence Typically in granite pegmatites and high-temperature quartz veins; also in cavities in granites and rhyolites. Also as rounded pebbles in alluvial deposits. As grains in granites which have been altered by fluorine-rich solutions and are accompanied by fluorite, tourmaline, apatite, beryl and cassiterite.

Uses As a gemstone.

Euclase

Composition $BeAlSiO_4(OH)$.

Crystal system Monoclinic.

Habit Commonly long prismatic.

Specific gravity 3·0–3·1.

Hardness $7\frac{1}{2}$.

Cleavage and fracture One perfect {010} cleavage; conchoidal fracture.

Colour and transparency Colourless to pale or dark blue-green; also yellowish-white; transparent to translucent.

Streak White.

Lustre Vitreous.

Distinguishing features Habit; cleavage distinct from that of topaz; hardness.

Formation and occurrence A rare mineral found in pegmatites in association with other beryllium minerals, notably beryl.

Uses Sometimes used as a gemstone.

Sphene (Titanite)

Composition $CaTiSiO_5$. Somewhat variable; calcium may be partly substituted by sodium or rare earths, and titanium by aluminium.

Crystal system Monoclinic.

Habit Crystals commonly flattened and wedge-shaped; also prismatic; also massive, compact, sometimes lamellar.

Twinning Common on twin plane {100}, giving contact twins or cruciform penetration twins.

Specific gravity 3·4–3·6.

Hardness $5–5\frac{1}{2}$.

Cleavage and fracture Prismatic {110} distinct cleavage; conchoidal fracture.

Colour and transparency Usually brown; sometimes yellow, green or grey; transparent to nearly opaque.

Streak White.

Lustre Resinous to adamantine.

Distinguishing features Habit; lustre; colour.

Formation and occurrence Widely distributed as an accessory mineral in intermediate and acid igneous rocks and associated pegmatites. In schists, gneisses and some metamorphosed limestones. Rarely a detrital mineral in sedimentary deposits.

Uses Transparent varieties cut as gemstones; faceted gems are fiery and brilliant.

Garnet:
rhombdo-
decahedron.

Garnet:
icositetrahedron.

Garnet:
combination of
rhombdo-
decahedron and
icositetrahedron.

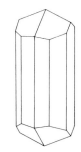

Idocrase.

Dark red rhomb-dodecahedral crystals of almandine in micaceous schist; from Var, France.

Garnet Group

Composition The garnets comprise a group of isomorphous minerals with the general formula $X_3Y_2Si_3O_{12}$ in which X may be Ca, Mn, Mg or Fe^{2+} and Y may be Al, Cr or Fe^{3+}. The compositions of naturally occurring garnets seldom approach the formulae for specific varieties because of extensive atomic substitution. The following names are in common use: almandine ($Fe_3Al_2Si_3O_{12}$) pyrope ($Mg_3Al_2Si_3O_{12}$) spessartine ($Mn_3Al_2Si_3O_{12}$) grossular ($Ca_3Al_2Si_3O_{12}$) andradite ($Ca_3Fe_2Si_3O_{12}$) uvarovite ($Ca_3Cr_2Si_3O_{12}$). There are, in effect, two main groups of garnets: the pyrope-almandine-spessartine group and the grossular-uvarovite-andradite group.

Crystal system Cubic.

Habit Crystals common; usually rhombdodecahedra or icositetrahedra or combinations of the two; other cubic forms occur but more rarely; sometimes massive or granular.

Specific gravity 3·6–4·3 (dependent on composition).

Hardness $6\frac{1}{2}$–$7\frac{1}{2}$.

Fracture Subconchoidal fracture.

Colour and transparency Colour varies with composition; pyrope, almandine and spessartine are usually shades of deep red and brown to nearly black; grossular is brown, pale green or white; andradite is yellow, brown or black; uvarovite is green; transparent to translucent.

Streak Normally white; may be pale shade of crystal colour.

Lustre Vitreous to resinous.

Distinguishing features Habit; colour; hardness. Colour and mode of occurrence often indicate the garnet species present.

Formation and occurrence Typically minerals of metamorphic rocks, although they are found in some igneous rocks. There is a link between composition and occurrence of garnets. Pyrope occurs in ultrabasic igneous rocks such as peridotite and associated serpentinites, and also in kimberlite. Found also in some high-grade, magnesium-rich metamorphic rocks. Almandine is the common garnet of schists and gneisses; also recorded in granites and pegmatites. Garnets from granite pegmatites and also metamorphosed manganese-

Fine tetragonal greenish-brown crystals of idocrase on quartz; from Eden Mills, Vermont.

bearing rocks are often spessartine. Uvarovite is rare; occurs in association with chromite in serpentinite. Grossular is typically formed by the contact or regional metamorphism of impure limestones. Andradite is commonly formed by the metasomatic alteration of limestones by iron-bearing solutions; and in metamorphosed limestones. The black variety, called melanite (titanian andradite), occurs in some feldspathoidal igneous rocks, such as phonolite. Garnet is also often found as a constituent of beach and river sands.

Uses Sometimes used as an abrasive. Transparent, unflawed garnets are cut into gemstones; many of the red garnet gemstones consist of pyrope from Czechoslovakia. Hessonite is a yellow to brownish red variety of grossular; demantoid is green andradite; rhodolite is a rose-coloured or purplish garnet of the pyrope-almandine series.

Idocrase (Vesuvianite)

Composition $Ca_{10}(Mg,Fe)_2Al_4(SiO_4)_5(Si_2O_7)_2$ $(OH,F)_4$. Considerable variation in composition; some calcium may be replaced by sodium and potassium, magnesium by ferrous iron and manganese, and aluminium by ferric iron and titanium.

Crystal system Tetragonal.

Habit Short prismatic; also pyramidal. Commonly massive, granular or columnar.

Specific gravity 3·3–3·5.

Hardness 6–7.

Cleavage and fracture Prismatic $\{110\}$ poor cleavage; uneven to conchoidal fracture.

Colour and transparency Various shade of green, dark green, brown, white-yellow; blue varieties called cyprine; transparent to translucent.

Streak White.

Lustre Vitreous to resinous.

Distinguishing features Habit; massive varieties may be mistaken for garnet, epidote or diopside.

Formation and occurrence By the contact metamorphism of impure limestones; commonly associated with calcite, grossularite or andradite garnet, and wollastonite. Also occurs in blocks of limestone erupted from Mount Vesuvius, hence its alternative name.

Hemimorphite

Composition $Zn_4Si_2O_7(OH)_2.H_2O$.
Crystal system Orthorhombic.
Habit Commonly thin tabular; striated; often in fan-shaped aggregates; also massive, granular or mamillary masses with fibrous structure.
Twinning Basal {001}, rare.
Specific gravity 3·4–3·5.
Hardness $4\frac{1}{2}$–5.
Cleavage Prismatic {110}, perfect cleavage.
Colour and transparency Usually white or colourless; also pale blue, greenish, yellowish, brown; transparent to translucent.
Streak White.
Lustre Vitreous.
Distinguishing features Habit.
Formation and occurrence Secondary mineral found in the oxidized zone of zinc-bearing ore bodies; and in limestones associated with such ore bodies. Common associated minerals are sphalerite, smithsonite, galena, cerussite and anglesite.

Epidote Group

The epidote group comprises several minerals with the general formula $X_2Y_3Si_3O_{12}(OH)$ in which X is commonly Ca, partly replaced by rare earth elements in allanite, and Y is Al and Fe^{3+}, partly replaced by Mg and Fe^{2+} in allanite, and by Mn^{3+} in piemontite.

Zoisite

Composition $Ca_2Al_3Si_3O_{12}(OH)$. Sometimes some substitution of ferric iron for aluminium.
Crystal system Orthorhombic.
Habit Aggregates of long prismatic crystals; often deeply striated; commonly massive.
Specific gravity 3·2–3·4 (increasing with iron content).
Hardness $6\frac{1}{2}$.
Cleavage and fracture Pinacoidal {100} perfect cleavage; uneven fracture.
Colour and transparency Grey, white, greenish brown, green; also pink (thulite), blue to purple (tanzanite); transparent to translucent.
Streak White to grey-white.
Lustre Vitreous, pearly on cleavage surfaces.
Distinguishing features Colour; cleavage.
Formation and occurrence Occurs in schists and gneisses and in metasomatic rocks, together with garnet, idocrase and actinolite. Occurs occasionally in hydrothermal veins.
Uses Up to 1967 the pink variety (thulite) was the only zoisite cut for gemstones but in 1967 violet-blue highly pleochroic zoisites from Tanga province, Tanzania were found; these have become valuable gemstones and are marketed under the name tanzanite.

Clinozoisite and Epidote

Composition $Ca_2Al_3Si_3O_{12}(OH)$ and $Ca_2(Al,Fe)_3Si_3O_{12}(OH)$. Clinozoisite and epidote are, in effect, a single species, with some replacement of aluminium by iron.
Crystal system Monoclinic.
Habit Crystals prismatic, often deeply striated parallel to their length. Commonly massive,

granular or fibrous.
Twinning On {100}, lamellar, not common.
Specific gravity 3·2–3·5 (increasing with iron content).
Hardness 6–7.
Cleavage and fracture Perfect {001} cleavage; uneven fracture.
Colour and transparency Pale green, or greenish grey (clinozoisite); yellowish green to black (epidote); transparent to nearly opaque.
Streak White, grey-white.
Lustre Vitreous.
Distinguishing features Colour; habit. Epidote is sometimes mistaken for tourmaline, but the latter lacks cleavage and often has a triangular cross-section.
Formation and occurrence Widely in medium- to low-grade metamorphic rocks, especially derived from igneous rocks such as basalt; or in acid igneous rocks contaminated with calc-silicate material.

Allanite (Orthite)

Composition $(Ca,Ce,Y,La,Th)_2(Al,Fe,)_3Si_3O_{12}(OH)$. Composition is variable; usually contains some thorium (up to 3 per cent recorded) also other rare earths such as Ce, Y, La. Allanite is commonly metamict, as a result of radiation damage caused by the radioactive decay of thorium.
Crystal system Monoclinic.
Habit Crystals usually tabular, long prismatic to acicular; commonly compact massive.
Specific gravity 3·4–4·2 (variable).
Hardness 5–$6\frac{1}{2}$.
Fracture Conchoidal to uneven fracture.
Colour Usually black, sometimes dark brown.
Streak Grey-brown.
Lustre Vitreous, sometimes pitchy.
Distinguishing features Colour; lustre. Radio-activity.
Formation and occurrence Widespread accessory mineral in many granites, syenites, pegmatites, gneisses and skarns.

Piemontite (Piedmontite)

Composition $Ca_2(Al,Fe,Mn)_3Si_3O_{12}(OH)$.
Crystal system Monoclinic.

Dark brownish-green, translucent, striated, prismatic crystals of epidote with quartz; from Prince of Wales Island, Alaska.

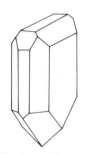

Hemimorphite.

Habit Crystals prismatic or acicular, similar to epidote; commonly massive.
Twinning On {100} lamellar; not common.
Specific gravity 3·4–3·5.
Hardness 6.
Cleavage and fracture Perfect {001} cleavage; uneven fracture.
Colour and transparency Red, reddish brown to reddish black; transparent to nearly opaque.
Lustre Vitreous.
Distinguishing features Colour; occurrence.
Formation and occurrence Rare; in some low-grade schists and also in metasomatic manganese ore deposits.

Pumpellyite

Composition $Ca_2MgAl_2(SiO_4).(Si_2O_7)(OH)_2.$ $2H_2O$.
Crystal system Monoclinic.
Habit Crystals usually fibrous, or as narrow, flattened plates; as stellate needle-like clusters or randomly orientated fibres.
Twinning On {001} and {100}, common.
Specific gravity 3·18–3·23.
Hardness 6.
Cleavage Distinct {001} and {100} cleavage.
Colour and transparency Green, bluish-green, brown; translucent.
Distinguishing features Habit; colour; occurrence.
Formation and occurrence Widespread and relatively common in rocks of different composition and geological environment; in amygdaloidal cavities in copper-bearing ores, in basalts, and with lawsonite in glaucophane schists.

White radiating globular masses of pectolite; from Paterson, New Jersey.

Wollastonite

Composition $CaSiO_3$.
Crystal system Triclinic.
Habit Crystals tabular or short prismatic; usually fibrous masses; sometimes granular and compact.
Twinning On {100}, common.
Specific gravity 2·8–3·1.
Hardness 4½–5.
Cleavage and fracture One perfect {100}, two other good cleavages; splintery fracture.
Colour and transparency White to grey; transparent to translucent.
Streak White.
Lustre Vitreous, somewhat silky in fibrous varieties.
Distinguishing features Colour; cleavages; association. Dissolves in hydrochloric acid with separation of silica.
Formation and occurrence By the metamorphism of siliceous limestones, either in contact aureoles or in high-grade regionally metamorphosed rocks; or in xenoliths in igneous rocks; associated minerals are calcite, epidote, grossular and tremolite.
Uses Some deposits of wollastonite are mined, the material being used in ceramics and paints.

Pectolite

Composition $NaCa_2Si_3O_8OH$. Most pectolite has a composition close to the above formula, but in some specimens the calcium is extensively replaced by manganese.
Crystal system Triclinic.
Habit As aggregates of fibrous or acicular crystals; usually radiating and forming globular masses.
Specific gravity 2·75–2·9.
Hardness 4½–5.
Cleavage and fracture {001} and {100} perfect cleavages; splintery fracture.
Colour and transparency Colourless, white; transparent to translucent.
Streak White.
Lustre Vitreous or silky.
Distinguishing features Habit; cleavage. Easily fusible to a white enamel.
Formation and occurrence Chiefly in cavities in basaltic rocks, often in association with zeolites; less commonly in calcium-rich metamorphic rocks, in alkaline igneous rocks and in mica peridotites.

Benitoite

Composition $BaTiSi_3O_9$.
Crystal system Hexagonal.
Habit Crystals rare, pyramidal or tabular; somewhat triangular in shape.
Specific gravity 3·6.
Hardness 6–6½.
Cleavage and fracture {10$\bar{1}$1}, indistinct cleavage; conchoidal to uneven fracture.
Colour and transparency Blue, purple, pink or white; transparent to translucent.
Streak White.
Lustre Vitreous.
Distinguishing features Habit; colour; association. Fluoresces under short-wave ultraviolet light.
Formation and occurrence As superb blue crystals in association with neptunite and natrolite

in serpentine in San Benito County, California; also occurs as detrital grains.
Uses Rarely cut as a gemstone.

Beryl

Composition $Be_3Al_2Si_6O_{18}$. Beryl often contains appreciable amounts of the alkali elements. Lithium can substitute for aluminium and aluminium for beryllium; to satisfy the resulting charge imbalance, sodium, potassium and caesium are introduced.
Crystal system Hexagonal.
Habit Short to long prismatic; often striated parallel to their length and etched.
Specific gravity 2·6–2·8.
Hardness $7\frac{1}{2}$–8.
Cleavage and fracture Basal {0001}, indistinct cleavage; conchoidal to uneven fracture.
Colour and transparency Colourless, white, light green, green, yellow, pink, red, pinkish orange, pale blue, blue; transparent to translucent.
Streak White.
Lustre Vitreous.
Distinguishing features Habit; colour; from apatite by hardness; from milky quartz by specific gravity.
Formation and occurrence Chiefly in granite pegmatites, biotite schists, gneisses, and pneumatolytic hydrothermal veins. Beryl crystals in some pegmatites grow to very large sizes [up to 5·4 metres (18 feet)] and crystal aggregates up to 100 tonnes in mass. The fine emeralds from Muzo, Colombia, occur in cavities in bituminous limestone.
Uses Commonest beryllium mineral, and the only commercial source of this element; some transparent varieties are prized as gemstones, dark green (emerald), pale blue (aquamarine), yellow (heliodor) and pink (morganite).

Cordierite

Composition $(Mg,Fe)_2Al_4Si_5O_{18}$.
Crystal system Orthorhombic.
Habit Rarely as prismatic or pseudohexagonal twinned crystals; generally massive or irregular grains.
Twinning On {110} or {130}, common, usually repeated, forming pseudohexagonal crystals.
Specific gravity 2·5–2·8 (increasing with iron content).
Hardness 7–$7\frac{1}{2}$.
Cleavage and fracture One poor {010} cleavage; subconchoidal fracture.
Colour and transparency Characteristically pale to dark blue or violet, also colourless, grey, yellow or brown; transparent to translucent.
Streak White.
Lustre Vitreous.
Distinguishing features Colour, but granular, colourless or grey cordierite is very like quartz, and has to be identified by optical or chemical tests. The gem variety (iolite) is recognized by its intense pleochroism. Alters readily to an aggregate of chlorite and muscovite called pinite.
Formation and occurrence By medium- to high-grade metamorphism of aluminium-rich rocks. Found in hornfelses, schists and gneisses in association with andalusite, spinel, quartz and biotite. Often formed by the assimilation of aluminous sediments in igneous rocks.
Uses Clear blue varieties of cordierite are cut as gemstones.

Dioptase

Composition $CuSiO_2(OH)_2$.
Crystal system Hexagonal (Trigonal).
Habit Crystals short to long prismatic, often terminated by rhombohedra; also massive.
Specific gravity 3·3
Hardness 5.
Cleavage and fracture Rhombohedral {$10\bar{1}1$} perfect cleavage; conchoidal to uneven fracture.
Colour and transparency Emerald green; transparent to translucent.
Streak Pale greenish blue.
Lustre Vitreous.
Distinguishing features Colour; habit; association with other copper minerals.
Formation and occurrence Not common but occurs in the oxidation zone of copper deposits; also as superb crystals with calcite from southwest Africa.

Gadolinite

Composition $Be_2FeY_2Si_2O_{10}$.
Crystal system Monoclinic.
Habit Crystals often prismatic, sometimes flattened; commonly massive, compact.
Specific gravity 4·0–4·7.
Hardness $6\frac{1}{2}$–7.
Fracture Conchoidal fracture.
Colour and transparency Black, greenish black, brown, rarely light green; transparent to translucent.
Streak Greenish grey.
Lustre Vitreous to greasy.
Distinguishing features Habit; fracture; hardness.
Formation and occurrence Chiefly in granites and granite pegmatites, usually in association with fluorite and allanite.

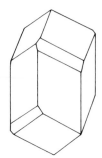

Beryl.

Dioptase.

Emerald-green modified rhombohedra of dioptase; from Tsumeb, southwest Africa.

Fine group of prismatic green diopside crystals with orange-brown grossular (variety hessonite) and dark green chlorite plates; from Ala, Piedmont, Italy.

Characteristic cleavages of pyroxenes.

Augite.

Pyroxene Group

The pyroxenes are an important and widely distributed group of rock-forming silicates. They have a general formula $X_2Si_2O_6$ in which X is usually magnesium, iron, manganese, lithium, titanium, aluminium, calcium or sodium. Of these, manganese and titanium are only present in minor amounts except in the rare pyroxene, johannsenite ($CaMnSi_2O_6$). Some aluminium may partly replace silicon, up to a Si:Al ratio of 3:1. The commonest pyroxenes are calcium, magnesium, iron silicates. Pyroxenes are characterized by two cleavages which intersect almost at right angles. Two main groups: the orthopyroxenes which crystallize in the orthorhombic system and contain very little calcium; and the clinopyroxenes which are monoclinic and contain either calcium, or sodium, aluminium, ferric iron or lithium. (For structure *see* page 101.)

Enstatite and Hypersthene (Orthopyroxenes)

Composition $MgSiO_3$ and $(Mg,Fe)SiO_3$. Enstatite and hypersthene form a continuous series. The boundary between them is arbitrarily fixed at about 10 per cent of the $FeSiO_3$ component. The series does not extend to pure $FeSiO_3$, this being unstable at high temperatures.
Crystal system Orthorhombic.
Habit Prismatic crystals rare; usually grains or massive.
Twinning On {100} simple and lamellar.
Specific gravity 3·2–4·0 (increasing with iron content).
Hardness 5–6.

Cleavage and fracture Prismatic {210} good cleavage; uneven fracture.
Colour and transparency Pale green to dark brownish green-black (darkening with iron content); bronzite is intermediate in composition between enstatite and hypersthene and has a bronze lustre; translucent to nearly opaque. Vitreous, pearly or bronzy.
Distinguishing features Cleavage; colour.
Formation and occurrence Typically in basic and ultrabasic rocks poor in calcium, such as pyroxenites, peridotites, norites and some basalts and andesites. Also, some high-grade metamorphic rocks. Compositions closest to $MgSiO_3$ are recorded for enstatite in some meteorites.

Diopside – Hedenbergite Series: Augite (Clinopyroxenes)

Composition $Ca(Mg,Fe)Si_2O_6$: $Ca(Mg,Fe,Al)(Al,Si)_2O_6$. Diopside, hedenbergite and augite are convenient species names for parts of a continuous series in chemical composition. Augite is the most common pyroxene.
Crystal system Monoclinic.
Habit Usually stout prisms of square or octagonal cross-section; also massive, granular.
Twinning Common on {100}.
Specific gravity 3·2–3·6 (increasing with iron content).
Hardness $5\frac{1}{2}$–$6\frac{1}{2}$.
Cleavage Prismatic {110} good cleavage; sometimes well-developed parting parallel to {100} (variety diallage).
Colour and transparency Usually dark green to black (augite); diopside is greyish white to light green; translucent to opaque, rarely transparent.

Streak White or grey.
Lustre Vitreous.
Distinguishing features Cleavage; habit; diopside colour is usually paler green than augite.
Formation and occurrence Augite is the most important ferromagnesian mineral of igneous rocks, abundant in basic and ultrabasic rocks, characteristic of gabbros and basalts, also occurs in many andesites. Diopside and hedenbergite occur in medium- and high-grade metamorphic rocks, especially those rich in calcium. White and light green diopside characteristically occur in metamorphosed dolomitic limestones and skarns; hedenbergite is often found associated with ore deposits formed at high temperatures.

Jadeite (Clinopyroxene)

Composition $NaAlSi_2O_6$. Sodium and aluminium may be partly replaced by calcium and magnesium, giving the variety diopside-jadeite; aluminium is sometimes partly replaced by ferric iron giving the dark green variety, chloromelanite.
Crystal system Monoclinic.
Habit Crystals very rare; normally fine granular or dense masses.
Specific gravity 3·2–3·4.
Hardness 6–6$\frac{1}{2}$.
Cleavage and fracture Prismatic $\{110\}$, good cleavage; fine-grained massive material extremely tough; splintery fracture.
Colour and transparency Various shades of light or dark green, sometimes white or lilac; translucent.
Streak White.
Lustre Vitreous, perhaps pearly.
Distinguishing features Colour; habit; tough nature of massive material. The name 'jade' is applied to two distinct minerals: jadeite, and nephrite (an amphibole); best distinguished by specific gravity.
Formation and occurrence At high pressures; occurs as grains in metamorphosed sodic sediments and volcanic rocks together with the amphibole, glaucophane. Found also in metamorphic rocks associated with serpentine.
Uses As a semiprecious and ornamental stone. Once known only as stream boulders from Burma; now found in situ in Burma, Japan, California and Guatemala.

Aegirine (Clinopyroxene)

Composition $NaFeSi_2O_6$. Solid solution produces a complete series between aegirine and augite.
Crystal system Monoclinic.
Habit Prismatic crystals; often elongated and terminated by steeply inclined faces giving a pointed appearance (acmite); also as discrete grains.
Twinning Common on $\{100\}$.
Specific gravity 3·5–3·6.
Hardness 6.
Cleavage and fracture Prismatic $\{110\}$, good cleavage; uneven fracture.
Colour and transparency Usually dark green to black; subtransparent to opaque.
Streak Grey.
Lustre Vitreous.
Distinguishing features From augite by habit.

Formation and occurrence In sodium-rich igneous rocks, especially those of the nepheline syenite family, and in associated pegmatites.

Spodumene (Clinopyroxene)

Composition $LiAlSi_2O_6$. Not subject to wide variations in composition, but some lithium may be partly replaced by sodium.
Crystal system Monoclinic.
Habit Crystals usually prismatic; often striated along their length; also massive and columnar often etched and corroded.
Twinning On $\{100\}$ common.
Specific gravity 3·0–3·2.
Hardness 6$\frac{1}{2}$–7.
Cleavage and fracture Prismatic $\{110\}$ perfect cleavage; usually with a well-developed parting parallel to $\{100\}$; uneven, splintery fracture.
Colour and transparency Commonly white or greyish white; some varieties transparent pink and violet (kunzite), or green (hiddenite); translucent.
Streak White.
Lustre Vitreous.
Distinguishing features Habit; cleavage. Presence of lithium (red flame test). Typical occurrence in complex granite pegmatites; alters readily to clay minerals.
Formation and occurrence Typically in lithium-bearing granite pegmatites, together with lepidolite, tourmaline and beryl. Some very large crystals have been reported, 12 metres (40 feet) or more in length and weighing up to 65 tonnes.
Uses Mined as a raw material for lithium compounds and ceramics; kunzite and hiddenite are cut as gemstones.

Amphibole Group

The amphibole group comprises a number of species, crystallizing in both the orthorhombic and monoclinic systems but closely related in structure and chemical composition. They can be grouped into a number of series within which extensive replacement of one ion by others of similar size can take place, giving very complex chemical compositions. The amphiboles contain essential hydroxyl groups in the structure. The angle between the prism faces and between two cleavages parallel to them is about 120° and is characteristic of the amphiboles.

Anthophyllite

Composition $(Mg,Fe)_7Si_8O_{22}(OH)_2$. The name gedrite has been applied to aluminium anthophyllites.
Crystal system Orthorhombic.
Habit Individual crystals rare; usually in aggregates of prismatic crystals; sometimes fibrous and asbestiform.
Specific gravity 2·9–3·3 (increasing with iron content).
Hardness 6.
Cleavage Prismatic $\{210\}$, perfect cleavage.
Colour and transparency White, grey, brown; translucent.
Streak White.
Lustre Vitreous, somewhat silky in fibrous varieties.

Characteristic cleavages of amphiboles.

Distinguishing features Colour and appearance. From cummingtonite, it can be readily distinguished only by optical, density or X-ray tests.
Formation and occurrence In magnesium-rich metamorphic rocks of medium grade, often associated with talc or cordierite.
Uses Asbestiform anthophyllite is used in industry, but fibres are brittle and of relatively low tensile strength, and are, therefore, used in asbestos cement and as an insulating material.

Cummingtonite

Composition $(Fe,Mg)_7Si_8O_{22}(OH)_2$. The iron to (iron + magnesium) ratio ranges from 1 to about 0·4; iron-rich varieties of cummingtonite are called grunerite. Manganese sometimes replaces part of the iron and magnesium.
Crystal system Monoclinic.
Habit Usually in aggregates of fibrous crystals; often radiating.
Specific gravity 3·2–3·6 (increasing with iron content).
Hardness 6.
Cleavage $\{110\}$, good cleavage.
Colour Pale to dark brown.
Streak White.
Lustre Vitreous to silky.
Distinguishing features Colour; specific gravity (to distinguish from anthophyllite).
Formation and occurrence In calcium-poor, iron-rich metamorphic rocks of medium grade, often in association with ore deposits. Also, in some igneous rocks such as certain rhyolites, and as a replacement product of pyroxene in diorites.

Tremolite-Actinolite Series

Composition $Ca_2(Mg,Fe)_5Si_8O_{22}(OH)_2$. The division between tremolite and actinolite is an arbitrary one, tremolite being the low-iron end of the series and actinolite the iron-rich member.
Crystal system Monoclinic.
Habit Usually in aggregates of long prismatic crystals; sometimes massive, fibrous.
Specific gravity 2·9–3·4.
Hardness 5–6.
Cleavage Prismatic $\{110\}$, good cleavage.
Colour and transparency White-grey (tremolite) becoming green with increasing iron content; transparent to translucent.
Streak White.
Lustre Vitreous.
Distinguishing features Colour; habit; fibrous, radiating tremolite resembles wollastonite, but can be distinguished optically and by the lack of reaction with hydrochloric acid.
Formation and occurrence Common in low- and medium-grade metamorphic rocks, tremolite being characteristic of thermally metamorphosed dolomitic limestones; actinolite generally occurring in schists produced by low- to medium-grade metamorphism of basalt and dolerite, or pelitic rocks.
Uses Nephrite, a dark green variety of 'jade', is usually an actinolitic or tremolitic amphibole, and is carved into ornaments; asbestiform varieties of tremolite are worked to some extent, but are less valuable industrially than serpentine asbestos.

Glaucophane-Riebeckite Series

Composition $Na_2(Mg,Fe,Al)_5Si_8O_{22}(OH)_2$. Glaucophane and riebeckite form a series varying in composition, mainly due to the replacement of magnesium and aluminium by ferrous and ferric iron.
Crystal system Monoclinic.
Habit Good crystals rare; prismatic to acicular crystals; sometimes fibrous or asbestiform.
Specific gravity 3·0–3·4 (increasing with iron content).
Hardness 5–6.
Cleavage and fracture Prismatic $\{110\}$, good cleavage; uneven fracture.
Colour and transparency Glaucophane is grey, grey-blue or lavender-blue; riebeckite is dark blue to black; translucent to subtranslucent.
Streak White to blue-grey.
Lustre Vitreous, silky in fibrous varieties.
Distinguishing features Colour; association.
Formation and occurrence Glaucophane typically in sodium-rich schists derived from geosynclinal sediments which have undergone low-temperature/high-pressure, regional metamorphism; in association with minerals such as jadeite, aragonite, epidote, chlorite and garnet. Riebeckite mainly in alkaline igneous rocks such as some granites, syenites and nepheline syenites; fibrous riebeckite, known as crocidolite or blue asbestos, as veins in some bedded ironstones.
Uses Crocidolite or blue asbestos extensively mined in South Africa and Western Australia; often crocidolite is partly replaced by quartz, and is then used as an ornamental stone under the name tiger-eye, the iron minerals having oxidized to a golden brown colour.

Richterite

Composition $Na_2Ca(Mg,Fe^{3+},Fe^{2+},Mn)_5Si_8O_{22}(OH,F)_2$.
Crystal system Monoclinic.
Habit Crystals long prismatic; rarely terminated.
Twinning On $\{100\}$.
Specific gravity 2·97–3·13.
Hardness 5–6.
Cleavage Prismatic $\{110\}$, perfect cleavage.
Colour and transparency Brown, yellow, red, pale to dark green; transparent to translucent.
Lustre Vitreous.
Distinguishing features Colour; habit; occurrence.
Formation and occurrence In contact metasomatic deposits, in igneous rocks and in thermally metamorphosed limestones.

Arfvedsonite

Composition $Na_3(Mg,Fe)_4AlSi_8O_{22}(OH,F)_2$.
Crystal system Monoclinic.
Habit Crystals long prismatic; often tabular $\{010\}$ but rarely well terminated; also in prismatic aggregates.
Twinning On $\{100\}$ simple lamellar.
Specific gravity 3·37–3·50.
Hardness 5–6.
Cleavage and fracture $\{110\}$, perfect cleavage; uneven, brittle fracture.

Colour and transparency Greenish black, black; nearly opaque.
Streak Dark bluish grey.
Lustre Vitreous.
Distinguishing features Habit; occurrence.
Formation and occurrence Characteristic of plutonic alkali igneous rocks such as nepheline syenite.

Hornblende

Composition $(Ca,Na)_{2-3}(Mg,Fe,Al)_5(Si,Al)_8O_{22}$ $(OH)_2$. Hornblende can vary greatly in composition; dark brown varieties occurring in basic igneous rocks often contain appreciable amounts of titanium (kaersutite), or they may be low in hydroxyl (basaltic hornblende).
Crystal system Monoclinic.
Habit Crystals usually of long or short prismatic habit often with six-sided cross section; also massive, granular or fibrous.
Twinning Common on {100}.
Specific gravity 3·0–3·5 (increasing with iron content).
Hardness 5–6.
Cleavage and fracture Prismatic {110}, good cleavage; uneven fracture.
Colour and transparency Light green to dark green to nearly black; translucent to nearly opaque.
Streak White to grey.
Lustre Vitreous.
Distinguishing features Cleavage; colour.
Formation and occurrence Important and widespread rock-forming mineral; in a wide variety of igneous rocks, being a common constituent of granodiorites, diorites, some syenites, gabbros and fine-grained equivalents. Common constituent of many medium-grade, regionally metamorphosed rocks; especially characteristic of amphibolites or hornblende schists.

Barkevikite

Composition $(Na,K)Ca_2(Fe,Mg,Mn)_5(Si,Al)_8O_{22}$ $(OH,F)_2$.
Crystal system Monoclinic.
Habit Crystals long prismatic; commonly well developed with terminal faces; also as prismatic aggregates.
Twinning On {100}.
Specific gravity 3·35–3·44.
Hardness 5–6.
Cleavage and fracture {110}, perfect cleavage; uneven, brittle fracture.
Colour and transparency Black; nearly opaque.
Lustre Vitreous.
Distinguishing features Habit; colour; occurrence.
Formation and occurrence In alkaline rocks such as trachytes, phonolites and nepheline and sodalite syenites.

Palygorskite (Attapulgite)

Composition $(Mg,Al)_2Si_4O_{10}(OH).4H_2O$.
Crystal system Monoclinic (possibly also orthorhombic).
Habit Crystals lath shaped; elongated in bundles; usually as thin flexible sheets, composed of inter-

Above
Black, short prismatic crystals of hornblende showing cleavage planes; from Hurd mine, New Jersey.

Left
Dull pink prismatic crystals of rhodonite in matrix; from Franklin, New Jersey.

laced fibres, resembling leather or parchment.
Specific gravity 2·2.
Hardness Soft.
Cleavage and fracture {110}, easy; tough.
Colour and transparency White, grey; translucent.
Lustre Dull.
Distinguishing features Habit.
Formation and occurrence Mainly in hydrothermal veins, in altered serpentinites or granitic rocks. Often referred to as 'mountain leather'.

Rhodonite

Composition $MnSiO_3$. Composition variable, part of the manganese being replaced by calcium, or iron.
Crystal system Triclinic.
Habit Crystals uncommon; prismatic or tabular; most commonly massive or granular.
Specific gravity 3·5–3·7.
Hardness $5\frac{1}{2}$–$6\frac{1}{2}$.
Cleavage and fracture {110} and {1$\bar{1}$0}, perfect, {001} good; conchoidal to uneven fracture.
Colour and transparency Pink to rose red, often veined by black alteration products.

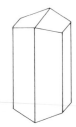

Hornblende.

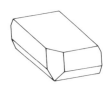

Rhodonite.

Streak White.
Lustre Vitreous.
Distinguishing features Colour; cleavage. From rhodochrosite by greater hardness and lack of effervescence in warm hydrochloric acid.
Formation and occurrence Commonly in association with manganese ore deposits in hydrothermal or metasomatic veins, or in regionally metamorphosed manganese-bearing sediments.
Uses Massive rhodonite is sometimes cut and polished as an ornamental stone.

Mica Group

The micas constitute an isomorphous group; compositions are variable within the mica group, but a general formula can be given as $W(X,Y)_{2-3}Z_4O_{10}(OH,F)_2$. In this W is generally potassium, X and Y represent aluminium, magnesium, ferrous and ferric iron and lithium; and Z is silicon and aluminium, the silicon : aluminium ratio being generally about 3 : 1. There are two main groups of micas: dark mica, rich in iron and magnesium, and white mica which is rich in aluminium.

Muscovite

Composition $KAl_2(AlSi_3O_{10})(OH)_2$. Some sodium is always present replacing potassium (replacement is greater the higher the temperature of formation). Aluminium in six coordination is often partly replaced by magnesium and ferrous iron or less commonly by chromium (bright green variety fuchsite) or vanadium (roscoelite); some fluorine may be present replacing hydroxyl.
Crystal system Monoclinic.
Habit Crystals tabular and hexagonal in outline; usually as lamellar masses or small flakes.
Specific gravity 2·8–2·9.
Hardness $2\frac{1}{2}$–4.
Cleavage Basal {001}, perfect; cleavage flakes flexible and elastic.
Colour and transparency Colourless, or pale shades of green, or brown; transparent to translucent.
Streak White.
Lustre Vitreous; pearly parallel to cleavage.

Distinguishing features Cleavage, and the flexible flakes, identify muscovite as belonging to the mica group, brown varieties can resemble phlogopite closely but associations are usually different.
Formation and occurrence Widely distributed. In igneous rocks it is confined to some granites and their pegmatites, in which it sometimes forms very large masses. Common and abundant in schists and gneisses of low- and medium-grade metamorphism. It also occurs as a secondary mineral resulting from the decomposition of feldspars; such muscovite is called sericite. Fairly resistant to weathering, and thus occurs as a common constituent of clastic sediments such as sandstones and siltstones.
Uses In the electrical industry for capacitors, insulating material, and heating elements; ground mica is used as a filler and dusting medium.

Illite

Composition Aluminosilicate of potassium. Illite is a clay mineral structurally related to the micas.
Crystal system Monoclinic.
Habit Massive, extremely fine-grained; usually with other clay minerals.
Specific gravity 2·6–2·9.
Hardness 1–2.
Cleavage {001}, perfect cleavage.
Colour Colourless in thin section; usually white or other pale colours.
Lustre Dull.
Distinguishing features Extremely fine-grained, necessitating study by electron microscope.
Formation and occurrence Dominant clay mineral in shales and other sedimentary deposits. As a hydrothermal mineral in some ore deposits and in alteration products around hot springs.

Phlogopite and Biotite

Composition $KMg_3AlSi_3O_{10}(OH)_2$ and $K(Mg,Fe)_3AlSi_3O_{10}(OH)_2$. Phlogopite and biotite are divisions of a single group with variable composition. Potassium can be replaced in part by sodium, calcium, barium, rubidium or caesium;

169

magnesium can be completely replaced by ferrous and ferric iron, and in part by titanium and manganese; the silicon-aluminium ratio is somewhat variable, and some of the hydroxyl may be replaced by fluorine.

Crystal system Monoclinic.

Habit Crystals tabular or short pseudohexagonal prisms; also as lamellar aggregates or disseminated flakes.

Specific gravity 2·7–3·4 (increasing with iron content).

Hardness 2–3.

Cleavage Basal {001}, perfect cleavage, giving flexible and elastic laminae.

Colour and transparency Phlogopite, pale yellow to brown, often with a distinctive coppery appearance; biotite, black, dark brown or greenish black; transparent to translucent.

Streak White or grey.

Lustre Vitreous; pearly on cleavage surfaces.

Distinguishing features Cleavage; colour. Biotite is the name commonly given to all dark, iron-rich micas. Phlogopite sometimes shows asterism – a point of light viewed through a cleavage sheet shows a six-rayed star.

Formation and occurrence A marked correlation exists between composition and geological environment; in igneous rocks the iron content of biotite increases with silica content of the rock, ultrabasic rocks containing phlogopite, and granites and granite pegmatites containing iron-rich biotites. Phlogopite commonly occurs in metamorphosed limestones and in magnesium-rich igneous rocks. It also occurs in kimberlite. Biotites are widely distributed in granites, syenites and diorites, and characteristic of mica lamprophyres. It is a common constituent of schists and gneisses, and of contact metamorphic rocks.

Uses Phlogopite is used industrially in the same way as muscovite; it has greater heat resistance and can withstand temperatures up to 1 000 C; therefore is used in electrical commutators.

Lepidolite

Composition $K(Li,Al)_3(Si,Al)_4O_{10}(OH)_2$. Varies considerably in composition; potassium is often partly replaced by sodium, rubidium and caesium; the relative amounts of lithium and aluminium in six-coordination varying widely; some hydroxyl may be replaced by fluorine.

Crystal system Monoclinic.

Habit Usually as small disseminated flakes; crystals tabular, pseudohexagonal.

Specific gravity 2·8–3·3.

Hardness 2½–3.

Cleavage Basal {001}, perfect cleavage, giving flexible and elastic laminae.

Colour and transparency Commonly pale lilac, also colourless, pale yellow or grey; transparent to translucent.

Lustre Vitreous; pearly on cleavage surfaces.

Distinguishing features Cleavage; colour.

Formation and occurrence In granite pegmatites, often in association with lithium-bearing tourmaline and spodumene.

Uses As an ore of lithium. Lepidolite is also used as a raw material in glass and ceramics.

Vermiculite

Composition Group name with general formula $(Mg,Fe,Al)_3(Al,Si)_4O_{10}(OH)_2.4H_2O$.

Crystal system Monoclinic.

Habit Platy.

Specific gravity About 2·3.

Hardness About 1½.

Cleavage Basal, {001}, perfect cleavage.

Colour and transparency Yellow, brown; translucent.

Lustre Pearly, sometimes bronzy.

Distinguishing features Expands greatly perpendicular to cleavage on heating.

Formation and occurrence As an alteration of magnesian micas, in association with carbonatites.

Uses Extensively as a thermal insulator.

Glauconite

Composition $K(Fe,Mg,Al)_2(Si_4O_{10})(OH)_2$.

Crystal system Monoclinic.

Habit Small, green, rounded aggregates.

Specific gravity 2·5–2·8.

Hardness 2.

Cleavage Perfect basal cleavage.

Colour Green to black.

Streak Green.

Lustre Earthy and dull.

Distinguishing features Colour; habit.

Formation and occurrence Usually in marine sedimentary rocks. The related mineral species, celadonite, is similar in structure and composition to glauconite, but its mode of occurrence is different being found as blue-green earthy material in vesicular cavities in basalt having been formed by hot solutions during late cooling stage of rock.

Stilpnomelane

Composition $K(Fe,Mg,Al)_3Si_4O_{10}(OH)_2.nH_2O$.

Crystal system Monoclinic.

Habit As thin foliated plates; also fibrous.

Specific gravity 2·6–3·0.

Hardness 3–4.

Cleavage and fracture Perfect, {001} cleavage. Brittle.

Colour and transparency Black, reddish brown, dark green; translucent to nearly opaque.

Lustre Pearly to subvitreous; sometimes bronze-toned metallic.

Distinguishing features Colour; habit; association.

Formation and occurrence Mainly as an abundant constituent of schists in association with chlorite, epidote, lawsonite. Also found in iron ore deposits associated with magnetite, hematite and pyrite.

Pyrophyllite

Composition $Al_2Si_4O_{10}(OH)_2$.

Crystal system Monoclinic.

Habit Sometimes in radiating, spherulitic aggregates of small platy crystals; otherwise in fine-grained masses.

Specific gravity 2·7–2·9.

Hardness 1–1½.

Cleavage {001}, perfect basal cleavage.

Colour and transparency White or pale yellow, frequently stained by iron oxides.

Lustre Pearly on cleavage surfaces.
Distinguishing features From talc by chemical tests for aluminium.
Formation and occurrence In low- and medium-grade metamorphic rocks rich in aluminium.
Uses As a carrier for insecticides; dusting powders generally contain the mineral.

Talc (Steatite, Soapstone)

Composition $Mg_3Si_4O_{10}(OH)_2$. Composition of talc is very uniform. The extreme softness results from the absence of bonding except by weak Van der Waals forces between the layers.
Crystal system Monoclinic.
Habit Crystals rare; usually as granular or foliated masses.
Specific gravity 2·6–2·8.
Hardness 1.
Cleavage {001}, perfect basal cleavage.
Colour and transparency White, grey or pale green, sometimes stained red-brown by iron oxides; translucent.
Streak White to very pale green.
Lustre Dull, pearly on cleavage surface.
Distinguishing features Extreme softness; soapy feel; colour.
Formation and occurrence As a secondary mineral formed as a result of the alteration of olivine, pyroxene and amphibole, being often derived from ultrabasic igneous rocks. Often in schists produced by low- or medium-grade metamorphism of magnesian rocks, in association with actinolite. Less commonly as a result of thermal metamorphism of dolomitic limestones.
Uses Well known as talcum and face powders; also in ceramics, electrical porcelain, filler in paint, paper and rubber; slabs of soapstone are used for electrical switchboards, acid-proof sinks.

Montmorillonite

Composition $Al_2Si_4O_{10}(OH)_2.nH_2O$. Composition of montmorillonite always deviates from the ideal formula through substitutions such as magnesium for aluminium, and aluminium for silicon.
Crystal system Monoclinic.
Habit Always in earthy masses; clay-like.
Specific gravity 2·0–3·0.
Hardness $1\frac{1}{2}$–$2\frac{1}{2}$.
Cleavage {001}, perfect basal cleavage.
Colour Usually grey, white, yellow, pink or brown.
Streak White.
Lustre Greasy or dull.
Distinguishing features Clay-like character and soapy feel; and the property of swelling and forming gel-like masses in water.
Formation and occurrence Commonly by the alteration of beds of volcanic ash. Bentonite is a rock consisting largely of montmorillonite.
Uses In drilling muds; also in moulding sands for foundries.

Chlorite Group

Composition $(Mg,Fe,Al)_6(Al,Si)_4O_{10}(OH)_8$. The chlorites form an extensive isomorphous series. In the general formula, magnesium and iron are mutually replaceable. Ferric iron is often present. Some varieties of chlorite contain appreciable chromium, nickel or manganese.
Crystal system Monoclinic.
Habit Crystals tabular pseudohexagonal, rarely prismatic; also as scaly aggregates, and massive, earthy.
Specific gravity 2·6–3·3 (increasing with iron content).
Hardness 2–3.
Cleavage and fracture {001}, perfect basal cleavage; flakes flexible but not elastic.
Colour Characteristically green; manganese-bearing varieties orange to brown; chromium-bearing varieties violet.
Streak White, pale green.
Lustre Vitreous to earthy.
Distinguishing features Colour; cleavage; non-elastic nature of cleavage flakes.
Formation and occurrence In igneous rocks as an alteration product of such minerals as pyroxenes, amphiboles and micas. Also occurs infilling amygdales in lavas. Characteristic of low-grade metamorphic rocks and is present in the clay mineral fraction of many sediments.
Uses Chamosite, an iron-rich chlorite, is an important constituent of some sedimentary iron ores.

Kaolinite

Composition $Al_2Si_2O_5(OH)_4$. Kaolinite is the commonest of four polymorphs, the others being dickite, nacrite and halloysite.
Crystal system Triclinic.
Habit Microscopic pseudohexagonal platy crystals, usually in earthy aggregates.
Specific gravity 2·6–2·7.
Hardness 2–$2\frac{1}{2}$.
Cleavage Basal {001}, perfect cleavage.
Colour White, sometimes stained brown or grey.
Streak White.
Lustre Dull; earthy crystalline plates pearly.
Distinguishing features Plastic feel; kaolinite cannot be distinguished from other clay minerals without optical or chemical tests.
Formation and occurrence Secondary mineral produced by the alteration of aluminous silicates, particularly of alkali feldspars.
Uses An important industrial mineral used as a filler in paper and an essential raw material in the manufacture of ceramics.

Serpentine Group

Serpentine, as a mineral name, applies to material containing one or more of the minerals chrysotile, antigorite, and lizardite. Their structures are similar to those of the kaolin minerals, antigorite being the analogue of kaolinite, and chrysotile the analogue of halloysite.
Composition $Mg_3Si_2O_5(OH)_4$. The composition of serpentine generally corresponds closely to the above formula; some iron may be present replacing magnesium and some aluminium.
Crystal system Monoclinic.
Habit Antigorite generally has a lamellar or platy structure; crystals virtually unknown. Chrysotile is fibrous.

Specific gravity 2·5–2·6.
Hardness Variable 2½–4.
Cleavage and fracture Basal, perfect (antigorite and lizardite); none in fibrous chrysotile.
Colour and transparency Various shades of green, also brownish; translucent to opaque.
Streak White.
Lustre Waxy or greasy in massive varieties, silky in fibrous material.
Distinguishing features Colour; lustre; smooth rather greasy feel; habit (chrysotile).
Formation and occurrence By the alteration of olivine and enstatite under conditions of low- and medium-grade metamorphism. In igneous rocks containing these minerals; but typically in serpentinites which have formed by the alteration of olivine-bearing rocks. The mines of south-east Quebec are some of the world's largest producers.
Uses Chrysotile is the most valued type of asbestos. Greatest amounts are used in asbestos-cement products. Massive serpentine is sometimes cut and polished as an ornamental stone. Garnierite, a nickel-bearing variety of serpentine, apple green in colour is an alteration product of nickel-rich peridotite mined as a residual nickel ore in New Caledonia, Pacific Ocean.

Apophyllite

Composition $KCa_4Si_8O_{20}(F,OH).8H_2O$. Analysis generally shows a little sodium replacing potassium.
Crystal system Tetragonal.
Habit Usually in good crystals of varied habit; combinations of prism, bipyramid, and pinacoid are most common.
Twinning On $\{111\}$, rare.
Specific gravity 2·3–2·4.
Hardness 4½–5.
Cleavage and fracture Basal $\{001\}$ perfect, prismatic, $\{110\}$ poor cleavage. Uneven fracture.
Colour and transparency Colourless, white or grey, sometimes pink to yellow; transparent to translucent.
Streak White.
Lustre Pearly parallel to basal cleavage; elsewhere vitreous.
Distinguishing features Habit; cleavage; lustre;

the basal pinacoid faces are often rough and pitted.
Formation and occurrence In association with zeolites in cavities in basalt; in association with prehnite, calcite, analcime, stilbite and other zeolites. Less commonly in cavities in granite, gneiss and limestone. Also in some hydrothermal veins.

Quartz

Composition SiO_2. Other elements are present in only trace amounts in the crystal structure; most quartz is close to pure silicon dioxide, although inclusions of other minerals may be common.
Crystal system Hexagonal (Trigonal).
Habit Crystals are usually six-sided prisms terminated at each end by six faces; prism faces may be striated at right angles to the length of the crystal; other faces may modify the form; misshapen crystals are common; shapes range from very elongate to equant; also massive.
Twinning Most crystals are twinned, but this may be difficult to detect; the three common types are the Dauphiné and the Brazil (penetration twins), and the Japan (contact twin); only the last type has visible re-entrant angles.
Specific gravity 2·65.
Hardness 7.
Fracture Conchoidal fracture.
Colour and transparency Usually colourless or white. Coloured varieties (due to a variety of factors including the presence of impurities and natural irradiation) are often used as semiprecious stones and are named according to colour: amethyst – purple; rose quartz – pink; citrine – yellow; smoky quartz – brown to almost black; milky quartz – white. Transparent to translucent.
Streak White.
Lustre Vitreous.
Distinguishing features Habit; lack of cleavage; hardness; lustre. Usually fresh and unaltered. Not attacked by acids, other than hydrofluoric.
Formation and occurrence Abundant and widely distributed, occurring in most igneous, metamorphic and sedimentary rocks, sometimes composing almost all of the rock, as in quartzites (metamorphic) and some sandstones (sedimentary). Also found abundantly as a gangue mineral in mineral veins, often as good crystals. May be iron-stained, or contain inclusions of iron oxides (as in yellow 'tiger-eye', a pseudomorph after fibrous crocidolite) or other minerals such as rutile, tourmaline or mica. 'Aventurine' is a variety containing brilliant scales of hematite or mica. Silicon dioxide that has crystallized at high temperatures and has kept the original crystal structure has properties different from those of quartz and constitutes a series of separate minerals. The two most common are: tridymite, which is orthorhombic; and cristobalite, which is tetragonal; they are found typically in volcanic rocks which have cooled too quickly for the crystal structure to be re-arranged.
Uses As a semiprecious stone or ornamental material. In the form of sand, in the construction industry and as a flux or abrasive in industrial processes. Pure forms are used in optical equipment such as microscopes, and as oscillators in electrical equipment (like some watches).

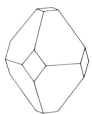

Apophyllite: combination of prism, bipyramid and pinacoid.

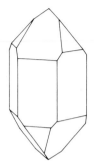

Quartz.

Quartz: showing striated prism faces.

Quartz: right-handed form.

Blue crystals of apophyllite on matrix; from Brazil.

Quartz: left-handed form.

Quartz: Japan twin.

Quartz: Dauphiné twin.

Quartz: Brazil twin.

Geode lined with agate and pale amethyst crystals; from Mexico.

Quartz (variety Chalcedony)

Composition SiO$_2$.
Habit Massive, as mamillated, botryoidal or stalactitic forms of fibrous cryptocrystalline silica with submicroscopic pores.
Specific gravity About 2·6.
Hardness About 6½.
Fracture Conchoidal fracture.
Colour and transparency Chalcedony is distinguished from agate by its lack of obvious colour banding, but exhibits the same range of colours, from colourless to white, grey, red, brown, green or yellow. Some coloured varieties have special names: carnelian – red; sard – brown; chrysoprase – apple green; heliotrope (bloodstone) – green with spots of red jasper. Transparent to translucent.
Lustre Vitreous to waxy.
Distinguishing features Occurrence; habit.
Formation and occurrence Lining or filling cavities or fissures in rocks, having been deposited from silica-bearing aqueous solutions. May also replace other substances, such as fossil shells or wood. Massive, cryptocrystalline quartz similar to chalcedony but with a granular rather than fibrous structure is very common, usually occurring in dull, often dark nodules or thin beds in sedimentary

sequences. It may be formed directly by deposition of silica on the sea-floor, or by redistribution and concentration of silica by percolating waters. Varieties are again named according to colour and appearance: flint—dark brown to black, translucent; chert – light in colour, often subtranslucent to opaque; jasper – red, opaque (due to inclusions of hematite); prase – dull green, opaque.
Uses As an ornamental material and semiprecious stone.

Quartz (variety Agate)

Composition SiO$_2$.
Habit Usually as concentric or irregular layers lining a cavity (and so forming a geode). The centre part may be partially or completely filled, sometimes by parallel layers of agate at an angle to the outer layers, sometimes by quartz crystals, and occasionally by the water-rich solutions from which the agate crystallized. The agate layers may be formed of chalcedony, granular quartz or opal.
Specific gravity About 2·6.
Hardness 6½–7.
Fracture Conchoidal fracture.
Colour Natural agate is variegated in shades from colourless, white or grey, to browns, reds, greens or black; much commercial agate is coloured artifi-

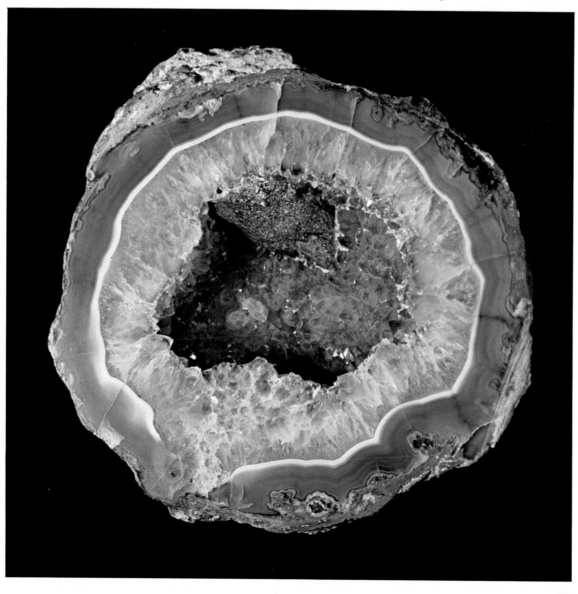

cially. Parallel-banded agate in shades of white with black or brown, or with brownish red, is named onyx and sardonyx respectively. Moss agate contains visible mineral impurities, such as manganese oxide and chlorite, in moss-like patterns.

Distinguishing features Habit; hardness.
Formation and occurrence Massive, crypto-crystalline form of quartz. Typically in volcanic lavas as a cavity filling, having been deposited by silica-rich aqueous solutions; the colour variation and banding may be due to slight changes in the composition of the solutions and of the physical conditions in the lava during deposition.
Uses As an ornamental and semiprecious stone. Also in some grinding apparatus.

Opal

Composition $SiO_2.nH_2O$. Opal is a hydrous submicrocrystalline form of cristobalite.
Habit Massive; as veinlets, or stalactitic, botryoidal or rounded forms; may replace other substances such as wood buried by volcanic tuff (hot ash falls).
Specific gravity Variable $1\cdot8-2\cdot3$.
Hardness $5\frac{1}{2}-6\frac{1}{2}$.
Fracture Conchoidal fracture.
Colour and transparency Variable, from colourless, through milky white, grey, red, brown, blue, green to nearly black. Much of the colour in 'precious' opal is not due to impurities but results from the splitting up of light into its component colours by the internal structures of the mineral. 'Common' opal does not have these internal reflections. Transparent to translucent.
Lustre Vitreous; often resinous.
Distinguishing features Lustre; low specific gravity; hardness.
Formation and occurrence Found lining and filling cavities and fissures in igneous and sedimentary rocks, where it has been deposited from silica-bearing waters, occurring especially in areas of geysers and hot springs. It may replace organic material such as wood, and forms the skeletons of some small sea organisms which may accumulate on their death to form a fine-grained sedimentary rock rather like chalk in appearance (diatomaceous earth). It may also form during the weathering and decomposition of rocks.
Uses As a gemstone; diatomaceous earth is used as an abrasive, filler, filtration powder, and in insulation products.

Feldspar Group

The feldspars are the most abundant of all minerals being widely distributed in igneous, metamorphic and sedimentary rocks. They are closely related in form and physical properties but they fall into two subgroups: the potassium and barium feldspars, which are monoclinic or nearly monoclinic in symmetry, and the sodium and calcium feldspars (the plagioclases) which are triclinic. The general formula is $X(Al,Si)_4O_8$ in which X is sodium, potassium, calcium or barium. In the plagioclases, sodium and calcium can substitute for one another. Twinning is common: Carlsbad, Manebach and Baveno twins are simple; albite and pericline twins

are repeated. Twinning shows as a difference in reflectivity of two halves of a crystal in the case of a single twin or as a series of parallel striations in a repeated twin. Repeated albite twinning is common in the plagioclases. Potassic feldspar alters readily to clay minerals, especially kaolinite, and the plagioclases usually alter to clay minerals or sericite.

Potassic Feldspars
Orthoclase and Sanidine

Composition $KAlSi_3O_8$.
Crystal system Monoclinic.
Habit Crystals usually short prismatic, somewhat flattened parallel to {010} or elongated with {010} and {001} prominent.
Twinning Common on the Carlsbad law (composition plane {010}); Baveno (composition plane {021}; and Manebach (composition plane {001}).
Specific gravity $2\cdot5-2\cdot6$.
Hardness $6-6\frac{1}{2}$.
Cleavage and fracture {001}, perfect, {010}, good cleavages; conchoidal to uneven fracture.
Colour and transparency Sanidine is colourless to grey; transparent. Orthoclase is white to flesh-pink, occasionally red; translucent to subtranslucent.
Streak White.
Lustre Vitreous; sometimes pearly on cleavage surfaces.
Distinguishing features Colour; cleavages; twinning; hardness. Sanidine and orthoclase can be distinguished from the plagioclase feldspars by the absence of twinning striations. The distinction from microcline is not easily made in hand specimen except by occurrence, because much of the potassic feldspar of igneous rocks is orthoclase, but practically all that of pegmatites and hydrothermal veins is microcline. Sanidine is distinguished by its colourless, transparent appearance, tabular habit, and occurrence.
Formation and occurrence Sanidine is the high-temperature form of $KAlSi_3O_8$ and it occurs as phenocrysts in volcanic rocks such as rhyolite and trachyte. Also occurs in rocks that have been thermally metamorphosed at high temperatures. Orthoclase is the common potassic feldspar of most igneous and metamorphic rocks. Also occurs as perthitic intergrowths with albite.

Adularia

Composition $KAlSi_3O_8$.
Crystal system Monoclinic.
Habit Distinctive simple crystals, usually a combination of prism terminated by two faces, which often have the appearance of rhombohedra.
Twinning Baveno twins, common.
Specific gravity $2\cdot6$.
Hardness 6.
Cleavage and fracture Two perfect cleavages; conchoidal to uneven fracture.
Colour and transparency Colourless or milky white, often with a pearly sheen or play of colours (moonstones); transparent to translucent.
Streak White.
Lustre Vitreous.

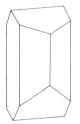

Orthoclase microcline: prismatic habit.

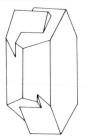

Orthoclase/ microcline: Carlsbad twin.

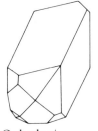

Orthoclase/ microcline: Baveno twin.

Orthoclase/ microcline: Manebach twin.

Green prismatic microcline crystals (variety amazonstone); from Florissant, Colorado.

Distinguishing features Habit; occurrence.
Formation and occurrence At low temperatures and occurs in hydrothermal veins.

Microcline

Composition KAlSi$_3$O$_8$.
Crystal system Triclinic.
Habit Usually short prismatic, similar to orthoclase.
Twinning Usually polysynthetically on the albite and pericline laws.
Specific gravity 2·5–2·6.
Hardness 6.
Cleavage {001}, perfect, {010} good cleavages.
Colour White, cream, pink; sometimes bright green (amazonstone).
Streak White.
Lustre Vitreous, sometimes pearly on cleavage surfaces.
Distinguishing features Distinguished from orthoclase only by its optical properties, although the variety amazonstone is easily identified by its colour.
Formation and occurrence Usually formed at lower temperatures than is orthoclase and is the common potassic feldspar in pegmatites and hydrothermal veins. Crystals a metre or more long are recorded from some pegmatites. Also occurs in some metamorphic rocks.
Uses Large quantities of microcline and microcline perthite are mined from granite pegmatites and used in the manufacture of glass, porcelain and enamel. Amazonstone is sometimes cut and polished as an ornamental stone.

Anorthoclase

Composition (Na,K)AlSi$_3$O$_8$.
Crystal system Triclinic.
Habit Crystals usually short prismatic, blocky;

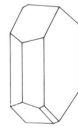

Plagioclase.

also tabular flattened; commonly massive, cleavable to granular.
Twinning Very common, on Carlsbad, Baveno and Manebach laws.
Specific gravity 2·56–2·62.
Hardness 6–6½.
Cleavage and fracture {001} and {010} perfect cleavages; uneven fracture.
Colour and transparency Colourless, white, grey, yellow and reddish; transparent to translucent.
Streak White.
Lustre Vitreous; pearly on cleavages.
Distinguishing features Occurrence.
Formation and occurrence Mainly in volcanic rocks such as andesite, phonolite and trachyte.

Plagioclase Feldspars

Composition NaAlSi$_3$O$_8$ – CaAl$_2$Si$_2$O$_8$. The chemistry changes progressively from albite (NaAlSi$_3$O$_8$) through oligoclase-andesine-labradorite and bytownite, to anorthite (CaAl$_2$Si$_2$O$_8$).
Crystal system Triclinic.
Habit Crystals prismatic or tabular; also massive, granular.
Twinning Repeated twinning is common on albite and pericline laws, as are simple twins on Carlsbad, Baveno and Manebach laws; both simple and repeated twinning may occur in one individual.
Specific gravity 2·6–2·8.
Hardness 6–6½.
Cleavage and fracture Good cleavages on {001}, {010}; uneven fracture.
Colour and transparency Usually white or off-white; sometimes pink, greenish or brownish; transparent to translucent.

Streak White.
Lustre Vitreous; sometimes pearly on cleavage surfaces.
Distinguishing features From potassic feldspars by the presence of repeated albite twin lamellae visible on one of the cleavage surfaces. Labradorite on cleavages commonly shows a play of colours in shades of blue and green.
Formation and occurrence The plagioclase feldspars are widely distributed minerals. They occur in many igneous rocks and are used as a basis of igneous rock classification. In general, the sodic plagioclases are characteristic of granitic igneous rocks, and calcic plagioclases of basalts and gabbros. Between potassic feldspar and albite there exists a continuous series as sodium substitutes for potassium; this series is called the alkali feldspar series. In places there are large masses of almost pure oligoclase-andesine rock called anorthosites. Albite is commonly found in pegmatites and in sodic lavas called spilites. Plagioclase is common in metamorphic rocks and occurs as detrital grains in sedimentary rocks.
Uses Albite and oligoclase are occasionally mined from pegmatites and used in ceramics.

Celsian

Composition $BaAl_2Si_2O_8$.
Crystal system Monoclinic.
Habit Short, prismatic crystals with prominent prism faces. Also massive.
Twinning Carlsbad, Manebach, Baveno, common.
Specific gravity 3·10–3·45.
Hardness 6–6½.
Cleavage and fracture {001}, perfect, {010} good cleavages; brittle, uneven fracture.
Colour and transparency Colourless, white, yellow; transparent.
Streak White.
Lustre Vitreous.
Distinguishing features Occurrence. (Celsian is dimorphous with paracelsian which occurs in a band in shale and sandstone at the Benallt manganese mine, Rhiw, Gwynedd, Wales.)
Formation and occurrence Chiefly in the contact zones of manganese deposits.

Feldspathoid Group

The feldspathoids are a group of sodium and potassium aluminosilicates which appear in place of the feldspars when an alkali-rich magma is deficient in silica. Therefore, they should never occur with free quartz.

Nepheline

Composition $NaAlSiO_4$. Natural nephelines always contain some potassium replacing sodium. A little calcium is generally present.
Crystal system Hexagonal.
Habit Crystals usually six-sided prisms; more commonly as shapeless or irregular grains.
Twinning On {10$\bar{1}$0}.
Specific gravity 2·6–2·7.
Hardness 5½–6.
Cleavage and fracture Prismatic {10$\bar{1}$0} and basal {0001}, indistinct cleavages; conchoidal fracture.
Colour and transparency Usually colourless, white or grey, but also brownish red or greenish; transparent to translucent.
Streak White.
Lustre Vitreous to greasy.
Distinguishing features Lustre; gelatinizes in hydrochloric acid.
Formation and occurrence Characteristic of silica-poor alkaline igneous rocks of both plutonic and volcanic associations; found, therefore, in nepheline syenites, ijolites and in lavas.

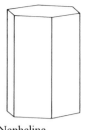

Nepheline.

Kalsilite

Composition $KAlSiO_4$.
Crystal system Hexagonal.
Habit Massive, compact.
Specific gravity 2·6.
Hardness 6.
Cleavage and fracture Prismatic, {10$\bar{1}$0}, and basal {0001} poor cleavages; subconchoidal fracture.
Colour and transparency Colourless, white, grey; transparent to translucent.
Streak White.
Lustre Vitreous to greasy.
Distinguishing features Occurrence.
Formation and occurrence As complex phenocrysts and as a constituent of the groundmass of certain alkali lavas.

Cancrinite

Composition $(Na,Ca)_{7-8}Al_6Si_6O_{24}(CO_3,SO_4, Cl)_{1.5-2.0} \cdot 1-5H_2O$. Composition is somewhat variable; cancrinite always contains some calcium, and often small amounts of sulphate and chloride.
Crystal system Hexagonal.
Habit Crystals rare but usually prismatic; generally massive or as discrete grains and veinlets.
Specific gravity 2·4–2·5.
Hardness 5–6.
Cleavage and fracture Prismatic {10$\bar{1}$0}, perfect cleavage; uneven fracture.
Colour and transparency White, grey, pink, yellow, blue; transparent to translucent.
Streak White.
Lustre Vitreous, pearly or greasy.
Distinguishing features Colour, occurrence. From other feldspathoids by effervescence in warm dilute hydrochloric acid.
Formation and occurrence As a primary mineral in certain alkali rocks, typically in nepheline-syenites, and is a common alteration product of nepheline. It occurs in some carbonatites and in some contact metamorphosed limestones.

Sodalite

Composition $Na_8Al_6Si_6O_{24}Cl_2$.
Crystal system Cubic.
Habit Crystals rare, sometimes as small dodecahedral crystals; commonly massive.
Twinning On {111}, common.
Specific gravity 2·3.
Hardness 5½–6.
Cleavage and fracture Dodecahedral {110}, poor cleavage, uneven to conchoidal fracture.
Colour and transparency Commonly azure blue, also pink, yellow, green or grey-white;

White
icositetrahedron of
leucite in volcanic
rock; from Ariccia,
Italy.

transparent to translucent.

Streak White.

Lustre Vitreous.

Distinguishing features Colour. From lazurite by occurrence and absence of associated pyrite.

Formation and occurrence With nepheline and cancrinite in alkaline igneous rocks such as nepheline syenites; also in some silica-poor dyke rocks and lavas.

Uses Cut and polished as an ornamental stone.

Haüyne and Nosean

Composition $(Na,Ca)_{4-8}Al_6Si_6O_{24}(SO_4)_{1-2}$ and $Na_8Al_6(SiO_4)_6SO_4$.

Crystal system Cubic.

Habit Crystals usually dodecahedral or octahedral; commonly as rounded grains.

Twinning On $\{111\}$ common, sometimes as penetration twins.

Specific gravity Haüyne 2·4–2·5; nosean 2·3–2·4.

Hardness $5\frac{1}{2}$–6.

Cleavage and fracture Dodecahedral, poor cleavage, uneven to conchoidal fracture.

Colour and transparency Often blue, also grey, brown, yellow-green; transparent to translucent.

Streak White.

Lustre Vitreous, to greasy.

Distinguishing features Colour; association.

Formation and occurrence In silica-poor lavas such as phonolites and related igneous rocks in association with leucite or nepheline.

Lazurite (Lapis-Lazuli)

Composition $(Na,Ca)_8(Al,Si)_{12}O_{24}(S,SO_4)$. Lazurite is isomorphous with sodalite but has sulphide ions in place of chloride ions in the structure.

Crystal system Cubic.

Habit Crystals rare, usually cubes or dodecahedra; commonly massive

Specific gravity 2·4.

Hardness 5–$5\frac{1}{2}$.

Fracture Uneven fracture.

Colour and transparency Azure blue; translucent.

Streak Bright blue.

Lustre Vitreous.

Distinguishing features Colour, association (c.f. sodalite).

Formation and occurrence As a contact metamorphic mineral in limestones, often in association with calcite and enclosing small grains of pyrite.

Uses Powdered lazurite is the source of the pigment ultramarine. Lapis-lazuli is a rock rich in lazurite, and is used in jewellery and as a decorative stone. Lapis has been mined since ancient times in the Badakshan region of Afghanistan.

Leucite

Composition $KAlSi_2O_6$. A small amount of potassium may be replaced by sodium.

Crystal system Tetragonal (pseudocubic) at ordinary temperatures; cubic above 625 °C.

Habit Crystals nearly always icositetrahedra.

Specific gravity 2·5.

Hardness $5\frac{1}{2}$–6.

Cleavage and fracture Very poor cleavage; conchoidal fracture.

Colour Colourless, white or grey.

Streak White.

Lustre Vitreous.

Distinguishing features Habit; occurrence. Analcime also crystallizes as icositetrahedra but occurs typically in cavities, not as embedded crystals. Leucite often alters to pseudoleucite, a pseudomorph consisting of nepheline, analcime and orthoclase.

Formation and occurrence Unstable at high pressures and, therefore, has a restricted occurrence. Typically in potassium-rich, silica-poor lavas such as certain trachytes. Fresh leucite does not occur in plutonic igneous rocks.

Scapolite Group

Composition $(Na,Ca,K)_4Al_3(Al,Si)_3Si_6O_{24}$ (Cl,F,OH,CO_3,SO_4). The scapolite series ranges in composition from sodium-rich (variety marialite) to calcium-rich (variety meionite).

Crystal system Tetragonal.

Habit Crystals usually prismatic, often with uneven faces; also granular, massive.

Specific gravity 2·5–2·8, increasing with calcium content.

Hardness 5–6.

Cleavage and fracture Good, prismatic cleavages $\{100\}$ $\{110\}$ impart a splintery appearance to massive scapolite; subconchoidal fracture.

Colour and transparency Usually white or grey, sometimes pink, yellow or brownish; transparent to translucent.

Streak White.

Lustre Vitreous to pearly.

Distinguishing features Blocky appearance; colour; cleavage.

Formation and occurrence Typically in metamorphosed limestones. The minerals also occur in skarns close to igneous contacts. Also in schists and gneisses, sometimes replacing plagioclase.

Uses Rarely cut as a gemstone. Fibrous inclusions in scapolite produce a fine chatoyancy, similar to that of moonstone.

Pollucite

Composition $(Cs,Na)AlSi_2O_6.nH_2O$.

Crystal system Cubic.

Habit Cubic crystals, sometimes dodecahedral; usually massive often deeply etched and corroded, fine-grained.

Specific gravity 2·9.

Hardness $6\frac{1}{2}$–7.

Fracture Conchoidal to uneven fracture.
Colour and transparency Colourless, white, grey, sometimes tinted pale pink, blue or violet; transparent.
Streak White.
Lustre Vitreous, slightly greasy.
Distinguishing features Association.
Formation and occurrence In granite pegmatites often as large segregations in association with microcline, quartz, amblygonite, spodumene and lepidolite.

Zeolite Group

Zeolites form a well-defined group of minerals closely related to one another in composition, in conditions of formation and mode of occurrence. They are all hydrated aluminosilicates, chiefly of sodium and calcium, less commonly potassium, barium and strontium. The zeolites, however, are not related in crystal structure but comprise a number of independent species, diverse in structure, and distinct in composition. Zeolites have a framework structure enclosing pores occupied by cations and water molecules, both of which have considerable freedom of movement permitting, within limits, reversible ion exchange and reversible dehydration. Natural zeolites show variability of composition in the ratio of silicon and aluminium and, with the exception of analcime, cation contents. The voids created by the loss of water have considerable specific surface area, which permits reversible absorption of molecules small enough to pass the apertures of the channels. Zeolites are, therefore, frequently used for removing water from other compounds, for example, drying natural gas, and in the petrochemical industry for hydrocarbon removal. They are also used as catalyst supports, providing large surface areas in chemical reactions.

Analcime (Analcite)

Composition $NaAlSi_2O_6.H_2O$. As much as a quarter of the sodium may be replaced by potassium in analcime formed at high temperatures, but little substitution takes place at low temperatures.
Crystal system Cubic.
Habit Crystals usually icositetrahedral; also granular and massive.
Specific gravity 2·2–2·3.
Hardness 5–5½.
Cleavage and fracture Cubic, very poor cleavage; subconchoidal fracture.
Colour and transparency Colourless, white or grey, sometimes tinged with pink, yellow or green; transparent to translucent.
Streak White.
Lustre Vitreous.
Distinguishing features Mode of occurrence.
Formation and occurrence Mainly with other zeolites as a secondary mineral in cavities in basaltic rocks and in sedimentary rocks as a secondary mineral. Occasionally as a primary mineral of silica-deficient igneous rocks.

Chabazite

Composition $CaAl_2Si_4O_{12}.6H_2O$. Often contains a little sodium and potassium replacing calcium.
Crystal system Hexagonal (Trigonal).
Habit Usually in simple rhombohedral crystals which look like cubes.
Twinning Common; interpenetrant.
Specific gravity 2·05–2·15.
Hardness 4–5.
Cleavage and fracture Rhombohedral $\{10\bar{1}1\}$, poor cleavage; uneven fracture.
Colour and transparency Usually white, yellow, often pinkish or red; transparent to translucent.
Streak White.
Lustre Vitreous.
Distinguishing features Habit; hardness; does not effervesce in acid.
Formation and occurrence Typically lining cavities in basalts and andesites.

Natrolite

Composition $Na_2Al_2Si_3O_{10}.2H_2O$.
Crystal system Orthorhombic (pseudotetragonal).

Analcime: icositetrahedron.

Chabazite: rhombohedral habit.

Natrolite.

Pinkish rhombohedral chabazite crystals in a cavity in basalt; from West Paterson, New Jersey.

Habit Prismatic crystals, commonly elongated and needle-like; also in radiating nodular forms and compact masses.
Twinning Rare.
Specific gravity 2·2–2·3.
Hardness 5–5½.
Cleavage Prismatic {110}, perfect cleavage.
Colour and transparency Colourless to white, grey, yellow, red; transparent to translucent.
Streak White.
Lustre Vitreous.
Distinguishing features Habit. (Mesolite and scolecite are fibrous zeolites of a similar composition and occurrence to natrolite; they are both monoclinic but fibrous in habit, and it is difficult to distinguish between them in hand specimen).
Formation and occurrence Typically in the cavities of basaltic rocks.

Radiating white, fibrous natrolite on green botryoidal prehnite; from Prospect Park, New Jersey.

Mesolite

Composition $Na_2Ca_2Al_6Si_9O_{30}.8H_2O$.
Crystal system Monoclinic.
Habit Crystals acicular or fibrous; usually as compact masses of radiating crystals.
Twinning Always twinned, on {100}.
Specific gravity 2·2–2·3.
Hardness 5.
Cleavage and fracture {101}, {10ī}, perfect cleavages; uneven fracture; compact masses, tough.
Colour and transparency Colourless, white; transparent.
Streak White.
Lustre Vitreous, or silky when fibrous.
Distinguishing features Habit. Optical determination usually has to be made to distinguish between the various fibrous zeolites.
Formation and occurrence In cavities in volcanic rocks, usually in association with other zeolites.

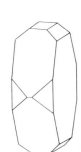

Heulandite.

Thomsonite

Composition $NaCa_2(Al,Si)_{10}O_{20}.6H_2O$.
Crystal system Orthorhombic (pseudo-tetragonal).
Habit Acicular crystals in radiating or divergent aggregates.

Twinning On {110}.
Specific gravity 2·1–2·4.
Hardness 5–5½.
Cleavage and fracture {010}, perfect, {100}, distinct cleavages; uneven fracture.
Colour and transparency White, sometimes tinged with red; transparent to translucent.
Streak White.
Lustre Vitreous to pearly.
Distinguishing features Similar to natrolite but usually slightly more coarsely crystalline.
Formation and occurrence In cavities in basalts and related igneous rocks. Also as a decomposition product of nepheline.

Laumontite

Composition $CaAl_2Si_4O_{12}.4H_2O$.
Crystal system Monoclinic.
Habit As small prismatic crystals often with oblique terminations; also massive, or as columnar and radiating aggregates.
Twinning Frequently on {100}, sometimes as 'swallow-tail' twins.
Specific gravity 2·2–2·4.
Hardness 3–4.
Cleavage and fracture Prismatic, {110}, pinacoidal {010}, perfect cleavages; uneven fracture.
Colour and transparency White, sometimes reddish; transparent to translucent.
Streak White.
Lustre Vitreous, pearly on cleavage surfaces.
Distinguishing features Habit. Characteristic alteration. Laumontite loses part of its water on exposure to dry air and becomes powdery, friable and chalky (variety leonhardite); most specimens in collections have altered to leonhardite.
Formation and occurrence With other zeolites in veins and amygdales in igneous rocks. It is produced as a result of very low-grade metamorphism of some sedimentary rocks and tuffs.

Heulandite

Composition $(Ca,Na_2)Al_2Si_7O_{18}.6H_2O$. Heulandite commonly contains some sodium substituting for calcium in the structure.
Crystal system Monoclinic.
Habit Crystals usually tabular parallel to {010}; trapezoidal.
Specific gravity 2·1–2·2.
Hardness 3½–4.
Cleavage and fracture {010}, perfect cleavage.
Colour and transparency Colourless, white, grey, pink, red or brown; transparent to translucent.
Streak White.
Lustre Vitreous, pearly on cleavage surfaces.
Distinguishing features Habit; lustre.
Formation and occurrence Common mineral associated with other zeolites, especially stilbite in cavities in basaltic rocks, and in sedimentary rocks as a secondary mineral.

Harmotome

Composition $BaAl_2Si_6O_{16}.6H_2O$.
Crystal system Monoclinic.
Habit Crystals usually twins, having a pseudo-orthorhombic or pseudotetragonal appearance.
Twinning Very common, interpenetrant.

Specific gravity 2·4–2·5.
Hardness 4½.
Cleavage and fracture One good cleavage; uneven to subconchoidal fracture.
Colour and transparency White, yellow, pink, reddish-brown; transparent to translucent.
Streak White.
Lustre Vitreous.
Distinguishing features Habit; occurrence.
Formation and occurrence Chiefly in cavities in basalts and related igneous rocks often in association with chabazite. Also in association with manganese mineralization and in lenses in gneiss. Found as fine crystals at Strontian, Argyllshire, Scotland.

Stilbite

Composition $NaCa_2Al_5Si_{13}O_{36}.14H_2O$.
Crystal system Monoclinic.
Habit Forms sheaf-like crystals which are aggregates of cruciform penetration twins; also massive or globular.
Twinning Common on {001} giving cruciform interpenetrant twins.
Specific gravity 2·1–2·2.
Hardness 3½–4.
Cleavage and fracture {010}, perfect cleavage; uneven fracture.
Colour and transparency White, sometimes yellowish or pink, occasionally brick red; transparent to translucent.
Streak White.
Lustre Vitreous, pearly on cleavage surfaces.
Distinguishing features Habit; lustre.
Formation and occurrence In cavities in basalts, commonly in association with heulandite.

Petalite

Composition $LiAlSi_4O_{10}$.
Crystal system Monoclinic.
Habit Crystals well-formed small and rare; usually massive, as large, cleavable, blocky segregations.

Twinning On {001}, polysynthetic.
Specific gravity 2·3–2·5.
Hardness 6–6½.
Cleavage and fracture {001}, perfect, {201} distinct; subconchoidal fracture.
Colour and transparency Colourless, white, grey or yellow; transparent to translucent.
Streak White.
Lustre Vitreous; pearly on cleavages.
Distinguishing features Cleavage; association.
Formation and occurrence In granite pegmatites in association with cleavelandite, quartz and lepidolite.

Prehnite

Composition $Ca_2Al_2Si_3O_{10}(OH)_2$.
Crystal system Orthorhombic.
Habit Crystals not common, usually tabular parallel to {001}; usually massive, botryoidal or stalactitic.
Specific gravity 2·90–2·95.
Hardness 6–6½.
Cleavage and fracture {001}, good cleavage; uneven fracture.
Colour Characteristically pale green, sometimes white or grey.
Streak White.
Lustre Vitreous to somewhat pearly.
Distinguishing features Colour; habit.
Formation and occurrence Prehnite occurs chiefly in cavities in basic igneous rocks, often associated with zeolites, also in low grade metamorphic rocks and after calcic plagioclase in altered igneous rocks.

Chrysocolla

Composition Near $CuSiO_3.2H_2O$. A group name with several members, hard to characterize.
Crystal system Possibly orthorhombic.
Habit Very finely fibrous or massive; sometimes botryoidal, earthy.
Specific gravity 2·0–2·5.
Hardness 2–4.
Fracture Conchoidal to uneven fracture.
Colour and transparency Various shades of blue, blue-green to green, sometimes brown to black when impure; translucent to nearly opaque.
Streak White (when pure).
Lustre Vitreous, earthy.
Distinguishing features Occurrence; colour.
Formation and occurrence As a common mineral in the oxidation zone of copper deposits.

Eudialyte

Composition $Na_4(Ca,Fe)_2\check{Z}rSi_6O_{17}(OH,Cl)_2$.
Crystal system Hexagonal (Trigonal).
Habit Crystals rhombohedral or tabular; also massive, granular.
Specific gravity 2·8–3·0.
Hardness 5–5½.
Cleavage and fracture Basal {0001} indistinct cleavage; uneven fracture.
Colour and transparency Red to brown; transparent to translucent.
Streak White.
Lustre Vitreous.
Distinguishing features Habit; colour; occurrence.

Sheaves of pale brown translucent stilbite crystals; from Paterson, New Jersey.

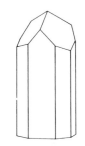

Tourmaline.

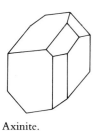

Axinite.

Formation and occurrence Typically in nepheline syenites and nepheline syenite pegmatites.

Tourmaline

Tourmaline is a general group term that applies to to several minerals with similar atomic arrangements and chemical compositions. Confusion sometimes arises between mineral species names, elbaite, schorl, dravite, and buergerite and names for coloured varieties – rubellite (pink and red), indicolite (blue), achroite (colourless), siberite (reddish violet).

Composition $Na(Mg,Fe)_3Al_6(BO_3)_3Si_6O_{18}(OH,F)_4$. Elbaite – $Na(Li,Al)_3Al_6B_3Si_6O_{27}(OH,F)_4$; schorl – $Na(Fe,Mn)_3Al_6B_3Si_6O_{27}(OH,F)_4$; dravite – $NaMg_3Al_6B_3Si_6O_{27}(OH,F)_4$. Tourmaline varies greatly in composition: magnesium and iron are interchangeable; aluminium may be partly replaced by ferric iron; sodium may be replaced by calcium; lithium is often present; and some hydroxyl may be replaced by fluorine.

Crystal system Hexagonal (Trigonal).

Habit Crystals usually prismatic, often with rounded triangular cross-sections; prism faces often strongly striated parallel to their length; the two ends of a crystal are often differently terminated; parallel or radiating crystal groups are common.

Specific gravity 3·0–3·2.

Hardness 7–7½.

Cleavage and fracture Very poor cleavage; conchoidal to uneven fracture.

Colour and transparency Usually black, especially schorl, also brown, dark blue, colourless (iron-free varieties), pink, green and blue; crystals commonly zoned, some are pink at one end, green at the other; transparent to nearly opaque.

Streak White.

Lustre Vitreous.

Distinguishing features Habit; striations; colour; triangular cross-section.

Formation and occurrence Commonly in granite pegmatites, or in granites which have undergone metasomatism by boron-rich fluids. Brown magnesium-rich tourmaline is found in metamorphosed limestones. Also common as an accessory mineral in metamorphic rocks, especially schists and gneisses.

Uses The piezoelectric properties of tourmaline are utilized industrially in the manufacture of pressure gauges. Coloured varieties as gemstones.

Axinite

Composition $(Ca,Mn,Fe,Mg)_3Al_2BSi_4O_{15}(OH)$. Composition of axinite varies considerably, as a result of varying proportions of calcium, manganese and iron. The proportion of calcium to manganese and iron is commonly about 2:1. A little magnesium may be present.

Crystal system Triclinic.

Habit Crystals usually tabular and wedge shaped; also massive, lamellar or granular.

Specific gravity 3·3–3·4.

Hardness 6½–7.

Cleavage {100}, good cleavage.

Colour and transparency Distinctive clove-brown colour, but sometimes yellow, grey or pink.

Streak White.

Lustre Vitreous.

Distinguishing features Colour; habit.

Formation and occurrence In calcareous rocks that have undergone contact metamorphism and metasomatism. Also occurs in cavities in granites and in hydrothermal veins.

Danburite

Composition $CaB_2Si_2O_8$.

Crystal system Orthorhombic.

Habit Prismatic crystals, with diamond-shaped cross-section; striated; also as granular masses.

Specific gravity 2·9–3·0.

Hardness 7.

Cleavage and fracture Basal, {001}, very indistinct; conchoidal fracture.

Colour and transparency Colourless, white, pale pink, yellow, brown; transparent to translucent.

Streak White.

Lustre Vitreous.

Distinguishing features Habit; hardness; indistinct cleavage.

Formation and occurrence With feldspar in dolomite at Danbury, Fairfield County, Connecticut, United States; also in Mexico and Japan.

Uses Occasionally cut as a gemstone.

Datolite

Composition $CaBSiO_4OH$.

Crystal system Monoclinic.

Habit Usually in short prismatic crystals, or as granular masses; crystals often show a rich variety of forms.

Specific gravity 2·8–3·0.

Hardness 5–5½.

Fracture Uneven to conchoidal fracture.

Colour and transparency Colourless, or pale shades of yellow and green, also white; transparent.

Streak White.

Lustre Vitreous.

Distinguishing features Colour, habit.

Formation and occurrence A secondary mineral, usually found in cavities in basic igneous rocks associated with zeolites, prehnite and calcite. Also in metallic veins and some granites.

Dumortierite

Composition $Al_7O_3(BO_3)(SiO_4)_3$.

Crystal system Orthorhombic.

Habit Prismatic crystals rare; usually in columnar or fibrous masses.

Specific gravity 3·3–3·4.

Hardness 8½.

Cleavage {100}, good, {110}, imperfect cleavages.

Colour and transparency Blue, violet or pink; transparent to translucent.

Streak White.

Lustre Vitreous or dull.

Distinguishing features Colour; habit.

Formation and occurrence Rare mineral which occurs in considerable quantities in a few localities. Found in aluminium-rich metamorphic rocks and occasionally in pegmatites.

Uses Dumortierite has been mined in Nevada, United States, for use in the manufacture of porcelain for spark plugs.

Allan Jobbins

Gemstones

There are a few gem materials of animal and vegetable origin such as ivory, coral, pearl, tortoiseshell and amber, but the vast majority belong to the mineral kingdom. However, by no means all minerals are suitable for use as gemstones. It is usually considered that a material must possess durability, beauty and rarity to be successfully used as a gem and few mineral specimens have all three attributes.

Durability The gemstone pebble that has survived in the alluvial gravels of a tropical river does so because it is at least as hard as the worthless quartz pebbles which surround it and it has the toughness to withstand the constant abrasion and percussion in the fluvial cycle. Thus, an ideal gemstone is harder (*see* page 79) than quartz (hardness 7) and does not cleave easily, not only because it needs to survive in nature, but because it needs also to resist the hard wear it will receive when set in a ring adorning a busy hand. Just as the abrading pebbles in a river system are largely quartz, so is the all-pervading dust so common in our everyday lives. Two examples of the fact that hardness is not equivalent to toughness may be quoted. Zircon (hardness $7\frac{1}{4}$) is harder than quartz but zircon does not wear as well in a ring because it is brittle and the edges of the facets tend to chip easily. Topaz is hard (hardness 8), but care must be taken to avoid physical or thermal shocks, which may induce fractures along the well-developed basal cleavages.

Beauty By no means all the pebbles of our tough and hard gem species will have apparent or even latent beauty. For most gemstones some degree of transparency is essential, as is freedom from flaws and disfiguring inclusions. This will reduce the usable gem material to possibly one per cent or less of the total amount recovered in mining operations. In the case of diamond, and only diamond, all material recovered can be utilized, due to the many uses of all types in industrial processes.

The complete absence of colour in diamonds (the 'white' category) produces the most desirable stone. However, in many other gemstones, for example, 'pigeon's blood' ruby, velvety green emerald and cornflower-blue sapphire, the possession of superb colour is the most sought after character.

The characters described so far may be readily apparent in the gem pebble, but there are other properties, mostly optical, which are best seen when the gemstone is cut to bring out the latent beauty.

The surface appearance of a mineral, known as the lustre, is related to its refractive index (*see* page 86) and the nature of the surface. Stones with refractive indices in the 1·5–1·6 range, such as aquamarine and quartz, tend to have a glassy or vitreous lustre, but diamond, with an index of around 2·4, has an intensely bright or adamantine lustre. Some stones with high index, such as zircon and sphalerite, have a resinous appearance, similar to that of amber, in addition to their

Well-terminated crystal of greyish-blue topaz in matrix, Alabaschka, USSR.

adamantine lustre. Stones with a low refractive index, such as fluorite, often have a watery or limpid appearance. The degree of polishing can greatly affect the surface brightness and a skilful lapidary can considerably enhance the external beauty of a stone. The surface brightness is controlled by lustre and polish but the internal brilliance of a stone, which is more obvious from a distance, is caused by internal reflections or flashes from suitably inclined surfaces or facets, the angles and arrangement of which are controlled by the lapidary when cutting the stone. By careful calculation of the facet angles it is possible to maximize the brilliance of the internal reflections in a gemstone. However, much cutting is done by the eye of experience and, in many cases, to retain maximum weight with much potential beauty being underdeveloped.

A prism will bend or refract an inclined beam of white light which strikes it and will split the light into its component colours. The resulting play of colours or fire (dispersion, *see* page 87) is commonly quoted as the B–G interval, B and G being the Fraunhofer lines near the red and violet ends of the visible spectrum. Diamond has a greater fire than any other natural colourless gemstone, but is surpassed by synthetic rutile and strontium titanate. Some coloured stones, such as demantoid garnet, sphene and cassiterite, have a greater fire than diamond.

Turning from the surface appearance, or the lustre, of a gemstone we now consider a group of varied effects caused by structures within the gemstone, which are known as sheen. A series of extremely fine alternating layers of ortho-clase and albite feldspar give rise to the beautiful soft sheen or schiller of moon-stones, the finer the layers the more bluish the effect. Schiller effects in minerals such as diopside may be caused by reflections from myriads of tiny iron ore crystals orientated in one plane. Chatoyancy or the cat's-eye effect (Fig. 6·1) is caused by point reflections building up to a single bright line from parallel microscopic fibres or hollow tubes orientated in one direction within the stone. From two or three sets of parallel fibres at angles of near 90° or 60° it is possible to get bright lines which cross to give four- and six-rayed stars respectively. Stones showing this effect (asterism) include diopside (four-rayed), corundum and quartz (six-rayed). Spinel and garnet occasionally show separate four- and six-rayed stars as the stone is rotated. Irregular inclusions of tiny shiny flakes or crystals of, for example, mica, may give reflected pin-points or spangles of bright light as in aventurine quartz and feldspar. A beautiful play of colour is also shown

Above left
Fine emerald crystal in calcite matrix, Muzo mine, Colombia.

Above
Parti-coloured tourmaline (elbaite) on matrix, Elba, Italy.

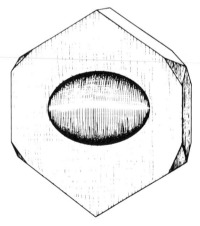

Fig. 6·1 In some chrysoberyls, extremely fine tubes run through the crystal in one direction. These produce chatoyance in cabochons cut parallel to the tubes. Stones cut in this way are called chatoyant stones or cat's-eyes.

Fig. 6·2 Some styles of cutting gemstones.

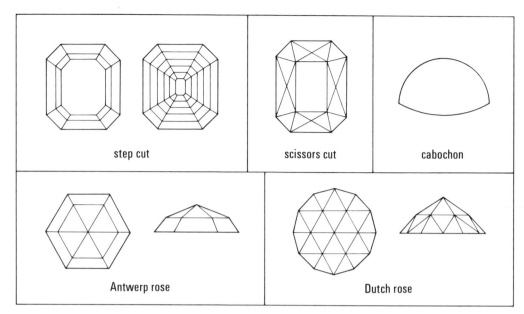

in minerals which develop fine planar cracks which may give rise to interference effects as a result of reflections from the surfaces of the thin films. Topaz may show this effect as does fire-marble and fire-agate.

Rarity This is possibly the most important factor among the gem attributes, but only when the gem already possesses the other necessary qualities. Thus, the very rare mineral painite, discovered as a deep red gem-quality material and with only two crystals known, does not compare in monetary value with, say, a fine pigeon's blood ruby. The ruby has an incomparably beautiful colour, but the painite is a less attractive brownish-red. There are other similar rubies, but there are also more buyers competing for them and the price remains high. Rubies of superb colour can be made synthetically to match the finest natural stone, but the value of the natural stone may be many thousands of times greater per carat then the synthetic stone.

Cutting and treatment

The aim in cutting a gem material is to bring out or enhance its latent beauty, whether it be colour, brilliance, fire or another desirable property, with the minimum loss of weight. Natural crystals of prismatic pale blue aquamarine or green tourmaline have beauty of form and colour but they lack brilliance. To achieve this the crystals are faceted to give internal reflections but in order to conserve weight the back facets on the shorter sides are often cut off quite sharply by the cutters. These cuts are generally known as step or trap cuts (Fig. 6·2). To avoid sharp and therefore vulnerable corners these are commonly eliminated to form an eight-sided or emerald cut, so called because this style of cutting is often used for emeralds. It has the advantage of giving some additional bright, internal reflection, while also transmitting light and, hence, giving full play to the fine body colours of emeralds, aquamarines, tourmalines and the like.

Care must be taken to avoid emphasizing the deeper or less pleasant pleochroic colours of, say, tourmaline which show when the crystal is viewed along its length, the optic axis (Fig. 6·3). Examples viewed across the crystal may appear an attractive yellowish-green but show a dirty green or brown colour down the length of the crystal. With any elongate crystal it is desirable to cut stones with their longer axes similarly extended – square or circular stones could be wasteful if the crystals are entirely of gem quality material. However, if the best colour is seen when the table facet is normal to the optic axis, as in ruby and sapphire,

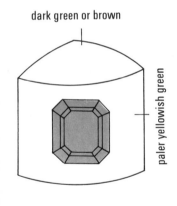

dark green or brown

paler yellowish green

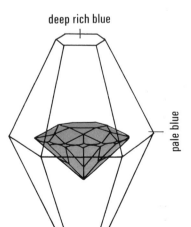

deep rich blue

pale blue

Fig. 6·3 The correct orientation of a mineral is important in cutting; tourmaline above, ruby/ sapphire below.

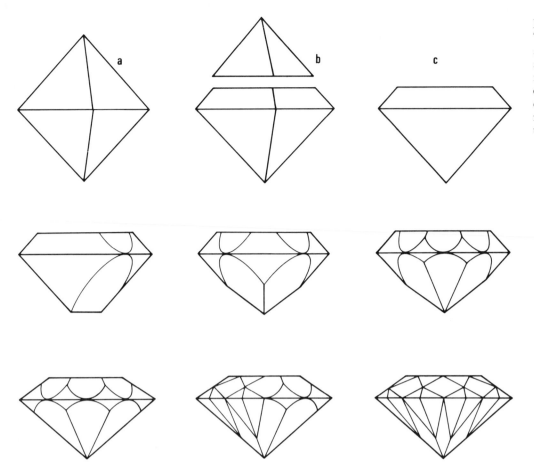

Fig. 6·4 Cutting a brilliant. Top row: (**a**) octahedron (**b**) after sawing (**c**) round shape after bruting. Middle row: the first facets are applied, in the order shown, by the cross-cutter. Bottom row: the remaining facets are added by the brillianteerer.

these circular or square stones may be more appropriate.

In the examples so far discussed, the gem material is not supremely hard and there is little trouble cutting with carborundum (SiC, hardness $9\frac{1}{2}$) or diamond tools. However, diamond cutting is different from normal lapidary work and two major factors have to be taken into account. The first is that diamond can only be cut effectively in certain directions and, therefore, the orientation of the stone is critical when cutting is planned (Fig. 6·4). Cutting in other directions can be difficult or almost impossible and could be even more wasteful of the diamond powder with which diamonds have to be cut. The second factor

Corundum group showing complete, fine Burma ruby and blue sapphire and range of other sapphires.

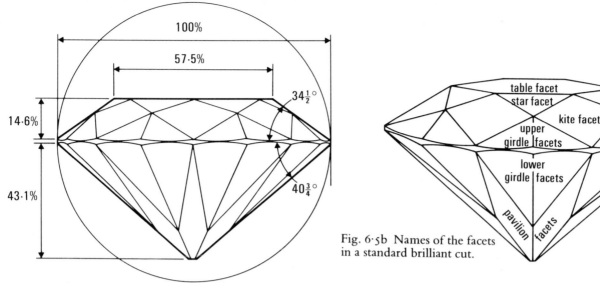

Fig. 6·5a Ideal proportions of a brilliant according to the Scandinavian standard.

Fig. 6·5b Names of the facets in a standard brilliant cut.

Fig. 6·6 Variations on the brilliant cut.

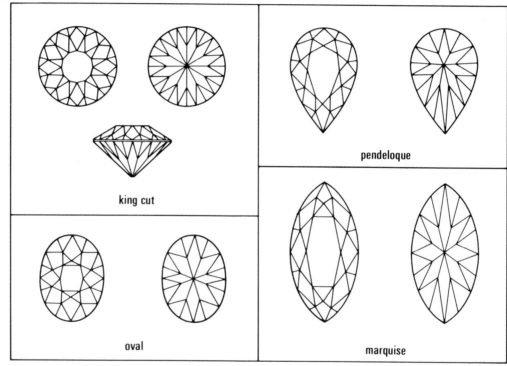

king cut

pendeloque

oval

marquise

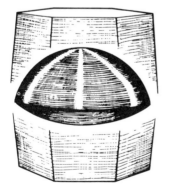

Fig. 6·7 In some rubies and sapphires, fine acicular crystals enclosed parallel to the three lateral crystal axes of the sapphire cause a star to appear in reflected light if a cabochon is cut in the manner illustrated.

relates to the angles at which the facets are arranged to give the best fire and brilliance. Much research has been carried out on the best proportions and modern diamonds adhere fairly well to the ideal cut (Fig. 6·5), but the stones are largely cut by eye, albeit very experienced ones.

In some gems, orientated inclusions within the mineral give rise to the beauty if the stone is cut in the correct way, usually the cabochon. Myriads of parallel minute fibres or hollow tubes will cause a series of point reflections if placed under suitable illumination (the sun, for instance) and this series of points will build up to a sharp line, the cat's-eye, when the stone is cabochon cut (Fig. 6·7). In chrysoberyl, the inclusions can be so fine as to be invisible to the naked eye and under the lens, in contrast to the very similar quartz cat's-eyes which often have much coarser fibres. With three sets of inclusions arranged at 60° to each other, as in ruby and sapphire, it is possible to get three sets of bright lines crossing each other – the six-rayed star of the star rubies and sapphires. In these cases the correctly orientated cabochon, with the star at the centre, is the normal style of cutting, the base frequently remaining unpolished to conserve weight.

Randomly orientated minute inclusions of reflective minerals such as mica and hematite may give rise to a series illuminated spangle effect as in aventurine quartz and feldspar. Glass with bright platy copper crystals, created in the glass

187

by chemical reaction during manufacture, is sometimes called 'goldstone' or aventurine glass.

The colours of gemstones may be altered in various ways including heating, dyeing and irradiation. Provided that such treatment is virtually implicit, as with blue zircon (heated brownish-red material), pale blue aquamarine (heated greenish material) and pink topaz (heated sherry coloured crystals), or is acknowledged at the time of sale, as with irradiated greenish-blue diamonds, many brownish-yellow diamonds and blue topaz (from colourless material) no harm is done. However, when large, yellow, irradiated diamonds are offered for sale as untreated natural material then some element of deception enters the transaction, especially since the difference in value may be very considerable indeed. The colours of agates have been changed by dyeing for centuries, but this is accepted practice and no deception is intended. Indeed, scrutiny with a lens will often reveal traces of dye in cracks.

Origin of gemstones

As with most other types of mining, it is preferable to work a concentration of gem minerals rather than attempt to mine them from their original source rock, where their distribution may be very sparse indeed. Among the commonest, and probably the best-known, type of concentration is that effected by river waters which have worn away the original source rocks and transported the debris, the heavier and more durable gem minerals being concentrated at suitable points such as meanders or potholes in the river-valley system. Thus, the rocks and minerals accompanying the gem minerals will be the quartz pebbles and sands, gravels, clays and muds of the alluvial cycle. However, there are other types of sedimentary origin. Precious opal, in which the play of colours is caused by the diffraction of light by submicroscopic, even-sized spheres of silica packed in a very orderly fashion, forms as a result of the precipitation of silica spheres in a tranquil environment over a very long period of time. Turquoise is a mineral of secondary origin formed in arid regions, such as Iran and the western United States, through the action of circulating water upon aluminous rocks in the vicinity of copper deposits. The turquoise is deposited in joints and veins or as concretionary masses.

The primary sources of many gem minerals are in the pegmatite vein systems often associated with the late-stage igneous activity of granite emplacement. Primary crystallization of the granite magma results in the formation of a rock composed of quartz, feldspar and magnesium- and/or iron-rich minerals such as mica and hornblende. Crystallization of these bulk minerals enriches the residual magma in the rarer constituents such as beryllium, lithium and fluorine and the consolidation of the granite allows the development of a well-structured pegmatite vein system in which large, euhedral, well-shaped crystals such as beryl, spodumene and many others are able to form, assisted by the increased concentration of aqueous fluids and fluxes after the primary crystallization has taken place. The subsequent weathering of the feldspar of these granitic pegmatite rocks under tropical conditions (and more slowly in temperate latitudes) gives rise to pale-coloured or iron-stained clays in which many of the gem minerals survive with little alteration or damage. Gem mining in many tropical areas consists of exploiting these clay deposits along the lines of veins or in accumulated pockets. In both cases there has often been little transport from the original source rock, and these deposits can be very rich locally.

However, granites and allied rocks and their satellite vein systems are not the only intrusive rocks to carry gems; eruptive basaltic rocks may carry ruby, sapphire, zircon and garnet in Indochina and kimberlite pipes are the main

Veins of lapis-lazuli in limestone; note the brassy pyrite marking the line of the veins and the alteration of the limestone to lapis-lazuli on each side of the veins, Afghanistan.

'primary' source of diamonds in Africa, Siberia and elsewhere. In these cases the eruptive rocks are thought to have a deep-seated origin and the gem minerals, probably also formed at depth, have been brought up with the basalts and kimberlites, but do not necessarily have the same parentage as these rocks. In some instances, sapphires have been formed nearer the surface such as those found in the engulfed aluminous country rocks (xenoliths) in basaltic rocks on the island of Mull in Scotland. Gem peridot has been recovered from basalts in Hawaii, and the principal source of this beautiful green gem is from ultrabasic rocks as in the northern part of the Mogok Stone Tract in Burma and on the island of St John (Zeberget) in the Red Sea.

The superb emeralds from Colombia occur in veins and pockets associated with calcite and other minerals in shales. In several other areas emerald occurs in pegmatite veins. In the old workings at Habachtal in Austria, emerald occurs in a mica-schist, a metamorphic rock, and the occurrences at Sandawana in Rhodesia, Pakistan and in India near Udaipur are also in schists. Metamorphic rocks also produce other gem material. Near Mogok in Burma, limestones containing clay as impurity have been thermally metamorphosed to produce a crystalline dolomitic limestone, with red spinel crystals resulting from the excess of magnesium and aluminium in the clay, and ruby, in lesser amounts, coming from the excess of aluminium after the spinel has been formed. In Afghanistan, limestones and other rocks have been invaded along joints by reactive fluids which have altered the limestone to blue lapis-lazuli for varying distances on each side of the joints. These joints are often filled by pyrite which commonly occurs in the lapis-lazuli itself. In this case the lapis-lazuli was formed by the movement of chemically active fluids, a type of metamorphism known as meta-somatism (*see* page 47), but in the case of the ruby formation at Mogok the materials were already present in the limestone before metamorphism.

The garnet group probably forms in a wider variety of geological environments than most other gem minerals and some examples are given below. Pyrope garnet occurs in South Africa in eclogites – rocks formed under great pressures and often associated with kimberlites and other deep-seated rocks. Almandine garnet may result from regional metamorphism of argillaceous (clay-rich) rocks and spessartine is often associated with granites and pegmatites. Grossular garnet may be formed by regional or thermal metamorphism of impure limestones, and andradite is commonly associated with the alteration of limestones by iron-rich mineralizing fluids.

Table 6·1

Igneous origin

beryl (emerald, aquamarine, morganite), ruby, sapphire, quartz (rock crystal, amethyst, citrine), chrysoberyl, spodumene, tourmaline, garnet, opal (some), chalcedony (agate), zircon, feldspars, diamond, peridot.

Metamorphic origin

emerald, ruby, the jades, lapis-lazuli, garnet, danburite, cordierite (iolite), axinite, spinel.

Sedimentary origin

opal, quartz.

Organic origin

amber, ivory, coral, jet, pearl.

A brief list of gemstones which are formed in the various geological environments is given in Table 6·1, but it should be realized that in many instances several types of environment will exist close by. Thus pegmatite veins (broadly igneous in origin) may be found in the same area as skarn (calc–silicate) rocks which have resulted from the alteration (metasomatism) of limestones by iron–rich fluids from the same granite body that produced the pegmatites.

Gem identification

The cardinal principle to bear in mind when identifying gemstones, especially those of high value, is that the gem must appear and weigh the same at the end of the investigation as it did at the beginning. This means that chemical methods and hardness tests are seldom practicable and physical and optical methods are preferable.

A times-ten ($\times$ 10) hand-lens, or loupe, will often enable the careful observer to learn a considerable amount about the stone or even identify it without further testing. For instance, a pale greenish blue stone might be heat-treated zircon, aquamarine, topaz or several other species. If the lens reveals doubling of the back facets, indicating a high birefringence, a high dispersion, and possibly chipped edges to the facets suggesting a brittle character then the stone is almost certainly zircon. Another example is the identification of diamond, where the bright, lustrous facets are a virtual hallmark to the initiated, but confirmation may be found by the triangular natural markings sometimes purposely left on the unpolished girdle by the diamond cutter. Careful examination with a lens will often reveal other characteristic internal structures, and the dealer or serious student well knows the value of this portable means of identification. Extension of the principle of magnification brings us to the microscope, probably the most useful single instrument (but unfortunately among the most expensive) in gem identification. A simple instrument will facilitate examination of the surface features better than the lens, but will also allow the internal features of the stone to be studied in some detail. Inclusions will frequently distinguish natural from synthetic stones and may quickly reveal the glass imitations with their bubbles and swirl marks. If the microscope has polarizing equipment and a Bertrand lens it will be possible to check for pleochroism and birefringence, thus distinguishing between isotropic (glass and cubic system species) and anisotropic stones (all other minerals), and in favourable circumstances to distinguish between uniaxial and biaxial stones and their optical sign (*see* pages 90–92). Refractive index may be measured by the real and apparent depth method and a concentrated beam of light directed through the stones from the sub-stage condensers may be used for the study of the absorption spectrum. There are other uses for the microscope in gemmology and it must be considered the most useful single instrument.

However, an experienced worker presented with a series of coloured stones, especially reds, blues and greens, might might well start his examination by using the spectroscope since these colours are usually caused by traces of transition elements such as chromium and iron which produce a very characteristic absorption spectrum when light is passed through the stone and examined by the spectroscope. Chromium, for instance, which colours many stones will often produce a series of fine lines in the red, and a band of partial absorption in the yellow-green. The position of these chromium lines will vary slightly, because of the differing effects of other constituent elements in the gem mineral, between, say, ruby and spinel and between emerald and alexandrite chrysoberyl. These differences are diagnostic to the experienced worker, who will not need to measure the exact positions of the bands, but will recognize the overall pattern. The spectroscope may be a complicated precision quartz-spectrograph, a metre (39 inches) or more in length, but very convenient portable instruments about 10 centimetres (3·9 inches) long are available.

For many faceted stones, the quickest method of identification is by use of the refractometer (*see* page 88). This normally has a range of from 1·3 to around 1·82, which is the convenient upper limit for reasonably durable, cheap, glass prisms for the instrument and for a relatively non-toxic fluid, essential for making contact between stone and prism. By using prisms of diamond or sphalerite and by using unpleasant contact liquids it is possible to get readings in the 2·1 or even higher ranges, but these are very expensive instruments, for use in the specialized gemmological laboratory only. It is possible to use a white light source with the refractometer, but sodium light is preferable since it eliminates the colour fringes to the shadow edges on the refractometer with consequent improvement in

Below
Detail of Mexican lace agate.

Below right
Fine example of the Mocha stone variety of chalcedony showing dendritic markings.

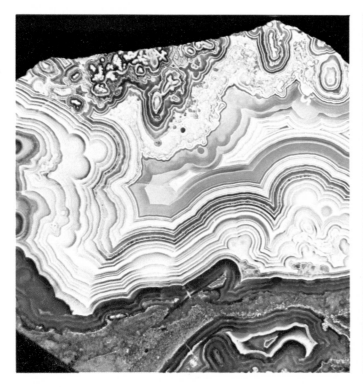

accuracy of reading. Unfortunately, both refractometer and light source are expensive.

The density or specific gravity (*see* page 80) may be very useful in identifying unmounted gem materials and is often the first line of attack when examining the jade minerals and their simulants in the form of carvings. In these cases, density results to one place of decimals will frequently separate a bowenite serpentine carving (2·6) from the jades (2·8–3·1 for nephrite and 3·3 for jadeite) and this determination may be carried out on a simple chemical balance, but more sophisticated balances may be necessary for small gemstones. For the highest accuracy (never easy to obtain in specific gravity measurements) it is often necessary to weigh in liquids such as toluene or ethylene dibromide which have low surface tension, rather than water which tends to 'cling' to supporting wires.

A very useful pocket aid is the Chelsea colour filter developed by the Gem Testing Laboratory in Hatton Garden, London. This filter transmits only deep red and yellow-green light and most emeralds (natural *and* synthetic, unfortunately) will appear distinctly red or pink when viewed through the filter, whereas most of the imitation emeralds and real stones resembling emeralds retain their green colour through the filter. The filter is also useful in detecting synthetic blue stones coloured by cobalt, and in other instances, but care must be exercised in drawing conclusions from the effects seen.

Both long- and short-wave ultraviolet lamps can be useful in gem-testing although they are normally used as a means of confirming opinions already reached by other methods of testing. Convenient 'boxes' containing both long- and short-wave tubes are available, but they are expensive.

For the budding gemmologist, the first acquisition must be the × 10 lens, followed probably by the Chelsea filter and a pair of stone tongs or tweezers. The next step depends upon opportunity, but careful study of advertisements in journals and the shelves of dealers might produce a reasonably priced microscope – new instruments are costly and the specialized gem-microscope particularly so.

Some interesting gems and ornamental minerals

In the space available it is not possible to deal systematically with all gem species. Physical, optical and other data on gem species will be found on pages 110–113. A number of gemmological topics, not always adequately dealt with elsewhere, will be considered here.

Chrysoberyl

The mineral chrysoberyl provides some of the most interesting and desirable of all gemstones. It has a hardness of $8\frac{1}{2}$ and when cut seems to possess a brightness superior to most other gems. From the extremely rare colourless stones found at Mogok in Burma it progresses through the pale yellow Brazilian stones which were so effectively used by the Spanish and Portuguese in seventeenth- and eighteenth-century jewellery, through shades of yellow, brown and green to deep bottle green. When the traces of iron responsible for some greens are replaced by chromium we get the rare alexandrite variety which, because of the absorption of light in the critical yellow-green part of the spectrum, appears green by daylight but red by tungsten light. The author recalls two pale green chrysoberyls which appeared identical to the eye. When photographed in daylight, however, the resulting transparency showed one as seen by the eye, but the other (an alexandrite) as pale pink, the film having reacted differently to the balance of light transmitted through each stone. Many minerals contain fine parallel fibres or hollow tubes, but in chrysoberyl they may be so fine that

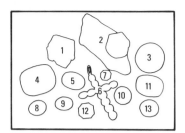

Chrysoberyl group: **1** twinned chrysoberyl, Brazil; **2** twinned alexandrite crystal in mica schist, Fort Victoria, Rhodesia; **3** 38·62 carat Burma; **4** 57·10 carat; **5** cat's-eye, 11·15 carat; **6** cat's-eye cross; **7** 1·97 carat Burma; **8** 7·25 carat; **9** 4·92 carat, Tiofilo Otoni, Minas Gerais, Brazil; **10** 8·55 carat Burma; **11** alexandrite 21·62 carat; **12** 5·35 carat; **13** 7·86 carat Brazil.

resolution is not possible with the ×10 lens, and the resulting bright bands are probably the sharpest found in any chatoyant gemstone. When these fibres occur in a honey-yellow matrix the resulting cat's-eyes are among the most beautiful of all gems and are priced accordingly.

Tourmaline

In Devon and Cornwall, tourmaline is usually a black or dark-brown acicular mineral of no outstanding beauty, but the superb red and green terminated crystals from the Pala district, San Diego, California, the zoned 'water-melon' crystals from Brazil and the magnificent red prisms from Mozambique and Malagasy are among the most spectacular of rough gem materials. With colour changes from one end of a crystal to the other because of its polarity, and with the zoning seen in the cross-sections of other crystals due to the changes in the chemical composition of the mother liquor during crystallization it is not surprising that tourmaline shows a greater range of colour than any other gemstone.

Spinel

Spinel ($MgAl_2O_4$) is colourless when pure, but such material is uncommon in nature, where chromium frequently substitutes for the aluminium in the very attractive range of red and pink stones from the Mogok Stone Tract in Burma. These deserve more appreciation, but are overshadowed by the magnificent reds of the Burma rubies with which they occur. In contrast, many of the stones from Sri Lanka are dull blue, greenish blue and purple in colour, caused by iron entering the lattice. Occasionally zinc may be present (up to 18 per cent ZnO), and these bluish green stones are known as gahnospinels.

Garnet

The garnets were mentioned in the Bible and doubtless were known very much earlier due to the attractive form of the crystals which are so easily weathered out of their schist matrix. In the past, the pyrope–almandine group has provided most of the cuttable material and the Bohemian garnets (approaching the pyrope end of the series) were especially well known in Victorian times. Today, beautiful rose pink and purple material comes from Tanzania and Sri Lanka. Sri Lanka has also provided the traditional orange-brown hessonite, a variety of grossular, but in the last few decades colourless and golden-yellow hessonite has come from East Africa, and in recent years we have seen the appearance of bright green vanadian grossulars which vie with the superb, fiery, demantoid garnets from the Urals area, USSR. Materials with the garnet structure have been synthesized and at least two of these, YAG (yttrium aluminium garnet) and GGG (gadolinium gallium garnet) provide very realistic simulants for diamond.

Fluorite

This is occasionally faceted for collectors, but is far too soft for use in jewellery. However, Blue John, the banded, purple and white or yellow variety which has come from at least fourteen differently patterned veins in Castleton in Derbyshire, England, has long been used for ornamental purposes, especially since the end of the eighteenth century when Matthew Boulton and others were so successful in fashioning vases and a variety of other objects. Green and cream material has been carved by the Chinese and attempts have been made to pass fluorite as jade, but the pronounced octahedral cleavage provides ready identification.

Jade and simulants

In many cities in the world, especially those geared to tourism, names such as 'new jade', 'soft jade', 'Korean jade', 'Transvaal jade' and even 'real jade' are applied to carved objects which resemble the true jades, nephrite and jadeite, and a gullible public buys them at true jade prices. In some countries there is relatively easy legal redress, but usually only in the case of objects bought in the country. Probably the commonest jade simulant is a massive serpentine of appropriate colour, usually of the antigorite variety. The hardness may vary from $2\frac{1}{2}$ to around 6 and the better material is not easy to distinguish by the use of a knife blade, which is bad practice anyway when dealing with valuable carved objects. However, the specific gravity of serpentine is usually in the region of 2·55–2·65 and such objects will float in pure bromoform (specific gravity 2·85–2·90) whereas the true jade will sink. Aventurine quartz, misleadingly called 'Indian jade', will also float in bromoform and is harder than the true jades. Saussurite, an altered gabbroic rock, has been used as a jade simulant in recent years. It may resemble both nephrite and jadeite, sometimes both in the same carving. The specific gravity may vary between 2·8 and 3·4 and since the hardness may be within the ranges of the true jades it can be a very difficult material to identify. In cases of doubt a laboratory with X-ray diffraction facilities should be consulted.

Blue John fluorite vase (Derbyshire) with cubes of purple fluorite on quartz (Durham).

Nephrite jade vase (left)
compared with saussurite
carving (right); note similarity
of lower bird to nephrite and
head of upper bird to jadeite
jade.

Carbonates

The fact that carbon dioxide dissolves in water to form carbonic acid has been
responsible for the production of a series of very attractive stalagmitic carbonate
minerals which are deposited from solution. Cream, green and brown banded
marbles (calcite) from Mexico, Algeria and Turkey are well known, and the
beautiful banded pink rhodochrosite from Capillitas, Argentina is very familiar
to the lapidary fraternity. However, an attractive blue stalagmitic aragonite is not
uncommon, and on rare occasions is accompanied by the colourful purple
cobalt-bearing calcite (cobalticalcite). In the upper zones of copper deposits,
circulating ground waters bring about the formation of vivid green banded
malachite which is often associated with deep blue azurite, both of which are
basic carbonates of copper. These materials have all been fashioned into ornaments of various kinds, usually of a smaller and more personal kind than those
fashioned from the large group of marbles (usually colourful sedimentary
limestones) which have been used for decoration and even building since prehistoric times. A notable fossiliferous marble with built-in optical effects is
fire-marble, in which fine hair-line cracks act as thin films for the production of
interference colours.

Arthur Notholt

Economic minerals

Economic minerals are of service to man, entering into nearly all aspects of modern industry and commerce. They are essential in sustaining the growth in world population, living standards and technological advances and, in this way, the history of the development of man is interwoven with their discovery and utilization. In spite of this and the fact that economic minerals are a natural resource which, once extracted and used, are not renewable, it is only in comparatively recent times that serious concern has been expressed over the long-term supply of man's mineral needs. Together with rocks of all kinds, unconsolidated materials such as clay, sand and gravel, as well as peat, coal, oil and natural gas, they constitute 'mineral resources'. These are usually regarded as second only to agricultural resources in promoting a nation's welfare and economic prosperity.

Primitive man at first used crude implements of wood and stone, chiefly flint, with which to feed, clothe and house himself and afford protection from his enemies. Ornamentation was provided by such pieces of copper, silver and gold as he might find in streams or in surface gravels while the search for pigments for decoration ultimately led to the discovery of ores of at least two metals. Yellow ochre or limonite and red ochre or hematite (also known as jeweller's rouge) are both ores of iron; the green mineral malachite and the blue mineral azurite are ores of copper. The pigments were ground to a fine powder and then probably mixed with animal fats to make rouge and eye-shadow.

Metal implements and weapons made of copper, bronze (an alloy of tin and copper) and iron gradually replaced those of stone, as successful metallurgical techniques to extract them from the ores in which they normally occur were developed. Indeed, working these metals into useful tools represents one of the outstanding achievements of early man, so that there is an association of the stages of man's civilization with different mineral materials – Copper Age, Bronze Age, Iron Age and, perhaps, now the Atomic or Nuclear Age. Flint, copper and other metals were also used in bartering, ultimately leading to the establishment of trade and trade routes. Another very important mineral commodity through the ages has been rock salt, which played a significant role, for example, in the development of trade between Syria and the Persian Gulf. It is well known that Roman soldiers were given money (salarium) with which to purchase salt.

Metals became of great economic importance after man learned to use fire for smelting ores as well as fashioning implements. The Phoenicians are reputed to have traded in tin, lead, iron, copper, silver and gold; tin had apparently brought these seafaring people to Cornwall, England, by 500 BC. Silver and gold have also played an important role in world history, first as highly prized media of exchange and, later, in their use in currency. It is thought that much of the expansion

Sulphur spring in the Danakil depression of Dallol, Tigre Province, Ethiopia.

of the Roman Empire depended on obtaining gold and silver, as well as other metals. The discovery of gold in California in the early part of the nineteenth century was largely responsible for industrial expansion in the American West as well as in Canada and Australia. There are many instances of the role played by minerals in world politics in more recent times. An often-quoted example is the invasion of Manchuria by Japan prior to World War II for the acquisition of much needed supplies of coal and iron ore.

In the geological and mining sense, economic minerals are defined as naturally occurring substances which can be extracted at a profit, although in some countries profitability may not be recognized as an overriding factor, and in such cases extraction is usually undertaken instead at the lowest possible cost. Another important factor is that since one or more economic minerals forming mineral deposits must be present in sufficiently large quantities to be of commercial interest, they represent relatively unusual geological phenomena occurring only where and when the processes responsible for their genesis have been particularly favourable. Thus, the concentration of elements in mineral deposits often greatly exceeds the average abundance of the element in the Earth's crust. For example, many of the copper deposits worked at present contain at least 100 times the crustal average of 0·005 per cent copper. Similarly, most of the deposits worked for iron, one of the most abundant elements, contain more than 25 per cent iron, or five times its average percentage in crustal rocks. The concentration of an element in a mineral deposit is termed its *grade*, expressed usually as a percentage or in parts per million (ppm). The lowest percentage at which a deposit can be economically worked is called its *cut-off grade*. If worked primarily for its metal content the mineral deposit is properly referred to as an *ore deposit* or an *ore-body*, although it has become common practice to extend the term ore to non-metalliferous substances such as sulphur, fluorite, potassium salts and phosphate rock.

Mode of occurrence

Economic mineral deposits have been formed in many ways, generally over appreciable periods of time by complex geological processes. In consequence they are characterized by many varied forms or shapes. They may be stratiform, occurring as sedimentary beds like regular seams of coal, or as lenses and irregular bodies, while others are vein-like, filling fractures or fissures in rocks or following other structural features. The general term 'lode' usually refers to several, often more or less parallel and steeply dipping, mineralized veins that are spaced closely enough for them to be mined as a unit, together with the intervening rock, rather than singly. Some form pipe-like as well as flat-lying tabular bodies or are present as disseminated masses in which the economic minerals occur as small particles or veinlets scattered irregularly through the surrounding rock (country rock). Such deposits are sometimes of great size and form important sources of, for example, copper. Sedimentary mineral deposits originate from surface or near-surface processes, such as evaporation (for example, deposits of rock-salt, gypsum and potassium salts); from biochemical extraction and precipitation, (for example, some deposits of iron ore and phosphate rock); or from the physical concentration of solid particles, as with many 'placer' or 'alluvial' deposits (for example, of tin, gold and diamond). These deposits are the erosional debris resulting from the destruction of the host rock in which they originally occurred.

Weathering, the precursor of sedimentation, may result in very large mineral deposits derived from otherwise valueless parent rock or *protore*. Of particular importance is chemical weathering, which takes place typically in equatorial regions of high temperature and moderate to heavy rainfall. In parts of Australia,

Top
Unrefined gold being cast into 31-kilogram (68-pound) bars in a cascade smelter at Johannesburg, South Africa. South African gold is usually 99·6 per cent pure; gold produced in the USSR is reported to be refined to 99·99 per cent purity.

for example, rocks are altered to depths of 60 metres (200 feet) or more below the surface. In this environment a soil is produced from which most or all soluble minerals have been dissolved. Such tropical soils are called laterites, and consist chiefly of hydrated iron and aluminium oxides, the most insoluble of minerals. When laterite consists of almost pure aluminium hydroxide, it is referred to as bauxite, the chief ore of aluminium. Important deposits occur in Jamaica, Guyana, Hungary, France, the United States and Australia. Similarly, igneous rocks of ultrabasic composition containing relatively large amounts of nickel, may on weathering in the tropics give rise to the development of nickel-rich laterites. Although relatively widespread, such deposits are mined in a few areas notably Cuba and New Caledonia. In this way, aluminium may be concentrated from 5 per cent to 20 or even 30 per cent and nickel from less than 0·2 per cent to about 1·5 per cent.

Another very important process is the chemical alteration brought about by circulating ground water, a common effect of which is to produce a *zone of secondary enrichment*. Valuable elements are carried downwards in solution and redeposited and concentrated within this zone. Metal sulphide ore deposits, notably copper sulphides, are particularly prone to this type of alteration, which may extend over a thickness ranging from 1–300 metres (3–1000 feet) or more, according to the climate, depth to ground-water level, composition and texture of the rocks and of the ore itself. Grades in the secondary ore produced may be ten times higher than those in the unaltered primary deposit beneath it which may, indeed, not be of economic grade. A common surface feature is a residual deposit consisting largely of iron oxides. Being resistant to erosion, it tends to stand above the surrounding countryside as a prominent reddish-brown capping, or *gossan*, and attracted the attention of early prospectors. Even today, a gossan is an extremely useful indication of underlying mineralization, although its size is not necessarily related directly to the size of the hidden deposit. The discovery of gossans containing up to 1 per cent nickel and 0·3 per cent copper led to the discovery of Western Australia's first nickel deposit at Kambalda in 1966.

Distribution

Deposits of economic minerals are unevenly distributed throughout the world and, generally, have been formed during specific geological times. Most present production and resources of iron are in sedimentary deposits chiefly of Precambrian age, notably in North and South America and Australia, many having been formed 2000 million to 3000 million years ago. Important iron-ore deposits also occur in Precambrian rocks in Sweden and USSR. Indeed, the Precambrian, which includes all rocks more than 600 million years old is one of the most important sources of economic minerals, particularly of metals. It accounts, for example, for most of the past production of nickel (chiefly from the Sudbury region of Ontario, Canada), lead and zinc (from the Mount Isa and Broken Hill districts in Australia and deposits in British Columbia in Canada), chromium (from the Bushveld Complex of South Africa and the Great Dyke of Rhodesia) as well as the bulk of the world's production of gold (from the Witwatersrand conglomerate in South Africa). In addition, it has provided a substantial proportion of the copper produced from the well-known Zambian copper belt, although most of the current production is derived from disseminated deposits which are situated mostly in the cordillera of North and South America and are generally of Jurassic to Pliocene age, that is, were formed from 190 million to about two million years ago.

Post-Precambrian rocks are important sources of many non-metallic and industrial minerals, particularly of sulphur, phosphate rock, potash, rock salt

and gypsum, the last three minerals being obtained largely from sediments of Permian and Triassic age (280 million to 190 million years ago) in the United States and western Europe, although important deposits also occur in older rocks of Devonian age in the Williston Basin of western Canada, which has emerged as a major source of potash since the 1960s.

Plate tectonics

The widely accepted plate tectonics theory of crustal evolution may be particularly relevant to the distribution of many mineral deposits and it is expected that a much better understanding of how and why mineral deposits form will be forthcoming before the end of the present century. The theory envisages segments of the Earth's crust slowly moving like a series of great rafts or 'plates'. One of the largest is the Pacific plate, around the margins of which are located deposits of chromite, nickel and copper associated with Mesozoic to Tertiary igneous activity. In contrast, many of the Precambrian iron ores, the Permian lead-zinc deposits of central Europe (in the famous Kupferschiefer), as well as the Permo-Triassic evaporites, have all been cited as examples of economic mineral deposits which have been formed within, rather than along, the margins of crustal plates.

Exploration

Exploring for mineral deposits is as old as mankind itself. Even primitive man, searching for specific flints for his stone tools, was a prospector, as were his successors when, a little later, they discovered the tin-bearing copper ores which made possible the manufacture of bronze. However, modern exploration does not depend on wandering prospectors who, with a 'nose for pay dirt' might stumble across a mineral deposit, but essentially on the systematic, detailed study of larger areas chosen either by geological analogy to known mineralized areas or by deduction based on theories of ore genesis. Although the geological hammer is still a primary exploration tool, a wealth of past experience and sophisticated techniques are available to the exploration geologist in his search for economic mineral deposits.

A successful mineral exploration programme usually begins with a general reconnaissance to decide from the geological evidence available which area offers the most promise of a discovery. In many of the developed countries, able to draw on many years of systematic geological mapping and research, this preliminary work frequently includes the use of geological maps and descriptive memoirs and reports that may already be available. Information from existing records stimulated exploration, for example, for low-grade copper deposits in Snowdonia, Wales, and lead-zinc deposits in the Irish Republic. One of the most intensely investigated areas is the British Isles, about four-fifths of which has been mapped on a scale of 6 inches to 1 mile (about 1:10 000), but in many other areas, geological, topographical and accompanying descriptive literature is meagre or not available and, in such situations, particularly in difficult terrain or where there is a covering of swamp or glacial material, various airborne photographic and geophysical or remote sensing techniques play an important part. Millions of square kilometres of Latin America, Africa and Asia have been mapped photo-geologically. A geological map depicts the type, age, distribution and structural relationships of rocks at the Earth's surface. With its aid, the exploration geologist can build up a three-dimensional picture of the geology of the area and select possible mineralized areas for closer scrutiny.

Geophysical exploration This generally involves the measurement of physical properties of rocks, such as magnetic intensity, density and electrical resistivity.

A particularly successful tool has been the aerial magnetometer. Although costs of airborne surveys are high, they are substantially less than those of corresponding ground surveys and some outstanding successes have been achieved. Important mineral deposits discovered largely by geophysical methods include the Marmora iron ores and the Kidd Creek massive sulphide deposits in Ontario, Canada, the iron ores of the Nimba Mountains in Liberia and large bauxite deposits in Surinam. Aerial geophysical surveys were responsible for confirming the existence of the Cerro Bolivar iron ore deposits in Venezuela, previously identified from aerial photographs as a promising site for exploration. Uranium and other elements can be sought with instruments that measure radioactivity. In general, γ-ray spectrometry is the most effective airborne tool for uranium exploration.

Over the last ten years or so electromagnetic radiation techniques using cameras and electronic recording devices operated from aircraft or space capsules have been used to obtain data. Airborne radar, for example, is capable of providing pictures of the Earth's surface even when there is dense cloud cover. This technique has been used principally in Venezuela and Brazil in the rapid evaluation of mineral and other resources. Thermal infrared imagery is produced by scanning devices operated from Earth satellites and has been instrumental in discovering nickel deposits in Canada and South Africa as well as copper in Pakistan, aided by modern computer techniques. Satellite imagery is being used to locate new mineral deposits in Yemen, Sudan, Venezuela and Bolivia. These remote sensing techniques have obvious advantages in exploring large, generally inaccessible areas of the Earth's surface. Although they cannot pinpoint underground mineral deposits with certainty, the techniques developed so far can identify surface geological features such as major faults and other structural features, together with a number of rock types. As a result, areas with economic mineral potential can be indicated.

Geochemical exploration This aims at finding areas which show considerable deviation from the normal concentration of the element of interest in an area under investigation. Such a deviation is referred to as a geochemical anomaly and may give clues to the presence of ore-bodies. Geochemical work involves analysing rocks, unconsolidated sediments, soils or stream water for certain elements, so that concentrations can be traced back to a workable mineral deposit.

One of the most valuable techniques, which is often used on regional surveys and has led to the discovery of many mineral deposits, consists of analysing stream sediments or waters, a development of the methods used by the old prospector who traced gold upstream to its source by panning alluvium in rivers and streams. Another method, using the mercury sniffer, has been developed to detect and measure minute traces of mercury in air, mercury being to a very small degree gaseous at ordinary temperatures. Apart from its obvious application in the search for mercury ores, it can assist in detecting certain lead-zinc deposits which contain relatively high concentrations of mercury as a minor constituent and also deposits that are buried beyond the range of geophysical methods. A similar tool, the helium sniffer, has been developed recently which may prove helpful in locating uranium-bearing ores. Arsenic is another useful indicator element in geochemical prospecting, being a widespread constituent of many types of mineral deposits, particularly those containing sulphides, while rhenium is of potential importance in the search for porphyry copper deposits. Rhenium occurs in small amounts (generally less than 2 000 ppm) with molybdenite, itself a common constituent of copper deposits of this type and very soluble in water.

Surveys involving the use of vegetation have been employed for some time as aids in geological mapping and in locating water supplies and mineral deposits. The type or colour of plants may indicate the presence of a particular element in

the soil in which they grow. For example, the calamine violet is found only over zinc deposits, and in North America the copper moss is considered an excellent indicator of copper mineralization, apparently growing principally on soils having a concentration of more than 100 ppm of copper. A characteristic of all indicator plants is that the metallic content of their ash is high. Much more recent is the potential application of remote sensing techniques from aircraft and Earth satellites and these are being tested in selected areas of known mineralization in Brazil, the Philippines, Thailand and the United States. Hydrogeochemical exploration based on rapid analysis of the water in lakes, streams and other features of surface drainage has been employed only to a limited extent. However, the technique has been known to detect uranium mineralization at depth, and the commercially important phosphate deposits in North Carolina were discovered around 1958 as a result of systematic ground-water studies. It has been found that artesian waters from the phosphate beds, which underlie an area of approximately 450 square kilometres (174 square miles), contain relatively high concentrations of iodides and bromides in solution. Such anomalies, if detected elsewhere in similar terrain, may indicate the presence of phosphate deposits.

Evaluation In general, only a few of the potentially interesting mineralized areas indicated by geological, geophysical or geochemical surveys qualify for this next, more detailed, stage of investigation, which invariably involves diamond drilling around the site of the prospective ore-body. The object is to determine as quickly and accurately as possible the size and shape of the deposit being investigated and to recover samples of mineralized rock for analysis and testing. As a result of this work, estimates can be made of the tonnage and average grade of ore proved by the drilling programme and also of the quantity of waste rock that will have to be removed in extracting, say, 1 tonne of ore. At this stage it is also usual to decide what kind of mineral processing techniques are likely to provide the most appropriate means of extracting the economic minerals that are present. With the exception of those employed in the search for oil and natural gas, most boreholes do not reach depths exceeding 300 metres (1 000 feet), and under favourable circumstances only a relatively small number may be needed to delineate a deposit. For example, in a systematic drilling programme on the Panguna copper deposit on Bougainville in the western Pacific, only 200 bore-holes were put down which drilled through an aggregate thickness of 84 000 metres (276 000 feet) of rock.

Investigating a mineral deposit generally involves much 'risk' capital and usually much less than 10 per cent of the mineral prospects on which drilling is begun eventually become mines. In addition, there may be two to three years or more between the time when investigation of an area begins and the point when the decision to develop a mine is taken, and an additional two to seven years may elapse before actual mineral production begins. This is clearly an expensive process and it has been estimated that the cost of exploration is not far short of that needed to develop and equip new mines. Thus, most modern exploration, as well as the construction of mining and processing facilities, is conducted by large, well-capitalized mining companies.

In addition to the many geological and mining factors which need to be taken into account, and which in some respects are relatively straightforward, there are other factors which determine whether individual mineral deposits become commercially valuable. These include price changes, technological developments and government policy. In addition, there is the impact of the growing restriction on environmental pollution. One of the most important considerations to a mining company is the political and fiscal stability of the territory where the mine is located.

A drilling platform extended over the side of *m.v. Heathergate* on charter to the British Institute of Geological Sciences in the northern Irish Sea. With equipment of this type it is possible to drill up to 100 metres (330 feet) of unconsolidated sediments into underlying solid rock.

The Jeffrey asbestos mine in the province of Quebec, eastern Canada, is reputed to be one of the largest of its kind in the world. The mine, which has been in operation since 1881, is an oval-shaped open pit, at present more than 1·6 kilometres (1 mile) across at its widest point and some 1 500 metres (4 950 feet) deep. It produces annually some 600 000 tonnes of asbestos fibre.

Mining and processing

Mineral deposits are found at all levels in the Earth's crust, ranging from surface outcrops to ore-bodies at considerable depths. In the early days of the mineral industry, mining was a relatively primitive procedure and it was possible only to work deposits at or very close to the surface. Today mining is carried on at depths of up to 3 kilometres (1·9 miles). Where a deposit occurs at or near the surface it is customary to operate an open-pit mine or quarry, or it may prove desirable to drive a tunnel or adit into the hillside. At greater depths shafts are sunk to reach the mineral deposit, the mine then consisting of a network of tunnels and larger openings known as stopes. Most mining operations take place at the relatively insignificant depths of less than 1 800 metres (5 900 feet) compared with the 15–50 kilometre (9–13 mile) thickness of the continental crust.

Open-pit mines, normally much less than 300 metres (1 000 feet) deep, are generally cheaper to operate, because they favour the extraction of ore from extensive, low-grade mineral deposits by means of large earth-moving equipment. On the other hand, underground mining is most profitable where the grades of ore are relatively high and the deposit comparatively small. Although surface mining accounts for most of the world's mineral production at present, it is believed that in the long-term underground mining will increase significantly due to the decline in the number of discoveries of surface deposits. Numerous technological advances have in recent years proved effective in reducing operating costs at some underground mines.

Another important underground method of extraction is solution mining. In this method, a liquid is pumped into the ground through a borehole to liquefy or dissolve the deposit, which is then pumped to the surface. The extraction of sulphur and rock salt are striking examples of solution mining; potassium salts are produced by this method near Regina, Saskatchewan, Canada, from deposits lying at a depth of about 1 500 metres (4 900 feet).

Placer deposits of tin or gold are mined by hydraulic mining, jets of water

breaking down the ore-bearing gravel so that it can be pumped to the treatment plant. The method is also used extensively in working relatively unconsolidated phosphate deposits in Florida.

It is evident that most economic minerals are obtained from the ground, despite the wide range of elements present in sea water. Only sodium chloride (as halite), magnesium, bromine and potassium salts are recovered commercially either by evaporation or chemical processing. Ocean or offshore mining by dredging in water up to 60 metres (200 feet) or more deep on continental shelf areas has been practised for many years in different parts of the world. The ocean floor and underlying strata hold potential as sources for future mineral production.

The task of converting mined ore into a marketable consumer product is termed mineral processing, and is one that is becoming increasingly sophisticated as progressively lower grades and more complex types of ore are worked. The various procedures vary widely but, basically, mineral processing involves

Salt is obtained by the solar evaporation of sea water in shallow ponds on the west side of Lanzarote in the Canary Islands. The installations cover an area of 8 hectares (20 acres). The sea contains vast resources of salt, most of which is used in the chemical industry in the manufacture of chlorine and soda ash.

crushing and grinding to liberate ore fragments from the enclosing barren rock, and the separation of the economic mineral particles from the ore matrix by methods that depend upon the different characteristics of the minerals concerned, for example, their density, magnetic or surface properties. In the concentrating process, flotation is widely used and has proved so successful that it is often referred to as the single most important development in mineral processing in the twentieth century. In this method, finely crushed ore is agitated in water containing one or more chemical 'frothing' agents; the unwanted material sinks and the economic mineral particles rise in the froth to the surface, where they are skimmed off into a collecting trough and eventually dried. The flotation process has enabled many low-grade ores, particularly of metal sulphides, to be utilized, but many other metallic and non-metallic ores are equally amenable.

The economic minerals so recovered are normally in the form of chemical compounds and further processing is often undertaken to provide products of still greater purity. For example, in copper smelting, a metal of 98–99 per cent purity (blister copper) is obtained. Even this is too impure for most applications of the metal and a refining process employed to obtain a product with a minimum purity of 99.9 per cent copper.

Important economic minerals

Excluding coal and bulk materials such as limestone, sand and gravel, it is estimated that some sixty minerals or mineral commodities are of economic value, of which somewhat less than half consist of metals. In terms of the quantities produced the most important, in order, are: iron, rock salt (halite), phosphate rock, bauxite, gypsum, sulphur, potash, copper, chromium, lead, zinc, asbestos, fluorspar and titanium. Such a listing inevitably does not include, for example, silver, gold, diamond, platinum, nickel, tin and tungsten, all of which are usually produced in much smaller amounts but represent high monetary value. Gemstones may be regarded as the ultimate in value for mineral resources. Of all economic minerals, by far the largest production is of iron ore, amounting to some 850 million tonnes a year.

Aluminium

This metal is often used as a substitute for steel, but in a much more important sense it is a substitute for copper. Its best-known characteristic is probably lightness, being only one-third the weight of copper or steel. Among all available substitutes only magnesium is lighter. Other outstanding properties include high thermal and electrical conductivity, high reflectivity to light and heat, and good workability. In addition it does not corrode easily and is unaffected by many chemical reactions.

Initially aluminium was used mainly in kitchen utensils, but the metal came to be used on a much larger scale during World War II, notably in aeroplane manufacture, with the development of such alloys as 'duralumin', which combines exceptional lightness with great strength and durability. Its principal applications now are in building and construction, and in various aspects of railway, road and air transport. On the basis of size the metal has about two-thirds the electrical conductivity of copper but weight for weight aluminium cable is a better conductor of electricity than copper cable of equal length. Aluminium came into use as cable around 1906, when a method was found of constructing an aluminium cable around a steel core. The application of aluminium in packaging as foil or cans is well known.

Almost all the aluminium produced to date has come from bauxite, of which the principal mineral constituents are the aluminium hydroxides gibbsite

[Al(OH)$_3$] and boehmite and diaspore [AlO(OH)]. Bauxite, so named after the village of Les Baux in southern France, where it was first recognized in 1821, is relatively widespread in the world. By far the largest producer is Australia, where there are major bauxite deposits at Weipa and Gove in northern Queensland, followed by Jamaica, Guinea and Surinam. These four countries account for more than half of the total world bauxite production. Most of the bauxite produced contains about 50 per cent aluminium oxide or more; depending on its richness, 4–6 tonnes of bauxite are needed to produce 1 tonne of aluminium metal which, when pure, normally contains 99·5 per cent aluminium. *See* distribution map, Fig. 7·1.

Chromite

Chromite [(Mg,Fe)$_2$CrO$_4$] is the only ore mineral of chromium, containing, theoretically, 68 per cent chromic oxide. Usually, however, impurities are present and commercial chromite deposits do not usually contain more than 50 per cent chromic oxide. Chromite is dark-brown to black in colour and occurs either in massive form as lenses and tabular bodies or is disseminated as granules and streaks in ultrabasic rocks. Large resources of chromite are present in southern Africa, in the remarkable Bushveld Complex of South Africa and the Great Dyke of Rhodesia, Rhodesia being the only major producer of high-chromium chromite. These are essentially layered igneous intrusions. Similar occurrences are to be found in several other parts of the world which represent potential low-grade sources. To date, the bulk of the world's supply of chromite has come from the USSR.

Chromium is primarily an alloying metal employed in the steel industry in the manufacture of high-speed tool steels and stainless steels. For this purpose high-chromium chromite is used. Low-chromium chromite is consumed chiefly in the manufacture of refractory materials which are used, for example, to line furnaces where chemically inert compounds with high melting temperatures are needed. Chrome-magnesite bricks are used in many steel furnaces. Before 1900 chromium was used only by the chemical industry, chiefly to produce point pigments. Pure chromium is used as electroplate. *See* distribution map, Fig. 7·2.

Copper

Together with gold and silver, copper is one of the metals known to early man, probably because it was found as a metal on the surface and could be shaped readily into ornaments and utensils. The present extensive industrial use of copper, however, depends mainly on its electrical and heat conductivity, corrosion resistance, ductility and alloying properties. Because of these remarkable properties, copper has come to be regarded as the 'work horse' of the electrical and allied industries. The metal is of paramount importance, for example, in the manufacture of transmission cables for power and lighting, motors, generators, telephone and telegraph wire and cable, electronic circuits and computers, and a wide variety of electrical appliances such as washing machines, refrigerators, air conditioning units, radio and television sets. It is also used extensively in complex weapon systems, and for electrical wiring in aeroplanes, motor cars, railways, while virtually all building wire for power and lighting is made of copper. Although aluminium has replaced copper for virtually all high-voltage overhead power cables, copper is still preferred for underground lines and dominates the smaller gauge wire market.

Being very ductile and malleable, pure copper can be rolled into extremely thin sheets, drawn into thin wire, and pressed, forged, beaten or spun into complicated shapes without cracking.

Copper corrodes and tarnishes readily, a sheet of copper exposed to the air very

The Morenzi open pit is one of the largest porphyry copper mines in the world, capable of producing 60 000 tonnes of ore per day.

quickly acquiring a thin and very attractive film of green copper compounds which subsequently remains immune to atmospheric attack. For this reason important buildings were often sheathed in copper, as were the wooden hulls of sailing ships, a practice which gave rise to the expression 'copper-bottomed'. Copper is widely used in the construction industry, narrow-gauge copper pipes in plumbing, for example, having almost entirely replaced the traditional lead or iron.

Copper is often alloyed with tin to form bronze, but it forms alloys with numerous other metals, notably zinc, aluminium, nickel and lead. Brasses, alloys of the copper-zinc group, are cheaper than either bronze or copper and used mainly as sheets, strips, wires, rods, pipes, tubes and castings.

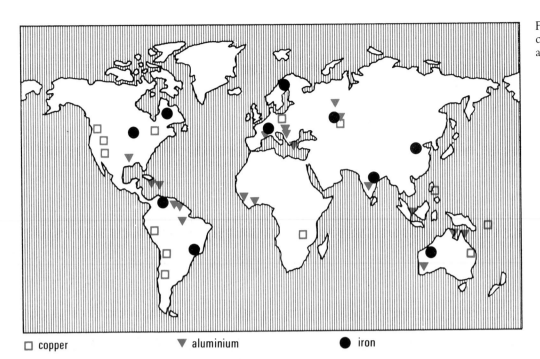

Fig. 7·1 Distribution of major commercial sources of copper, aluminium (bauxite) and iron.

□ copper   ▼ aluminium   ● iron

More than 160 copper-bearing minerals are known, but only a relatively small number are of economic importance. The principal copper minerals are – chalcopyrite ($CuFeS_2$), bornite (Cu_5FeS_4), chalcosine (Cu_2S), malachite [$Cu_2.CO_3(OH)_2$], azurite [$Cu_3(CO_3)_2(OH)_2$], chrysocolla ($CuSiO_3.2H_2O$).

The United States has been the world's leading copper producer since 1883, except for 1934 when adverse economic conditions affected domestic production. Other major producers are, in order of importance, Chile, Canada, USSR, Zambia and Zaire. More than 50 per cent of all the copper ore produced is derived from porphyry copper deposits and an estimated 20 per cent from stratiform deposits, principally those making up the Zambian Copper Belt. *See* distribution map, Fig. 7·1.

Gold

Gold and silver were both known to man in prehistoric times, gold being long considered as the 'king of metals'. Its untarnishable yellow glitter resulted in gold being a much-sought-after ornamental material, while its relative scarcity and physical properties have since made the metal an acknowledged international measure of wealth. The discovery of goldfields has often prompted men to search frantically for quick fortunes; among famous gold rushes are those to California in 1848; to the 'Comstock Lode' of Nevada in 1858, to Ballarat in Victoria, Australia, in 1851 and to Klondike in Alaska in 1896.

The principal uses of gold are as coinage and bullion. At present, much of the world's gold is held in the United States, mostly as ingots in well-guarded vaults. The next most important use is for ornamentation, especially in jewellery, for which purpose gold, as the most ductile and malleable metal, is hardened by alloying it with copper, silver, palladium, or nickel, the process giving the various 'carat' qualities in everyday use. The term carat indicates a twenty-fourth part; thus, twenty-two carat gold has twenty-two parts by weight of pure gold and two parts of alloying elements. Gold plating has a variety of decorative uses, as does gold leaf made by beating gold by hand. Gold thread is used in making gold lace and various fibres. Most of the remainder goes largely into electronic circuitry, dentistry, pen nibs, and some chemical and photographic preparations.

Gold does not usually combine with other elements but occurs as native gold which, because it is dense and indestructible, is ideally suited for concentration in placer deposits, the source of much of the world's production. Among the most remarkable is the Precambrian Witwatersrand sedimentary basin in South Africa, commonly called the 'Rand', which has provided almost one-third of the gold so far recovered. Besides South Africa, other important producers of gold are the USSR and, to a much lesser extent, Canada, the United States and Australia. *See* distribution map, Fig. 7·3.

Iron

Next to aluminium, iron is the most abundant metal in the Earth's crust. It is also the most indispensable of metals because of its many useful properties, including hardness, strength and durability, and is therefore rightly regarded as the backbone of modern industry. Iron ore is consumed in very large quantities in the manufacture of iron and steel, the uses of which are too numerous to mention. Unlike aluminium, however, which has become economically important only since the beginning of the present century, iron has been used by man for at least 3 000 years.

A large number of iron-bearing minerals are known, but only four are important commercial sources of the metal. These are magnetite (Fe_3O_4) containing theoretically 72·4 per cent iron; hematite (Fe_2O_3) with 70 per cent iron; goethite $(Fe_2O_3.H_2O)$ containing about 60 per cent iron; and, to a lesser extent, siderite $(FeCO_3)$, sometimes known as spathic iron ore, which when pure contains 48 per cent iron. In some iron ores, notably the oolitic ironstones mined in the United Kingdom and France, the greenish, fine-grained complex hydrous iron silicate, chamosite, is a major constituent, together with siderite and goethite. Few if any of the iron-ore deposits worked have an iron content approaching the theoretical percentages found in the minerals mentioned. Rich deposits of magnetite may contain up to 68 per cent iron and under favourable circumstances ores with as little as 25 per cent iron are mined profitably. Among the impurities normally present, titanium and manganese may enhance the value of iron ore, but sulphur, phosphorus and arsenic are undesirable constituents, although they can be removed by suitable treatment.

Iron ore deposits are reasonably well distributed throughout the world and are mined in most countries. However, the USSR, Australia, the United States, Brazil, China, Canada, France, India, Sweden and Venezuela account for the bulk of the world's output. The largest single source of iron now being mined is the extensive sedimentary banded iron ore of Precambrian age, also known as the Lake Superior type ore, after the area in North America where it was first studied. Its mineralogy varies but economically the most significant rock type consists of magnetite (or hematite) and chert (silica) in alternating thin layers, often less than 1 millimetre (0·04 inch) thick, with an iron content commonly ranging from 25–40 per cent.

Since World War II the underlying hard, unleached and lower grade ores, called taconites, have been mined and ways found to concentrate them successfully. A major technological achievement has been the development of a process for the large-scale treatment of such low-grade ores to produce high-grade pellets. Iron ore pellets, which contain at least 60 per cent iron, are an ideal furnace feed and pelletizing is now common practice for both low-grade and high-grade iron ores. *See* distribution map, Fig. 7·1.

Lead and zinc

Lead and zinc invariably occur together in nature, and galena (PbS) and sphalerite (ZnS) are the principal minerals accounting for most of the world's production. The United States, the USSR, Australia, Canada and Peru are the leading producing countries of both lead and zinc ores.

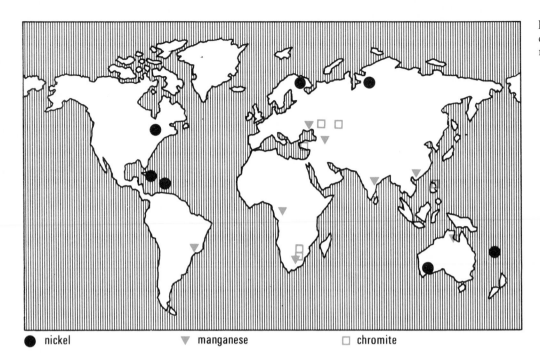

● nickel   ▼ manganese   ☐ chromite

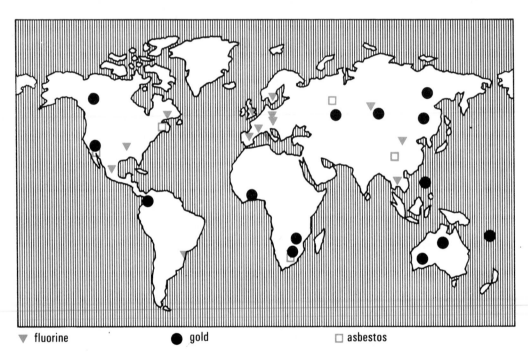

Fig. 7·3 Distribution of major commercial sources of fluorine, gold and asbestos. The only major source of fluorine is fluorspar, which consists essentially of the mineral fluorite and variable amounts of other minerals. Asbestos is the name given to the fibrous form of many minerals, but chrysotile accounts for more than 90 per cent of the world production of asbestos.

▼ fluorine   ● gold   ☐ asbestos

Lead is among the metals to have been used by early man. The Romans, for example, used the metal extensively in pipes for conveying water and, later, lead alloyed with tin and antimony proved the most satisfactory type-metal when movable type was invented in Europe in the fifteenth century. At the present time, lead is the fifth-ranking metal of trade and consumption after iron, aluminium, copper and zinc. It is used principally for storage batteries, tetraethyl lead, cable covering, paint pigments, building construction, ammunition and various alloys, chiefly solder, bearing metal and type-metal. Tetraethyl lead, the most important of the chemicals containing lead, is an active ingredient of 'antiknock' compounds added to petrol to improve its performance. All these industrial uses take advantage of lead's softness and extreme workability, high specific gravity, alloying properties, and good corrosion resistance. Because lead cannot be penetrated by short-wave radiation, it finds an important use also in radiation shields. Contrary to popular belief, lead is much lighter in weight than gold, platinum or mercury.

For many years the only important use of zinc was for alloying with copper to make brass, indeed, sphalerite was frequently considered a nuisance in lead mines and was dumped as worthless material. The metal came into industrial prominence with the discovery that a thin coating of zinc protected iron sheets from rusting for long periods (galvanizing). This technique has widespread applications. Other important uses are in zinc-base alloy die castings, the manufacture of brass and zinc oxide in rubber and paints. Zinc is employed in the chemical industry as a reducing agent and in dye making.

Well-defined generally steeply dipping vein deposits of lead and zinc ore are common and have been worked in many parts of the world but much more important are the stratiform ore-bodies found chiefly as replacements in limestone. Examples are the almost horizontal or low-dipping 'flats' in Carboniferous limestone which formerly yielded large tonnages of galena in the United Kingdom in Derbyshire and Durham, and the highly folded and faulted metamorphosed Precambrian deposits. The latter include the Broken Hill and Mount Isa districts in Australia, which are among the richest in the world. Some of the ores at Broken Hill, for example, contain more than 20 per cent each of lead and zinc. Geologically exceptional are the deposits of the Mississippi Valley type, which occur as replacements in limestones of various ages. In the type area, they extend from Oklahoma and Missouri to southern Wisconsin, over much of the drainage system of the Mississippi River. Similar deposits have been located in many other parts of the world.

In tropical climates, weathering has led to the development of enriched ore zones by processes akin to lateritization. In such cases the primary or original minerals are oxidized to form cerussite ($PbCO_3$) and, to a much lesser extent, anglesite ($PbSO_4$) in the case of galena, while sphalerite similarly undergoes alteration to smithsonite ($ZnCO_3$) and hemimorphite [$Zn_4Si_2O_7(OH)_2 . H_2O$]. The deposits of the Franklin Furnace - Sterling Hill district of New Jersey are almost unique among the major zinc deposits in that the ore minerals consist predominantly of the red zinc oxide zincite, the generally greenish-yellow zinc silicate, willemite, and the black complex zinc oxide, franklinite. *See* distribution map, Fig. 7·4.

Fig. 7·4 Distribution of major commercial sources of lead and zinc, tin and tungsten. The most important sources of tungsten, which is used principally in the steel industry, are the minerals wolframite and scheelite.

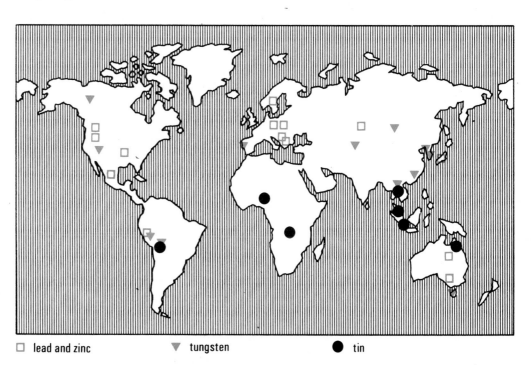

☐ lead and zinc   ▼ tungsten   ● tin

Manganese

Manganese is essential in steelmaking to avoid the formation of deleterious amounts of iron oxide and iron sulphide. It is normally added as an alloy, ferromanganese, which averages about 80 per cent manganese, the remainder being iron and carbon. Very small amounts in steel itself are desirable to increase its strength. These uses account for most of the manganese ore produced, but the metal has important uses also in the manufacture of dry cells, varnishes, paints and inks. Sodium permanganate (as in Condy's fluid) and potassium permanganate are disinfectants.

Manganese does not occur alone in nature in the metallic form, but chiefly as two black minerals, pyrolusite (MnO_2) and psilomelane, an impure hydrated oxide (Mn_2O_3. $2H_2O$). Commercial deposits normally contain between 25 and 40 per cent of manganese. Large sedimentary manganese deposits are found in the USSR at Chiaturi in Georgia and Nikopol in the Ukraine. The USSR has been by far the most important producer since 1929, followed by India, South Africa, Ghana and Gabon, the manganese deposits of which are chiefly of residual origin.

Ferromanganese nodules which have been found on the ocean floors, notably of the Pacific Ocean, constitute a large potential source of manganese. In certain areas, they are of added interest because of their copper and nickel contents. *See* distribution map, Fig. 7·2.

Nickel

Nickel, like manganese and molybdenum, finds its greatest single use in the steel industry to provide steel and cast iron with added strength and hardness, and resistance to corrosion and heat. Nickel ranks second only to manganese as an additive for ferro-alloys. Pure nickel is used principally in electroplating and also in the chemical, electrical, and petroleum industries.

There have been relatively few sources of supply of nickel ore. Important deposits were discovered in New Caledonia in the South Pacific in 1854 and were the world's chief source of the metal until the early part of the present century, when Canada became the largest producer. The famous deposits in the Sudbury district of Ontario were developed mainly because of the growing realization of the value of nickel as an alloy in armour plate for naval ships. The Sudbury deposits are Precambrian nickel-copper-iron sulphide ores in which nickel is present as the mineral pentlandite [(FeNi)S] in intimate association with pyrrhotite, and chalcopyrite. Sulphide ores are also worked in the USSR, the second largest world producer, and in Western Australia where nickel was first produced in 1967. Some of the Australian ores contain up to 6 per cent nickel, compared with an average of about 1·5 per cent in the Sudbury deposits.

In complete contrast, the deposits of New Caledonia are nickeliferous laterites which contain relatively insoluble nickel silicates, notably garnierite ($H_4Ni_3Si_2O_9$). Such low-grade ores, which contain only up to about 1 per cent nickel, are worked in Cuba and have been discovered in the Philippines, Greece and elsewhere. *See* distribution map, Fig. 7·2.

Phosphate rock

Phosphate rock forms the basis of a large mining industry of worldwide importance which ranks second to iron ore among economic mineral commodities in terms of gross production and international trade. More than four-fifths of the world production of rock is consumed each year in making phosphoric acid and a wide variety of phosphate fertilizers, notably superphosphates and ammonium phosphates; most of the remainder is used in the growing production of detergents and animal feed supplements. Only to a relatively small extent is phosphate rock

Fig. 7·5 Distribution of major
commercial sources of
phosphate and potash. The
soluble mineral sylvine is by
far the most important of the
potassium-bearing minerals.
Like phosphate rock, potash
is used chiefly in the
manufacture of fertilizers.

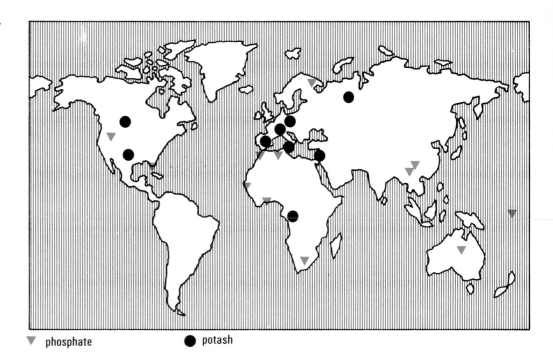

▽ phosphate   ● potash

used for direct application to the soil as fertilizer when finely ground.

Commercial production of phosphate rock began in Suffolk, England, in 1847, when phosphate nodules (mistakenly referred to as 'coprolites') were produced and used in superphosphate manufacture. At present, phosphate rock production and trade is dominated by the United States, Morocco and the USSR, although other major producers are Tunisia, Nauru, South Africa, Togo, Senegal, China, Christmas Island (in the Indian Ocean), Jordan and North Vietnam. Morocco is the world's largest exporter. Western Europe has been for many years the major market outlet, accounting in recent years for almost one-half of the total world trade in phosphate rock.

Most commercial deposits are of marine sedimentary origin, but a significant output is obtained also from alkaline igneous complexes and, to a much smaller

Electric draglines are typical of
many open-pit mining
operations. Machines equipped
with buckets of up to 37
cubic metre (49 cubic yard)
capacity are used in Florida, to
mine phosphate ore ('matrix')
and dump it into a pump-
sump from where it is pumped
as a slurry to beneficiation
plants.

extent, from residual deposits. Their grade is expressed in terms of the percentage of phosphorus pentoxide (P_2O_5) determined by chemical analysis. Phosphate rock containing as little as 5 per cent phosphorus pentoxide is mined and processed, but the bulk of the world's output contains 14 per cent or more. Nearly all these varied sources of supply contain one or more of the apatite group of minerals. By far the most common mineral of the group is fluorapatite [$Ca_5(PO_4)_3F$], which occurs as a frequent accessory mineral in almost all types of igneous rock and is found also in many metamorphic rocks. In sedimentary deposits, the phosphate mineral is usually apatite, whose chemical composition may be represented approximately by the formula $Ca_5(PO_4)_3(F,Cl,OH)$.

Particularly extensive phosphate deposits were formed in the western United States during Permian times in a marine basin covering what is now parts of Idaho, Nevada, Utah, Colorado, Wyoming and Montana. Large resources of phosphate rock are available within this area. *See* distribution map, Fig. 7·5.

Sulphur

Sulphur (S) has been known and used since earliest times, the name being derived from the Latin word for brimstone (burning stone). It is an abundant element and occurs both in the free state as native sulphur and in direct combination with many other elements, notably as metal sulphides, hydrogen sulphide gas (H_2S) and the calcium sulphate minerals gypsum and anhydrite.

Because of its diverse uses, sulphur is not generally thought of as a fertilizer raw material. Nevertheless, the bulk of the production goes into the manufacture of sulphuric acid much of which, in turn, is utilized in the preparation of various fertilizers. The element is a vital raw material in many other processes, being used extensively in the rayon, pulp, paper, dye stuffs, pharmaceutical and rubber industries. It has an important use, for example, in 'vulcanizing' rubber, a process based on the discovery by Goodyear in 1839, that by heating rubber with sulphur, the rubber lost part of its natural stickiness and many new applications could be found for it. Sulphur is used in agriculture as a fungicide and insecticide and as a plant nutrient.

The United States, Canada, Poland, France, USSR, Mexico and Japan are all major producers of elemental sulphur, the United States having been the most important exporter for many years until superseded by Canada in 1968. By far

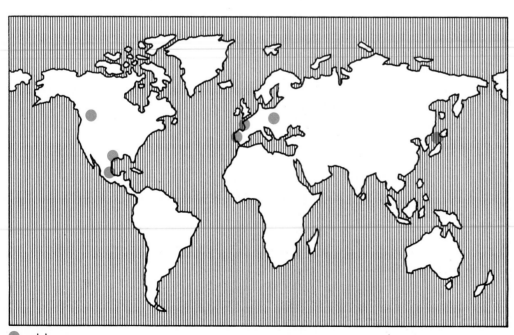

Fig. 7·6 Distribution of major commercial sources of sulphur.

● sulphur

the most important deposits of native sulphur occur as crystal aggregates, veinlets and disseminated particles in limestone and gypsum which, together with anhydrite, form the 'cap rock' of many salt domes scattered along the Gulf of Mexico from Louisiana and Texas to southern Mexico. The sulphur produced from these deposits is often known as *Frasch sulphur*, after Herman Frasch who developed the process by which this sulphur is mined. A typical commercial deposit may be up to 30 metres (100 feet) thick and contain 20–40 per cent sulphur. Some thirty-two domes have been mined for sulphur since production first began in Louisiana in 1894.

Another extremely important source is sulphur recovered from hydrogen sulphide (H_2S) gas associated with natural gas, commonly referred to as 'sour' gas. The very large natural gas fields in the western Canadian sedimentary basin and at Lacq in south-western France are striking examples. In Canada most of the producing gas fields contain up to 20 per cent hydrogen sulphide. The natural gas found to date in the North Sea has a low sulphur content.

Native sulphur is also found in most volcanic regions of the world, particularly in the Andes Mountains of South America and in Japan, but production from these sources is relatively small. Declining in importance as a commercial sulphur source is pyrites, a general trade name for pyrite or iron pyrites (FeS_2) and other iron sulphide minerals. Pyrite is one of the most abundant of all sulphide minerals, occurring in lenses, veins and disseminated crystals. The USSR, Japan and Spain are important producers. *See* distribution map, Fig. 7·6.

Tin

Tin was undoubtedly one of the first metals to be used by man, although it was not until the nineteenth century that tin was employed in industry on a large scale. The metal continues to be used in the manufacture of various bronzes containing up to 25 per cent tin and of other non-ferrous alloys, but the largest uses are now in the tin-plate industry and in the production of soft solders. Perhaps the most important of tin's properties is its non-toxicity, and its resistance to corrosion is superior to that of both copper and nickel. Tin is also very soft and has a low melting point (231·9° Centigrade) and when molten has the ability to wet and remain firmly coated on other metals.

Tin plate, the manufacture of which was a virtual monopoly of the United Kingdom until 1891, consists of mild steel with a protective coating of tin which, in the case of the 'tin' can, is generally less than 0·0004 millimetres (0·00001 inch) thick. Besides its use in the manufacture of metal containers of various kinds, tin plate is also used for making bottle tops, kitchen utensils and light engineering parts. The so-called 'silver paper' is commonly a very thin sheet of lead covered by a film of tin. Other alloys, in addition to bronze and brass, include the well-known pewter which is a high-tin alloy containing about 20 per cent lead, although numerous compositions have been employed. Britannia metal, a modern pewter, is essentially a tin-antimony alloy.

The only important ore mineral of tin is the oxide *cassiterite* (SnO_2), which when pure contains 78·6 per cent tin. However, the best commercial concentrates average only about 70 per cent tin. Small quantities of tin have been recovered as complex tin sulphides, such as stannite and cylindrite. Cassiterite is obtained mainly from alluvial placer deposits worked in Malaysia, Thailand and Indonesia and from vein deposits or lodes mined underground, notably in Bolivia. These countries are the leading world producers of tin concentrate. Tin lodes are also worked in south-west England, where there has been a long history of tin mining, although production has fluctuated greatly. The region was the world's major source of supply during the latter half of the nineteenth century. *See* distribution map, Fig. 7·4.

Dr Bob King

Building a collection

Collections of minerals are built in ways which often characterize the individual. Inevitably, as experience develops, the collector will modify ideas as to the kind of collection he or she wishes to build. The collection will develop in parallel to changes of ideas, accumulation of knowledge, limitations of accommodation, and so on.

A collector, due to fortuitous circumstances, may specialize in some aspect of mineralogy or chemical group of minerals. He may, for example, live in an area where zeolites are particularly abundant and his collection may become outstanding in examples of this group. Other collectors may shun the presence of silicates in their collections, perhaps due to their complexity, and may instead concentrate on adding metallic sulphides or some other apparently more easily understood class of species. Others, due to ready access to a local mine, may concentrate their attention on species found there.

In the early stages, however, it is not advisable to specialize too stringently. A broad spectrum of knowledge of minerals and how to handle them is highly desirable, augmented, if possible, by attendance at some institution which offers courses in mineralogy.

Unfortunately the great abundance and ready access to sites of mineralogical potential which existed in Europe 100 years ago, can not be matched today. The collector has, therefore, a great responsibility to preserve any mineralogical material which comes his or her way, and not only the specimens themselves, but, of equal importance, the data which accompanies them. The would-be collector should have these facts strongly in mind in the early stages of his or her interest. The new collection should be built in such a way that research workers, knowing of its existence, may use it to promote their research.

In the early stages, the collector may have no intention of specializing. It must be realized, however, that there are approximately 2 500 known naturally occurring species of minerals. Though it is highly unlikely that a collector would acquire anywhere near that number, even during a lifetime, the acquisition of a general collection of this kind does present a formidable task which may eventually prove beyond a collector's means, especially with regard to storage. He or she may have to decide whether to increase the accommodation, specialize in some limiting aspect, or drastically reduce the size of the specimens. Today the latter two alternatives are enjoying most popularity.

Specialization is to be strongly recommended after the collector has attained experience in the general field. The collector may become an expert in the small branch of mineralogy chosen. Today there are few professional mineralogists who can command an advanced knowledge of the whole mineral kingdom. The large majority can only hope to possess such knowledge in some limited aspect of mineralogy.

Lemon-yellow transparent radiating group of mimetite crystals; from Tsumeb, south-west Africa.

The third alternative has achieved great popularity in the United States, and is of increasing popularity elsewhere. In the States, there are collections made up entirely of very small specimens, known as micromounts. These are specimens, mounted on pedestals in small plastic boxes, which show association of species invisible to the naked eye and which must be examined by the use of a binocular microscope. A collection, consisting of several hundreds of specimens, may be housed in a modest cabinet. The serious disadvantages of this method are the necessity of acquiring a microscope (Fig. 8·1), which can be expensive, and the reduction in exhibition value of the collection.

As a collector advances, in whichever alternative he or she chooses, three basic collections will develop. If the material is collected in the field, it will need to be stored until convenience allows its examination, identification and so on. This collection should be stored under a system which allows ready retrieval, the specimens being numbered in numbered boxes. After examination, some of this material may be transferred to the main collection, but some will be excess to requirements, and may be placed in a third collection in a similar manner to the first where it may be retrieved quickly for exchange with other collectors, gifts to local museums or teaching institutions, or perhaps sold to a local dealer.

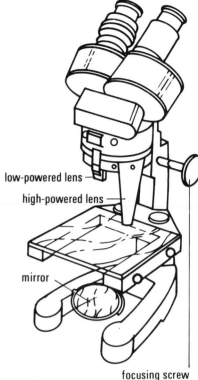

Fig. 8·1 A binocular microscope showing its ready adaptability for the use of reflected and transmitted light. In this model the eye-piece may be focused independently.

Collecting minerals

Where to collect

As mentioned previously, compared to the mineral scene a hundred years ago, minerals are now more difficult to obtain in the field. Gone are the days when a collector could visit a working mine or comparatively young mine dump and there select, from an abundance of good quality material, that which attracted him most.

Today the young collector is faced with a minor range of alternatives. Obviously it is most desirable to collect material in the field, but he or she may be faced with severe limitations. In some countries, access to mines, quarries and so on, is seriously restricted or even prohibited, and the value of mine dumps as collectors' hunting grounds decreases annually due to the enormous increase of interest in mineralogy and easier access to the once remote mining areas.

In addition, conservation of geological and mineralogical sites has become a vital issue. Due to their rarity, certain mineral associations, still visible in the field, now need guarding from collectors in order to retain their unique value for posterity. Access, even restricted to visual examination, is permissible only under supervision.

It is highly recommended that young collectors should join a local geological or mineralogical society. They will, thus, be given access to interesting sites and receive guidance from observing the activity of more experienced collectors.

In spite of this apparently gloomy picture there is much to be done in the field. Working quarries, where access is permitted, produce excellent mineralogical material from time to time. Coastal sections, especially where erosion is taking place, are profitable mineral hunting grounds. Stream sections, where sedimentary minerals may be found, and road works are often surprisingly rewarding.

A visit to a museum possessing a good mineral collection may depress the young collector since there might seem to be little hope of matching the quality of specimens on display. This is not necessarily the case, however. High-quality material may be purchased. There are now many mineral dealers who have obtained old collections and made their contents available for purchase. Mineral specimens are expensive. Indeed, there is a growing trend for non-mineralogists to purchase mineral specimens as an investment for the future.

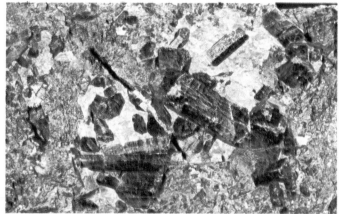

Above
Rosettes of platy specular hematite on quartz; from Cleator, Cumberland, England.

Above right
Dark greenish-brown augite in matrix; from Mount Hope, New Jersey.

Right
Group of 'smoky' quartz crystals dusted with chlorite crystals, on matrix; from St. Gotthard, Switzerland.

Below right
Brilliant red prismatic crystals of crocoite; from Dundas, Tasmania.

Some dealers will undertake to exchange specimens in their hands for something the collector has obtained from the field in bulk.

As the young collector grows in experience, an exchange arrangement may often be made with other collectors to their mutual advantage. This can usefully be extended to exchange or trade internationally since minerals from one country may be greatly sought after in another.

Should the collector be the recipient of a bequest in the form of a mineral collection, advice should be sought immediately on its value to science. Before parting with any of its contents a careful and experienced examination should be made. Material which, to the young collector, may have little appeal, may nevertheless be important and his or her responsibility for its conservation must be made obvious.

Techniques and tools

Field techniques in mineral collecting are based on experience of geological situations, together with an acquired skill in the use and handling of quarrying tools.

The collector should, however, first consider his or her safety in the field. Indeed, permission to enter sites may demand the wearing of safety equipment. A comfortable hard hat of regulation type, a pair of strong and waterproof boots and a pair of safety goggles are essential. Serious accidents involving head, feet and eyes may occur in apparently safe situations.

A good hand lens is an essential piece of equipment. Its magnification need not exceed times ten. Beyond that point of magnification there are problems of depth of focus. Its use is an acquired skill. For field use it is convenient to hang it round the neck by a cord or ribbon, thus saving the difficulty of its extraction from pockets in awkward situations.

A good webbing pack should be obtained, which can accommodate 15 kilograms (33 pounds) or so in weight and, when full, sit comfortably on the wearer's shoulders. It should possess a number of external pockets with cover flaps in which to carry the smaller tools, notebooks, pencils, maps and so on. Internal pockets are not recommended. Ex-military equipment is hard wearing and practical.

The collector will need a range of hammers. A general-purpose hammer of about 1 kilogram (2·2 pounds) in weight is highly recommended. It should possess a flattened head and a chisel edge (Fig. 8·2). A rubber hand-grip is a great asset. The use of a bonded steel shaft as opposed to a wooden one is a matter of preference. A 'lump hammer' of approximately 1·5 kilograms (3·3 pounds) is useful when driving in chisels or wedges. For heavy work a sledge-hammer is required. Its weight need not exceed 3 kilograms (6·6 pounds). Skill, not brawn, is required. A skilled operative using the lighter tool will achieve more than an unskilled operative using heavy equipment, with much less risk to the specimens. Furthermore, it should be remembered that tools have to be carried often over long distances. Cold chisels are essential to lever off slabs of rock, or to open up a crack to provide access to a vug or joint. (Fig. 8·3). If much rock has to be moved in the uncovering of a cavity, steel wedges are a great asset. They are produced for forestry work and are readily available from agricultural suppliers under the name of timber wedges. Their effectiveness may be increased by the use of 'feathers' which are placed on either side of the wedge, producing much-increased leverage in joints or cracks (Fig. 8·4). Much effective work may be done by the use of a crowbar. It is awkward to transport, but its use as a lever to open up or move fissured blocks is highly recommended. It should be of well-tempered steel and need not exceed 1·5 metres (5 feet) in length. Offset points greatly increase leverage, though the offset bend should not exceed 30° from the length of the bar.

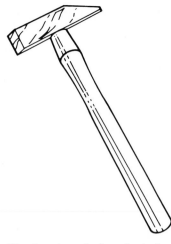

Fig. 8·2 A typical geological hammer showing the chisel end and flat face. The shaft should be in balance with the weight of the steel head, as shown here.

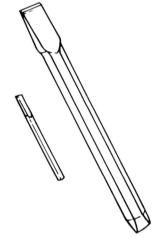

Fig. 8·3 Cold chisels used in mineralogy. The largest is 30 centimetres (12 inches) in length, the shortest 10 centimetres (4 inches).

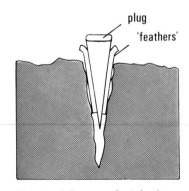

Fig. 8·4 The use of a 'plug' and 'feathers' to split open a piece of rock down a natural joint or fissure.

Above
Parallel veins of fibrous asbestos in matrix; from South Africa.

Right
Coated rhombohedra of ankerite associated with blende and chalcopyrite; from Eagle mine, Gilman, Colorado.

Below right
Green massive zoisite enclosing red corundum crystals (ruby); from Tanzania.

If a geological site is heavily overgrown it may have to be dug out, and a pick and spade are valuable additional tools.

A gold pan is a useful tool if the collector is examing alluvial deposits, not only for gold, but for other dense, resistant minerals, such as cassiterite and scheelite. Pans vary in size, but the most popular diameter is 0·44 metre (1·4 feet), with a depth of 8·5 centimetres (3·3 inches). Panning is an acquired skill, best demonstrated in the field. Essentially, the pan is filled with sand, gravel and water and is stirred by hand to loosen clay and vegetable matter. Large pebbles are removed. About half the water is poured off and, by skilful manipulation of the pan, the remaining water is made to run round the pan in a circular motion, eventually spilling over the edge to remove the lighter minerals present, such as quartz and feldspar. The heavy minerals, including gold, are concentrated as a 'tail' towards the bottom rim of the pan.

If the collector intends to work underground, a good reliable source of illumination is essential. For reconnaissance work in most situations an electrical system is quite adequate. A grave disadvantage is the inability of electric lamps to detect the presence of the greatest enemy of the unsuspecting underground worker – carbon dioxide gas or 'black damp', so common in mines operating in limestone terrain. For excellent overall illumination, and for the detection of 'black damp', acetylene or carbide lamps are thoroughly recommended. In low concentrations of carbon dioxide gas, the acetylene flame turns blue but in strong concentrations it is extinguished. Two main types are available, a hand and a hat model. The former (Fig. 8·5) is the more practical, for it may be hung or stood adjacent to working areas and the operator's head need not be kept aligned to the work as is the case with hat lamps. In addition, the larger the lamp, the longer the period of illumination, and the size of a hat lamp is obviously limited. Water is usually available in underground situations, but spare carbide, from which acetylene gas is produced by the addition of water, should be carried in a well-sealed can. Care should be taken to remove spent carbide from the lower container of the lamp and to maintain the cleanliness of the burner jet. It is advisable to carry a small electric hand torch when underground in case of lamp failure or the necessity to recharge a carbide lamp. It should be remembered that most light sources are predominantly yellow. The inexperienced collector may be mistaken in his discrimination of colours under such conditions.

Certain minerals reveal their presence under the influence of ultraviolet light. Minerals, such as scheelite, fluoresce strongly in short-wave ultraviolet light, but are inconspicuous in ordinary light. This provides a valuable prospecting aid underground or at night. There are battery powered mobile units on the market, but a model should be selected which provides both long-wave and short-wave light.

Although a rather formidable list of equipment has been outlined, it must be appreciated that much of it is specialized and only required in particular situations. The beginner only really requires a hammer, one with a chisel edge or pick, a lens, notebook and pencil, labelling and wrapping materials, and some sort of haversack. Other equipment can be acquired as needed.

Data collection

A good collector should make data collection a top priority. The accumulation techniques should be standardized so that the data are understandable to anyone who may wish to re-examine or promote research into the mineralogical situation described. The young collector may find making field notes and sketches an onerous task, especially under inclement weather conditions. If a determined attempt is made at an early stage, however, then the wisdom of it will become apparent later.

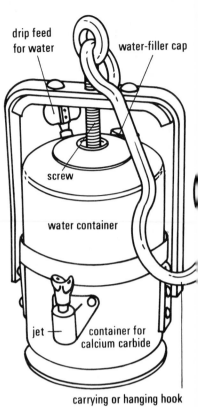

drip feed for water

water-filler cap

screw

water container

jet

container for calcium carbide

carrying or hanging hook

Fig. 8·5 A hand-model acetylene lamp. The lower half carries the calcium carbide container and porcelain jet. The upper half, screwed on to the lower by the carrying hook, holds water controlled by a drip feed tap.

A good notebook, preferably one with waterproof covers, is essential. Surveyors and architects use an ideal type. The left-hand pages bear horizontal rules, the right-hand pages being square-ruled. Squared paper is ideal for the production of field sketches. Each square is taken as a unit of measure and may represent a metre or a centimetre, thus aiding the production of an accurate sketch of, for example, a quarry face many metres in height and width, or a small section of a mineral vein only square centimetres in area (Fig. 8·6).

A soft pencil is recommended for taking field notes. If conditions are damp it will still be usable, whereas other writing implements may cease to function under similar conditions. Pencil marks tend to fade in time and the data should be inked in at a later and drier date.

A long tape measure is a necessity. Estimates of distance, and especially heights, vary greatly from person to person. Though not essential, a good-quality, black-and-white photograph of the area of the sketch greatly helps in its subsequent interpretation.

The data should include exact details of the locality observed from an up-to-date map if available, accompanied by the nearest town and the region. If access to the site has been physically difficult, notes should be taken describing how

Fig. 8·6 A typical page from a field notebook showing the use of squared paper.

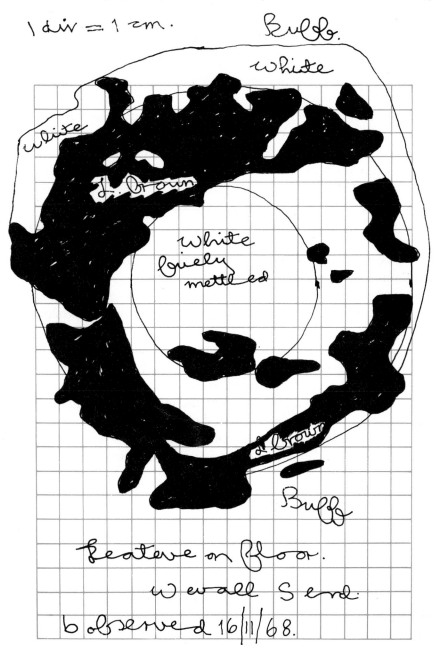

223

access was made. After visiting many localities, a collector's memory may become confused. In abandoned mining terrain, great care should be taken to avoid open shafts and open cuts on veins, and their position should be carefully marked on the map.

A good compass, prismatic if possible, is a valuable aid to orientation especially where the exact site worked on is inconspicuous on the map. A number of bearings to prominent landmarks greatly aids its location on the map.

If the site is open to daylight, for example, a quarry, a sketch of it is often required. Upon this all observed field data are plotted, including the location of all collected material. It is impossible to stress sufficiently the importance of adding as much detail as possible to field sketches. If the site is a working mine, then access to its plans is desirable. Coordinates may be made available which can determine the exact point from which specimens were collected. If the mine is abandoned then the acquisition of old plans is highly desirable, but if this is not possible, and the collector intends to carry out detailed work, a surveyed sketch plan, sufficiently detailed to allow the plotting of site data, should be produced.

It may be necessary to make a collection from a sedimentary sequence where, for example, septarian nodules occur in a succession of shales. A measured section should then be prepared marking the thickness of each bed or lithological unit and plotting the position from which each specimen was taken.

Brownish-black crystals, some twinned, of staurolite in schist; from Brittany, France.

Field labelling

The labelling of specimens collected in the field is essential. A collector may think it impossible to forget the exact site from which each specimen was obtained, but it is only too easy.

Labelling techniques may be a matter of personal preference, but the situation should be a fluid one and be dependent on weather and field conditions. It should also be remembered that field labelling is only a temporary measure and labels should remain on the specimens only long enough to ensure ready retrieval from temporary storage prior to assimilation into a permanent collection.

If conditions are dry, then the use of white adhesive tape to label specimens is recommended. Strips of tape should be cut prior to field work and pressed on to a sheet of thin cellulose, mounted in the back cover of the field notebook, and numbered. As specimens are collected, the relevant numbered piece of tape can be pressed firmly on to the specimen. Lack of adhesion may occur if the specimen is wet, but too much adhesion may occur if the label is left on the specimen too long. A solvent, such as acetone, may be required to release it.

If collecting conditions are wet, then a series of printed or handwritten numbers coinciding with the notes taken may be wrapped with individual specimens in paper or polythene bags. On returning to base, the collector should, of course, allow the specimens to dry (immediately removing material from polythene bags) and apply a more permanent method of labelling.

The use of felt-tipped pens for field marking is not recommended. Many are water solvent, the ink spreading on damp material. Marks made by spirit-based pens are difficult to remove, unsightly and may effect future chemical analysis.

Styles of field numbering are a matter of personal preference. One system which can be recommended is given here. The field number commences with the initial letter of the collector's surname, continues with the year of collecting, and finally a number, for example, K77-147. This will remind the collector when the specimen, or group of specimens, was collected, and will indicate the relevant field notebook. If the collector is working as a member of a team, the desirability of using the initial prefix is obvious. Others may prefer to be reminded of localities and may add additional letters, for example, K77-SG4, indicating that the specimen is the fourth collected from Shap Granite Quarry in 1977.

Fig. 8·7 The stages of
wrapping a mineralogical
specimen.

a The specimen, delicate side
upwards, is placed in the
corner of a sheet of newspaper.

b The left- and then the
right-hand sides of the
newspaper sheet are folded
on to the specimen.

c The specimen is then
rolled up along the length of
the folded newspaper.

d The final package with the
bulk of the folded newspaper
over the most delicate part of
the specimen.

Transportation

Many a good specimen has been destroyed or badly bruised by bad packing and transportation out of the field. A young collector usually learns quickly from previous mistakes, but there are a few basic techniques which should always be observed.

If the material is reasonably robust, adequate protection is afforded by the use of newspaper. The specimen should be wrapped in such a way that the bulk of the paper wrapping is immediately over the most delicate part of the specimen (Fig. 8·7).

The use of cellulose wadding is not recommended for field use. If material so wrapped is allowed to remain damp for long, the cellulose may be difficult to remove.

In some geological situations minerals may be covered in the mud in which they have grown. It is advisable to leave this on the specimen during transportation, as it is a natural protective medium. Its removal is a matter of urgency, however, as growth of the underlying material may continue if the specimen remains damp. The additional growth is usually not crystallographically continuous over the whole of large crystals but takes the form of orientated overgrowths producing a frosted appearance. Carbonates are particularly susceptible to this activity.

If specimens are to be carried out of the field in a large pack (the best way), then layers of specimens and creased up paper prevent damage due to rubbing caused by movement of the pack.

The use of a motor vehicle in the field is a mixed blessing. Specimens may more safely be extracted from field situations and be carried in prepared padded boxes. The great danger, permitted by greater carrying capacity, is to collect indiscriminately. Selectivity is a good thing to observe. If everything had to be carried from the field by pack, much less collecting would be done. Sites would survive and much of the material thrown away from the boot of a car would be available for someone else.

If material has to be transported by public transport or by post, more careful packing is required. Specimens should be completely dry and the field data, identification, and so on should be enclosed against the specimen in the same wrapping. Cellulose tissue is an admirable medium to use for this purpose provided the specimen is first covered by a piece of light-weight tissue paper. Cottonwool should never be used, nor should it ever touch a specimen, because the fibres, especially after long periods of time, are almost impossible to remove. The cellulose and tissue-paper wrapped specimen may now be boxed, surrounded by a tightly fitting mass of rolled-up newspaper.

The use of expanded polystyrene masses or vermiculite is often recommended, but rolled-up newspaper, providing it is tightly packed, is perfectly adequate and, over long distances, is superior.

Extremely delicate specimens, such as crocoite from Tasmania, are almost impossible to transport by conventional methods without damage. The use of low-melting-point wax is a practical way of overcoming the difficulty.

A layer of wax is poured into a metal can of sufficient size and allowed to set. The specimen is placed on it making sure no part of it touches the sides of the can. The can is then completely filled by wax. The specimen is thus sealed in a solid block which will stand a great deal of hard treatment in transportation. On arrival at its destination the can, with its lid removed, is placed upside down in a deep dish in a low temperature oven and the wax allowed to melt off the specimen. Any excess wax remaining on the specimen may be removed by flushing with solvents.

Curation

Classification of accommodation

There are two main headings under which minerals might logically be housed. The first is geographical. It is perfectly reasonable to store minerals under such headings as individual mine sites, regional collections, British minerals, minerals from the Bohemian Massif, and so on. There are, however, often limitations to the number of species known under each heading and problems arise if the collector's scope should widen. An alternative system is, therefore, required.

The commonly accepted chemical subdivision of minerals by anions into sulphides, oxides, silicates, sulphates, and so on, provides a convenient method of classification which is flexible and adaptable and is the one used in The Mineral Kingdom. Older methods using the presence of a dominant element are impracticable, the dominance of any one element often being unknown.

As mentioned earlier, a collector may develop a marked preference for the species of a particular chemical group. He or she may restrict acquisition of material to that group, disposing of other species found simultaneously in the field by exchange or other means.

Cleaning minerals

Before embarking on the cleaning of specimens, the collector should consider carefully the ethics behind the action. What is meant by mineral cleanliness? To some people, especially in the United States, this includes chemical cleanliness. Stains on normally lustrous material are considered undesirable and are removed often in a vicious manner, using highly corrosive mineral acids, for example. It should be remembered that the stain is part of the paragenesis of the mineral and should be preserved as such. The green films on sheets of native copper belong there, as do the iron or manganese stains and dendrites on quartz crystals.

Running tepid water is usually perfectly adequate to remove mud or loose material from the surface of insoluble specimens. If the water supply can be hosed and brought to a fine jet, so much the better, since the additional force dislodges more efficiently clays and particles trapped between crystals. The use of wooden splints in conjunction with the jet also helps.

Brushes should be selected and used with caution. Brushes of man-made fibres may be particularly abrasive. A brush composed of natural fibres, preferably animal hair, should be selected. Copious water should be used in conjunction with the brush to avoid the abrasive action of trapped particles of matrix.

If the use of soap is required to remove fingermarks, dust, and so on, soft or domestic soaps, never toilet soaps, should be used. A few drops of ammonia added to a solution of soft soap greatly assists the cleaning action and has no chemical effect.

Gypsum crystals (variety selenite) should never be washed in soapy water or detergents. The perfect cleavage readily allows the infiltration of soapy solutions which, on evaporation, leave films on the cleavage planes producing an opalescent effect and destroying the true lustre. Clean water only should be used to remove any adhering clays.

The use of ultrasonic cleaning devices is not recommended. Even low sound intensities can cause great damage, especially if the material is well cleaved or heterogeneous.

Should the collector be forced to lower his or her ethical standards due to the need to remove stains from a specimen for display purposes, the removal techniques should be as mild as possible. Weak solutions of chemicals should be tried first. The use of mineral acids is undesirable and should be avoided unless absolutely necessary. Their presence may remain long after application, in spite

Top
Pearly white, translucent tabular crystals of gypsum; from Spain.

of steps taken to remove them. The careful application of an acid to a bruise on a carbonate mineral, however, will greatly improve the appearance of the specimen.

Iron staining may be removed by the use of oxalic acid in the presence of aluminium. The specimen is placed in a glass vessel and covered with a solution of 5 per cent oxalic acid in water. A piece of aluminium foil is added and the whole gently boiled. If the solution becomes strongly discoloured, it should be discarded and the operation repeated until the stain has been removed. The specimen should then be thoroughly washed in running water.

Stains caused by manganese oxides can only be removed by mineral acids, preferably hydrochloric acid, though its strength should not exceed 30 per cent in water.

No attempt should be made to neutralize acids by adding an alkali to either the solution or the specimen. Hydroxides of iron are far more difficult to remove as their mobility as solutions forces them into cleavages and cracks from which they are impossible to remove.

A common cause of staining of minerals is man-made oily compounds, though natural associations of hydrocarbons do occur. This kind of staining is common on material from coastal sections. A range of organic solvents should be tried, commencing with acetone.

Some minerals are poor conductors of heat and great care should be taken to wash them in water at room temperature. Sulphur crystals may disintegrate completely in warm water.

Preservation of minerals

Minerals are subject to a host of physico-chemical effects which, unless controlled, may destroy them completely. The most common effects are deliquescence, efflorescence, effects of light (decomposition, colour change) and pyrite decay. The competence of a curator is soon apparent from an examination of his or her collection, the changes brought about by these effects being only too apparent.

Deliquescence This term is applied to the absorption of atmospheric water into the structure of the material, dissolving it. Halide minerals are particularly subject to deliquescence which in damp conditions may result in a specimen becoming a puddle of liquid. The effect may be so difficult to control that many experienced collectors refuse such species a place in their collections. Among common halides subject to this effect are halite ($NaCl$), especially if magnesium is present; carnallite ($KMgCl_3.6H_2O$); and sylvine (KCl). Effective control necessitates establishing dry air conditions. This may be achieved by placing the specimen in a sealed polythene bag containing a little self-indicating silica gel in a small open-weave canvas bag. A rapid and regular examination of the specimen may thus be made.

Efflorescence This is the loss of water of crystallization from the material. It is a much less serious condition than deliquescence, though far more common. Centrally heated storage conditions are the greatest cause of efflorescence. Minerals such as chalcanthite ($CuSO_4.5H_2O$), laumontite ($CaAl_2Si_4O_{12}.4H_2O$) and melanterite ($FeSO_4.7H_2O$) may be quoted as good examples of efflorescent minerals. The normally translucent crystallized material may change to a white powder.

The only way to control the effect is to store the specimens prone to it in humid air. An excellent situation is a damp cellar, provided the specimens are protected from the effects of damp on timber and metal, and they may be readily retrieved for examination. Open polythene trays on racks are ideal.

Above
Pale pink, hopper-faced cubic crystals of halite; from south-west Africa.

227

Effects of light *Decomposition*. Many minerals are strongly affected by the action of light, especially daylight, in the presence of oxygen. Certain species should never be subjected to long periods of lighting. Silver salts, being photosensitive, come first to mind. On badly curated material the effects are only too apparent. The beauty of the crimson crystals of proustite (Ag_3AsS_3) or pyrargyrite (Ag_3SbS_3) is completely lost, the colour changing to dull black. The effect of light on realgar (As_4S_4) is equally unfortunate. The strikingly beautiful bright red colour, changes to bright golden-yellow powdery orpiment (As_2S_3). Many of the normally brightly metallic-lustred sulphides are affected by light, developing a tarnish which may, at times, be highly iridescent.

This phenomenon presents a display problem to the collector who may wish to exhibit material under strong lighting to enhance its beauty. Experience will indicate which species cannot be exhibited under such conditions, but advice should always be sort. Much valuable material has been lost due to inexperience.

Colour changes. Several minerals are noted for their ability to change colour under strong lighting. Many are restricted to single occurrences for the phenomenon is not universal to a species.

Fluorite, especially the green varieties from the Weardale area of County Durham, England, may change their unique colour to the more common purple and blue.

Miners of the now abandoned Mowbray Mine, near Frizington in west Cumbria, England, found they could sell the normally yellowish brown baryte crystals to collectors far more readily after they had been subjected to strong sunlight, which caused them to turn to the more attractive zoned green and blue colours.

Complete loss of colour may take place after subjection to light. Topaz from Transbaikal in the USSR may lose its attractive sherry colour and become colourless. Anhydrite, especially the fine blue crystals from Japan, may also become colourless. These processes are irreversible, but there are others in which the colour may gradually return under darkened storage conditions. Hackmanite from the Kola Peninsula in the USSR loses its pink colour in light, but regains it after a period in darkness.

Pyrite decay Of all conditions the collector has to contend with, pyrite decay is, without doubt, the most difficult. Though research into its control has advanced in the last decade, the complete answer is still to be found.

Without any warning, under conditions little understood, such minerals as pyrite and marcasite develop acidic surface encrustations which, if left unattended, progress until the specimen is converted to a heap of white or greenish powder. Simultaneously there is a release of sulphuric acid which plays havoc with storage arrangements and with adjacent specimens. All sulphides are liable to be affected and careful supervision should be maintained to detect the first stages of attack. It is thought to be promoted by the activity of sulphur-reducing bacteria and the techniques used to combat the condition involve their control.

Palaeontologists have this problem to contend with also, but the techniques they use, including basifying, oven drying and soaking in heavy films of lacquer or polyvinyl acetate, are unacceptable to mineralogists. The use of a transparent film of any compound over the specimen destroys its natural lustre and shade of

A group of fine honey-yellow tabular crystals of baryte with hematite and dolomite; from Cumberland, England.

colour, and the resultant specimen may as well be thrown away.

As with all problems connected with bacterial control, prevention is better than cure. Once the attack has been strongly promoted, there is little hope for the specimen, especially if it is polycrystalline, as most sedimentary sulphides are. Recently acquired material should, therefore, be dealt with immediately. The use of cationic inhibitors has met with a degree of success using such antiseptics as Savlon and Cetrimide (Cetyltrimethyl ammonium bromide). Savlon should be used as a 10 per cent by volume dilution, and Cetrimide as a 0·1 per cent by volume solution, both in industrial methylated spirit.

The first stage of the technique is to basify the material susceptible to attack by immersion in a weak solution of ammonia in water, and allow it to dry. The specimen is then immersed in a solution of either Savlon or Cetrimide for 30 minutes and again allowed to dry. It is pointless to increase the concentration of either solution, as excess solid may be left on evaporation. It should be noted that Cetrimide is highly reactive to ultraviolet light.

Although a measure of success has been achieved by using this technique, a regular routine of application should be established, three monthly intervals being recommended.

More recent work has achieved additional success by the use of volatile inhibitors, though their use by amateurs is difficult, as they can function properly only in air-tight show cases. Meanwhile research continues on this most unpleasant of curatorial problems.

Labelling

The creation of good labels demands five basic essentials – clarity; a unique accession number; as much data as is practically possible; waterproofing; and a high degree of permanency.

Every specimen should be given a unique accession number when placed in the main collection. There may be, for example, eight specimens of an unidentified sulphide housed under the field number K77-147. After its identification, two specimens may be given away, three placed in an exchange collection, and the remaining three selected for inclusion in the main collection. Each of the three specimens should now be labelled and bear a unique accession number, preferably in consecutive order. If the specimen is large enough, the species name, locality, and so on, should be added, in as much detail as size will allow, together with the accession number. If the specimen is small, the accession number may have to suffice, but its presence is essential. The data on the specimen should correspond to a catalogue which should contain the complete history of the specimen, including its name, associates, exact locality, date of collection and/or name of donor.

There are numerous techniques in label production. An examination of private collections demonstrates that label production is a personal matter. Indeed, old labels are now achieving value as collectors' items and are much sought after (Fig. 8·8). Some people advocate writing the data in waterproof ink on the specimen itself, but this may be impracticable. Others apply an area of white paint to the specimen on which to write the data. The technique recommended here is to use a good white bond paper with waterproof ink which is then cut out and affixed to the specimen by a spirit-based adhesive. If water-based glues are used, labels soon come adrift under damp conditions. After the adhesive has dried, the label, including an overlap area all round the label, is covered with the same adhesive to seal it completely from damp. The specimen may even be washed without fear of data loss, and under heavy handling this type of labelling survives best.

Some collectors create labels and affix them in a way in which they hope will deter thieves. A determined mineralogical thief will not be deterred, however, since any kind of label may be removed. Labels should be produced for in-

Fig. 8·8 A selection of specimen labels. Some are from famous mineral collections and are themselves collectors' items.

formation and maximum security against natural hazards.

The character of a collection is enhanced if the labels are handwritten, but some people find it difficult to produce small print legibly. Photography may then be of help. Each label is typed on to a piece of white paper. It is then photographed on to a line film. From the negative thus obtained it is possible to print labels on a light-weight document paper reduced to any size to suit the size of specimen. Labels may be produced which need a hand lens to read them.

Housing

As mentioned previously, the size range of material collected must determine the kind of housing required. If the collector has decided to invest in a binocular microscope and limit his or her collecting to micromounts, then a cabinet of shallow drawers is perfectly adequate. If the collector has placed no limitation on the size of specimens, then there are practical problems which must be faced, especially if the collector intends to store and exhibit his or her material in cabinets or drawers.

Well-made drawers are much to be preferred as a method of storage. Numerous devices are available to ensure relatively dust-free conditions and the cabinets may be locked to prevent access to younger members of the family. They are, however, expensive to have made and the collector should watch carefully for auction sales of furniture where actual specimen cabinets sometimes come on to the market.

A refinement of storage in drawer accommodation is the use of cardboard trays. Each specimen is placed in an individual tray which avoids the possibility of 'rolling' and consequent bruising. The trays should be size multiples to provide maximum economy of room in a drawer. They are expensive, however, and a collector who opens the drawers of the cabinet with care should not need to use them.

For economy of space, excess exchange and field collections may be wrapped and stored in crates or boxes. The accession, exchange collection number, or field reference number should be marked boldly on the wrapped specimens. The crates or boxes used should not be too large. Not only are large filled crates difficult to move, but the weight of the upper layers of specimens may damage the lower.

Against each accession number in the several collections housed, there should be a 'finding number' which will indicate in which drawer or crate the specimen is stored, and for simplicity and ease of 'finding', each drawer and crate should bear a number.

Meteorites

If the collector had the good fortune to witness the rare occurrence of a 'fall' and acquire a 'stone' or an 'iron' from it, he or she should inform the department of mineralogy of the national museum of the acquisition and be prepared to donate it it to that institution for the benefit of research (*see also* page 11).

Portions of meteorites are available for purchase from such firms as the American Meteorite Laboratory. Their curation demands a high standard of vigilance to ensure stability for they possess mineral associations rare on Earth. Many are prone to active deliquescence and conditions of absolutely dry air are essential.

Spraying with a lacquer (hair spray is ideal) is often recommended, but does present problems if the specimen is eventually used in research. It is better, therefore, to seal the specimen in a polythene bag accompanied by self-indicating silica gel. The development of ferric salts from such minerals as lawrencite can rapidly destroy a specimen unless great care is taken.

Dr Alan Woolley

A guide to the literature on minerals

This encyclopedia presents a broad and relatively comprehensive coverage of many aspects of minerals and rocks. However, a further two particular aspects may be of considerable interest, particularly to the beginner, namely how to go about finding out more about the subject, and how to discover the whereabouts of collecting localities in your particular area. These few pages are intended to try to answer these questions, bearing in mind that the main need is not necessarily to have a vast encyclopedic knowledge, but to know where to find the information required.

There are two principal ways of answering these questions, namely by joining a society or club and by reading. In most countries there are national geological and mineralogical societies. These hold meetings, conduct field trips, and often publish journals, but these societies tend to cater mainly for professionals and their interests are very academic. Of more general interest are local societies and clubs and there are many of these in most countries. They hold meetings, usually with a talk on some topic of common interest, organize field trips, and sometimes hold shows at which members exhibit and exchange their specimens. Contacting these clubs can be difficult, but the local library may be able to help while some journals and magazines give the names and addresses of many of them.

Joining a society is undoubtedly the best way for the beginner to feel his way into the subject. This is particularly true with regard to collecting. No book can make you a good and knowledgeable collector. However, field excursions with others who are more experienced will soon reveal the best techniques, essential equipment and the better localities. Members of the society will also know where equipment can be obtained locally, and this is particularly important with regard to a geological hammer without which little or no collecting can be done. The society may also have a small library of books and magazines from the pages of which further information may be gleaned. There may also be a collection of geological maps, which can be a most valuable source of information, as will be described later.

Turning now to printed information, the wealth of knowledge that comprises the geological and mineralogical literature is contained in a great volume of books, journals and magazines, and geological maps. The books range from the comprehensive mineralogies, perhaps in several volumes, to the lavishly illustrated books of more general interest. The serious collector of minerals will probably require at least one comprehensive mineral text, in addition to this encyclopedia, for although this book covers a wide range of minerals it does not claim to include them all. The encyclopedia by Roberts, Rapp and Weber is probably the most comprehensive text available, giving just about all the mineral species known up to 1974, but it is expensive. The various books by J. D. and E. S. Dana, which have gone through many editions and have been revised by later authors, are some of the great classics of the mineralogical literature and one or more of them is a must for the bookshelf of the more serious collector. These and a number of other mineral books are listed in the bibliography.

There are many books, pamphlets and articles describing the minerals to be found in particular countries, areas, or even particular mines, quarries or veins. Information of this sort is obviously of great value to the collector, but too dispersed and vast to be summarized here. However, this information can be sought from other collectors or through a club or library.

There are only a few general texts which are devoted to describing rocks as such, because these are usually considered in a broader geological context, and the reader is probably best advised to turn to such books rather than trying to learn more about rocks in isolation. However, one book that does describe and illustrate in colour a wide range of rocks, as well as minerals and fossils, is *The Hamlyn Guide to Minerals, Rocks and Fossils*. Of the geology books perhaps the most comprehensive, beautifully illustrated, and well written is *Principles of Physical Geology* by Arthur Holmes.

Journals and magazines on mineralogy and geology are basically of two kinds. Firstly there are those such as the *Mineralogical Magazine*, the *American Mineralogist*, the *Canadian Mineralogist*, and the *Journal of the Geological Society of Australia*, which are devoted almost wholly to scientific research papers. These publications are of little help to the collector, although they do contain articles on new minerals and new mineral occurrences. The second group of magazines are orientated specifically to the collector and it is recommended that the more serious collector either takes one of these or gets access to copies. Some of these are listed in the bibliography.

A further very important category of literature which can be of considerable value to the collector is the publications of the various national geological surveys. Just about every country in the world has a department of some kind devoted to exploring, mapping, and describing the geology of the country. Many of these geological surveys are vast organizations publishing considerable numbers of books, periodicals, pamphlets, and maps. In large countries such as the United States, Canada and Australia, the individual states also have their own geological survey departments. It is well worth while obtaining a list of publications of the survey, for many publish literature orientated particularly to the collector and of a general nature; for instance, the Institute of Geological Sciences in London publish a series of eighteen guides called *British Regional Geology* which describe the geology of Great Britain area by area in general terms. Other survey publications, however, particularly bulletins, although primarily devoted to describing the geology of particular areas, often have economic sections from which considerable information can be gleaned.

Nobody interested in geology can get very far before he or she is confronted with a geological map of some kind, and the collector only interested in minerals is also well advised to know something about geological maps to pursue the interest efficiently. Geological maps show the distribution of rocks at the surface of the Earth, either by some form of shading, or by colour. They are of two main kinds. 'Drift' maps show superficial deposits such as alluvium and glacial deposits, and sometimes also the bedrock lying beneath these. 'Solid' geology maps on the other hand show only the distribution of the bedrock as it is presumed to be after the superficial deposits are stripped away. For the collector, the 'solid' maps will usually be the most useful because most collecting sites will be situated in bedrock.

The scale of geological maps varies considerably from world and continental maps on a scale of one to many millions, to very detailed local maps. For most collectors it is probably advisable to have a copy of your national geological map, and such maps are available from most geological surveys. With this it will be possible to become familiar with the range and distribution of rock types in the country and hence have some idea of where particular minerals are more likely to be found. It may be possible to consult larger scale maps at a library or borrow them from a geological or mineralogical society or club.

On most geological maps the rocks are arranged into three groups: sedimentary, igneous and metamorphic. Sometimes the sedimentary and metamorphic rocks are grouped together. The sedimentary rocks are arranged stratigraphically, that is, they are placed in order of geological age with the youngest at the top. Sometimes rocks deposited during past geological periods are lumped together, such as Ordovician or Jurassic, whereas on more detailed maps these will be broken down into smaller units. Igneous rocks are usually divided according to rock type such as granite, gabbro and basalt, and may be further subdivided according to age. Metamorphic rocks may be arranged in a stratigraphical sequence like sedimentary rocks, because many are sedimentary rocks which have been metamorphosed. In other cases, particularly for areas of very old schists and gneisses, they may simply be shown in a general way. The extent of metamorphic aureoles around igneous intrusions are sometimes indicated and can be a useful guide to the collector.

For some old mining areas in particular, large scale and very detailed maps may be obtained indicating the positions of old workings and perhaps the distribution of veins and pegmatites. Such information is obviously invaluable. Other kinds of maps such as mineral maps and maps showing the distribution of particular rocks such as limestone are sometimes available, and it pays to consult any map lists that can be obtained either directly from the survey or from a bookseller.

Bibliography

Anderson, B. W., *Gem Testing*, 8th edition, Butterworths, London, 1971.
A manual for the identification of gemstones.
Anderson, B. W., *Gemstones for Everyman*, Faber & Faber, London, 1976.
A fine, very readable book on all aspects of gemstones.
Chalmers, R. O., *Australian Rocks, Minerals and Gemstones*, Angus and Robertson, Sydney, 1967.
Dana, E. S., *A Textbook of Mineralogy*, 4th edition revised and enlarged by W. E. Ford, John Wiley & Sons, New York, 1932.
Although feeling its years perhaps still the most useful single volume describing the great majority of minerals, and with introductory chapters on the science of minerals.
Dana, J. D., *Manual of Mineralogy*. John Wiley & Sons, New York, numerous editions.
A very useful general text with introductory chapters then systematic mineral descriptions.
Dana, J. D. & Dana E. S., *The System of Mineralogy*, 7th edition by Palache, Berman and Frondel, John Wiley & Sons, New York, 1944.
A series which when completed will describe all minerals in detail, but only the first three volumes so far published.
Desautels, P. E., *The Mineral Kingdom*, Hamlyn, London, 1969.
A superbly illustrated and fascinating account of all aspects of minerals.
Hamilton, W. R., Woolley, A. R., & Bishop, A. C., *The Hamlyn Guide to Minerals, Rocks and Fossils*, Hamlyn, London, 1974.
A pocket guide in colour covering numerous minerals, rocks and fossils.
Hey, M. H., & Embrey, P., *Index of Mineral*

Species and Varieties Arranged Chemically, British Museum (Natural History), London, 1955–74.

Holmes, A., *Principles of Physical Geology*, Nelson, London, new and fully revised edition 1965.
A superbly written and illustrated account of all aspects of geology. No background knowledge whatsoever is required to understand this splendid book.

Hurlbut, C. S., *Minerals and Man*, Thames and Hudson, London, 1969.
Concerned with the use of minerals.

Institute of Geological Sciences, London, *British Regional Geology*. Often revised.
A series of eighteen guides covering the geology of Britain.

Jones, W. R. *Minerals in Industry*, Penguin Books, Harmondsworth, 1963.
An account of how minerals are used in industry.

Kesler, S. E., *Our Finite Mineral Resources*, McGraw-Hill Book Company, London, 1976.

Lamey, C. A., *Metallic and Industrial Mineral Deposits*, McGraw-Hill Book Company, New York, 1966.

Liddicoat, R. T., *Handbook of Gemstone Identification*, 10th edition, Gemmological Institute of America, Los Angeles, 1975.

O'Donoghue, M. (Ed), *The Encyclopedia of Minerals and Gemstones*, Orbis, London, 1976.

Plough, F. H., *A Field Guide to Rocks and Minerals*, Houghton Mifflin Company, Boston, 1960.
A most useful field guide that will slip easily into the pocket.

Roberts, W. L., Rapp, G. R., & Weber, J., *Encyclopedia of Minerals*, Van Nostrand Reinhold Co., New York and London, 1974.
The most comprehensive of all mineral texts. Gives physical and chemical details of all minerals and references for further reading. Many coloured illustrations.

Rutland, E. H., *An Introduction to the World's Gemstones*, Country Life, London, 1974.

Sinkankas, J., *Gemstones and Mineral Data Book*, Winchester Press, New York, 1972.
A mine of information on minerals and gems, including preservation hints and problems.

Sinkankas, J., *Gemstones of North America*, Van Nostrand Company, Princeton and London, 1959.
A guide to gem materials to be found in North America.

Sinkankas, J., *Prospecting for Gemstones and Minerals*, Van Nostrand Reinhold, New York and London, 1970.

Traill, R. J., *A Catalogue of Canadian Minerals*, Geological Survey of Canada, 1970. Paper 69–45.

Webster, R., *Gems*, Newnes-Butterworth, London, 1975.
A comprehensive and readable account of all gem materials.

Magazines and Journals
Lapidary Journal, P.O. Box 80937, San Diego, California 92138, United States.

Mineralogical Record, John S. White Jr., P.O. Box 783, Bowie, Maryland 20715, United States.
Contains articles on all aspects of minerals, usually with splendid black and white and colour illustrations.

Gems: The British Lapidary Magazine, Lapidary Publications, 84 High Street, Broadstairs, Kent, England.
Devoted to the amateur collector and packed with information on collecting localities, societies and clubs, dealers, books etc.

Index

Page numbers in italic refer to illustrations. Page numbers in bold indicate the main description of the mineral.

Acknowledgements

Professor N. H. Ambraseys, Imperial College
of Science and Technology, London 16;
Ardea – I. R. Beames 20–21; Ardea – K. W.
Fink 11; Ardea – P. J. Green 115 top;
Ardea – C. K. Mylne 35; F. B. Atkins,
Oxford 16–17, 21, 27 top left, 66 bottom,
67 bottom, 90, 95 bottom, 103 left, 108–109,
160; R. Boardman, Kew 72, 198–199 top,
229; Rob Bock, Johannesburg 25, 33 left,
51, 62, 74–75, 141, 153; California Institute
of Technology and Carnegie Institution of
Washington 6, 12; Canadian Johns-Manville
Company, Quebec 203; Hervé Chaumeton,
Chamalières 32, 76 bottom, 96, 99 right, 127
top, 151, 164, 172, 226–227 top; Bruce
Coleman – Jane Burton 37, 46, 70, 104, 173,
191 left; Bruce Coleman – NASA 9 right;
Bruce Coleman – G. D. Plage 196; Bruce
Coleman – Toni Schneiders 22–23; Bruce
Coleman – Werner Stoy 40; Hamlyn Group

Picture Library 2–3, 27 bottom, 105, 177,
224; Institute of Geological Sciences, London
186, 189, 193, 195, 202; Jacana – Hervé
Chaumeton 97, 103 right, 143, 161 top;
Jacana – Jean Philippe Varin 65; Breck P.
Kent, High Bridge, New Jersey endpapers,
4–5, 19, 39, 43, 44–45, 49 left, 56, 63, 66 top,
67 top, 73, 77, 84, 85 left, 85 right, 95 top,
100, 106–107, 115 bottom, 116, 118, 119,
120, 121, 123, 124 top, 124 bottom, 125, 127
bottom, 129, 135, 136, 137, 139, 142, 147,
148, 149, 150 left, 150 right, 152, 154, 155,
159, 161 bottom, 162, 163, 165, 168 top,
168 bottom, 169, 175, 178, 179, 180, 182,
184 left, 184 right, 216, 219 top left, 219 top
right, 219 centre, 219 bottom, 221 top right,
228; A. Notholt, London 213; Phelps Dodge
Corporation, New York 206–207; RIDA
Photo Library – D. Bayliss 76 top, 99 left,
107, 117, 221 left, 221 bottom; R. Symes,
London 27 top right; Peter H. Truckle,

London 36; University Museum, Oxford
12–13, 191 right, 194; ZEFA – W. Backhaus
204; ZEFA – P. de Prins 49 right; ZEFA –
R. Halin 41, 198–199 bottom; ZEFA – Photo
Researchers 10–11; ZEFA – Photri 9 left;
ZEFA – M. Serban 34–35; ZEFA – G.
Sirena 30; ZEFA – Heins Steenmans 33
right; ZEFA – I. Steinhoff 226–227 bottom.

The following photographs are reproduced
by courtesy of the Smithsonian Institution,
Washington D.C.: 4–5, 19, 56, 63, 67 top,
77, 84, 95 top, 100, 116, 118, 119, 120, 121,
123, 124 bottom, 125, 127 bottom, 137, 139,
147, 148, 149, 150 left, 150 right, 152, 159,
161 bottom, 162, 165, 182, 184 left, 184
right, 216, 219 bottom, 221 top right, 228.

The photographs on pages 6 and 12 are
reproduced by permission from the Hale
Observatories.